Evolutionary Ecology of Amphibians

Editors

Gregorio Moreno-Rueda

Dpto. de Zoología, Facultad de Ciencias
Universidad de Granada
Granada, Spain

Mar Comas

Department of Biological Sciences
Dartmouth College
Hanover, NH, USA

CRC Press
Taylor & Francis Group
Boca Raton London New York

CRC Press is an imprint of the
Taylor & Francis Group, an **informa** business

A SCIENCE PUBLISHERS BOOK

Cover credit: Male Mediterranean tree frog *Hyla meridionalis* climbing an almond tree (Catalonia, Spain). Photo by Mar Comas.

First edition published 2023
by CRC Press
6000 Broken Sound Parkway NW, Suite 300, Boca Raton, FL 33487-2742

and by CRC Press
4 Park Square, Milton Park, Abingdon, Oxon, OX14 4RN

Library of Congress Cataloging-in-Publication Data (applied for)

ISBN: 978-0-367-55396-8 (hbk)
ISBN: 978-0-367-55397-5 (pbk)
ISBN: 978-1-003-09331-2 (ebk)

DOI: 10.1201/9781003093312

Typeset in Times New Roman
by Radiant Productions

Contents

CHAPTER 1

Introduction to the Book

Gregorio Moreno-Rueda[1,*] and *Mar Comas*[1,2]

INTRODUCTION

Natural selection is considered the main driver of evolution (Darwin 1859). Accordingly, animals face a determinate environment, which subjects the animals to a series of selective pressures. The organisms capable of leaving more descendants in the next generation, those best adapted to the environment, have the alleles that last the longest (higher fitness; Maynard Smith 1989). The fittest animal in one environment does not have to be the fittest in another. Similarly, as the environment changes, selective pressures change and hence the traits that make an organism adapt better varies. Considering how natural selection operates, we must understand how the organisms relate to their environment to understand how they evolve. This is the role of Evolutionary Ecology, the branch of Ecology or the branch of Evolutionary Biology that tries to understand how the environment affects evolutionary processes. This discipline is at the intersection between Ecology and Evolutionary Biology. The environment will contribute to the selective pressures that will determine which phenotypes reproduce and which do not. If these phenotypes are genetically determined, the environment will eventually shape the change in allele frequencies in subsequent generations, giving rise to microevolutionary processes. Evolutionary Ecology, therefore, allows us to understand how animals evolve and even predict how they will evolve.

Amphibians constitute an ancient group of vertebrates, in fact, the oldest of the terrestrial vertebrates (Carroll 2009). Despite their apparent morphological simplicity, which can be reduced to three *blauplan*, they present enormous physiological and behavioral diversity. There are amphibians with internal and external fertilization, viviparous, ovoviviparous and oviparous, with parental care of different lengths (from simply releasing the eggs to extensive parental care of the young). Among

---
[1] Dpto. de Zoología, Facultad de Ciencias, Universidad de Granada, E-18071, Granada, Spain.
[2] Department of Biological Sciences, Dartmouth College, Hanover, NH, USA.
  Email: marcomas@ugr.es
* Corresponding author: gmr@ugr.es

those with extensive parental care, there is enormous variety in the mode of that care, including amphibians that feed their offspring. In addition, amphibians also have complex cycles, with typically aquatic larvae whereas adults can live in different environments, from aquatic adults with gills to adults that have adapted to desert environments. Therefore, the same individual must adapt to two very different environments (aquatic and aerial), undergoing an important metamorphosis in between. Once they are adults, there are aquatic, fossorial, terrestrial, and arboreal amphibians (even glider amphibians; Emerson and Koehl 1990). They also show enormous phenotypic plasticity, being able to greatly modify their morphology depending on whether they are exposed to terrestrial or aquatic environments (Denoël et al. 2005).

Due to their complex life cycles and their moist, highly permeable skin (to the extent that some amphibians breathe through their skin and lack lungs; Jørgensen 2000), amphibians have specific and unique requirements that differ from those of other vertebrates. Most amphibians need water points in which their larvae develop. In addition, they usually have high humidity requirements and, as ectotherms, they depend on the external temperature for the development of their activity.

In sum, amphibians present a considerably greater diversity than might be thought at first. Studying the evolutionary ecology of such a diversity of strategies is of enormous interest *per se*. However, the evolutionary ecology of amphibians is not only interesting in itself but also from an applied point of view. After living more than 300 million years on Earth, amphibians are currently facing a global decline (Collins et al. 2009). The causes of this decline seem to be multiple (habitat destruction, diseases, invasive species, water contaminants, climate change, etc.), but in any case, understanding the evolutionary ecology of amphibians is essential to understand their decline and predict the response that amphibians will have in the future to imminent changes in climate, landscape and water availability. Inevitably, the conservation of amphibians involves understanding their evolutionary ecology.

The Book

Given the importance of the study of evolutionary ecology for amphibians' conservation, and the speed at which scientific knowledge is advancing, we consider it is pertinent to compile the existing knowledge. In this sense, this book aims to establish a compendium of what is known so far about the evolutionary ecology of amphibians. In this volume, we have tried to deal with the most important aspects of the amphibian evolutionary ecology. This volume provides an excellent starting point for the study of the evolutionary ecology of amphibians.

The book is structured in several sections, in general going from basic biological aspects to more complex traits. Chapter 2 is dedicated to amphibian evolution. Here, San Mauro et al. present to us the extant amphibians (Caudata, Anura and Gymnophiona) and their evolutionary history. They explain to us that amphibians evolved from lungfishes 385–360 million years ago. Extant amphibians comprise a monophyletic group, called Lissamphibia. Still, although the origin of amphibians seems to be clear, some debate exists about the origin of Lissamphibia and its possible phylogenetic relationships with extinct groups. Extant amphibians seem to have originated during the Carboniferous, whereas Anura (frogs and toads) and Caudata (salamanders and newts) diverged during the Permian, diversifying from Triassic to the end of the Cretaceous. San Mauro, Agorreta and Garcia-Porta end the chapter by explaining the evolutionary mechanism by which amphibians diversified into more than 8,000 species.

Chapters 3, 4, and 6 deal with different aspects of amphibian physiology. Chapter 5 deals specifically with diseases. Thus, this section deals with basic aspects of amphibian biology, which will be necessary for a better understanding of more complex traits that are studied in later chapters.

In Chapter 3, Burraco and Gabor describe the basic aspects of amphibian physiology. They discuss the importance of phenotypic plasticity for amphibians to cope with environmental variation and how physiological pathways regulate the plastic responses to environmental cues. The physiology of amphibians has received considerable attention because it is unique among Tetrapods, given that most of amphibians suffer a metamorphosis implying a drastic change in the environment where they live. Amphibian larvae may respond to stressful conditions such as pond drying, predation or competition by accelerating their development. This plasticity, however, is not free of trade-offs and individuals developing faster often show reduced body size and fat reserves at metamorphosis. The authors describe the physiological pathways implied in metamorphosis and in the plastic response to environmental change, and then they discuss the implications of these physiological pathways for amphibians to cope with human-derived environmental change. The review made by Burraco and Gabor concludes with a list of recommendations for future studies on eco-physiology in amphibians which, undoubtedly, will stimulate new studies on the topic.

In Chapter 4, Comas and Moreno-Rueda talk about Ecoimmunology. Parasites and pathogens are probably the main selective forces in nature (Schmid-Hempel 2011). In the case of amphibians, moreover, diseases constitute one of their main threats (Daszak et al. 2003). In this context, understanding the immune system is essential for amphibian conservation. However, the immune capacity cannot be maximized, since a series of trade-offs are established between the immune system and other components of fitness. In this chapter, Comas and Moreno-Rueda review the trade-offs between the immune system and other components of fitness such as locomotion, oxidative stress or breeding success in amphibians.

As a continuation of Chapter 4, in Chapter 5, Thumsova, Bosch and Rosa discuss the diseases that affect amphibians. Infectious diseases are serious threats to amphibian conservation (Rachowicz et al. 2006). An array of fungi, bacteria, protozoa and viruses have been described as pathogens that cause significant mortality in amphibians. Still, the impact of pathogens seems to vary according to environmental conditions. The authors of this chapter review the different factors that may affect the susceptibility of amphibians to different pathogens, such as climate, habitat, pollutants, or intrinsic factors such as the own immune system, life history, or other factors affecting pathogen transmissions such as behavior or population or community structure. Understanding the factors affecting infection is essential for amphibian conservation, so the chapter ends with a section dedicated to strategies useful for mitigating the impact of infectious diseases on amphibian populations.

In Chapter 6, Ortega and coauthors fathom other important aspects of amphibian physiology: thermoregulation and hydric balance. Amphibians are ectotherms so they depend on external sources of heat. Moreover, adults have highly permeable skin. Therefore, in amphibians, heating may be strongly compromised with water balance because of evaporation, making amphibian ecophysiology unique. In this chapter, Ortega, Garci and Rivas review and synthesize the knowledge of thermal ecology and hydric balance of amphibians. They summarize the morphological, physiological, and behavioral mechanisms that amphibians use for thermoregulating and maintaining water balance. Obviously, the topics treated in this chapter are thickly related to the response capacity of amphibians against climate change, so Ortega et al. close the chapter by discussing extensively the importance of thermal and hydric ecology for amphibian conservation in a changing world.

From Chapter 7 onwards, the book deals with complex traits, an understanding of which requires an understanding of the underlying physiology of the animal. Olalla-Tárraga describes in Chapter 7, the evolution of body size and body morphology in amphibians, which cannot be understood without understanding Chapter 6 first. Body size is one of the more relevant traits in ecology and evolution. Undoubtedly, body size may determine several fitness-related parameters in amphibians, such as detectability by predators, locomotor performance (see Chapter 8), competition ability in sexual contests, fecundity in females, acoustic performance (see Chapter 11), etcetera. Chapter 7 reviews the interspecific variation in body size, showing that this variation follows determinate climatic gradients. The analysis of Olalla-Tárraga suggests that geographical patterns in body size are the result of a trade-off between thermoregulation and hydro regulation mediated by surface area to volume ratios. Understanding how climatic conditions affect amphibian body size and morphology is fundamental in order to forecast how future climatic change will affect amphibian populations.

Continuing with the study of complex traits in amphibians, in Chapter 8, Zamora-Camacho makes an extensive review of amphibians' locomotion. Locomotion in amphibians is essential to escape predators, chase prey, patrol their territories, disperse in the environment, or search for a mate. It is therefore a feature of paramount importance in ecology and evolution. Amphibians, which inhabit different types of environments, even throughout the same life, present numerous means of locomotion, from fossorial (caecilians), burrowers (e.g., spadefoot toads; Zamora-Camacho et al. 2019), jumpers (frogs), runners and of course swimmers, with different styles of swimming, from swimming based on the movement of the tail (tadpoles), the body (newts) or propelled by the hind legs (frogs). There are even wing flying frogs (e.g., genus *Rhacophorus*). Zamora-Camacho provides an overview of the forms of locomotion across amphibian groups, but also explains the possible physiological and anatomical trade-offs between locomotor performance and other fitness-related traits, such as oxidative stress or immune system.

In Chapter 9, Vági and Szekely present an analysis of the amphibian reproduction modes and their evolution. Amphibians show the highest diversity in reproductive modes among the terrestrial vertebrates. Typical amphibian life history implies an aquatic phase and adult terrestrial phase. The analysis of Vági and Szekely throws light on the amphibian reproductive adaptations in response to life-history constraints and the abiotic and the social environments. They conclude that, in the three amphibian groups, the reproductive innovations are selected especially when they help in gaining independence from the aquatic habitats. Still, in some amphibians, like paedomorphic salamanders, reproductive innovations contributed to a complete return to the water.

The last three chapters of the book review the information concerning the three main communication channels used by amphibians: chemical (olfaction), acoustic (sound) and visual (coloration) channels. In Chapter 10, Iglesias-Carrasco and Wong speak about the most widespread and most ancient form of communication, the chemical communication. Olfaction is used by amphibians for conspecific communication, food and predator detection as well as predator deterrence. The authors explain how the environments which amphibians inhabit (aquatic and terrestrial) have shaped the types of molecules used for chemical communication. They review the contexts in which amphibians use chemical compounds. Lastly, Iglesias-Carrasco and Wong integrate the information on chemical communication with the threat of environmental changes, by which several human activities can impinge on chemical communication, negatively affecting amphibian conservation. The understanding of amphibian chemical communication may be relevant for amphibian conservation in a changing world.

Sandra Goutte presents an overview of acoustic communication in anurans in Chapter 11. The chapter describes the different types of anuran vocalizations and the context in which they are produced, formulating the hypotheses explaining such a huge variation. The chapter also describes the morphological and physiological mechanisms underlying call production and hearing. Goutte explains the main causes of call inter- and intraspecific variation, discussing their evolutionary causes, mainly regarding the plastic capacity of anurans to respond to environmental cues. Finally, the chapter discusses how anthropogenic changes, with emphasis on urbanization, may affect anuran acoustic communication and the consequences of human activities on anuran evolution and conservation.

Finally, in Chapter 12, Rojas, Lawrence and Márquez present an extensive and well-illustrated review of the evolution of coloration in amphibians. Their review begins with a detailed description of how the coloration is formed in the skin of amphibians, where chromatophores acquire special relevance. Rojas and coauthors describe the physiology and genetics underlying color formation and end this first part by discussing color changes in amphibians. The chapter's authors then describe the selective pressures that affect color evolution, beginning with predation, which encourages colors that favor camouflage, aposematism, mimicry, deimatic displays, or visual illusions. In the following section, they describe the latest findings in relation to the evolution of color in amphibians under sexual selection. Although sexual selection has been considered a powerful force driving color evolution in other vertebrates (fish, reptiles and birds), in amphibians it had gone unnoticed until recently. The chapter ends with an interesting section of suggestions for future research in the area.

References

Carroll, R.L. 2009. The Rise of Amphibians: 365 Million Years of Evolution. Johns Hopkins University Press, Baltimore, MD, USA.

Collins, J.P., M.L. Crump and T.E. Lovejoy III. 2009. Extinction in our Times: Global Amphibian Decline. Oxford University Press, Oxford, UK.

Darwin, C. 1859. On the Origin of Species. John Murray, London.

Daszak, P., A.A. Cunningham and A.D. Hyatt. 2003. Infectious disease and amphibian population declines. Divers. Distrib. 9: 141–150.

Denoël, M., P. Joly and H.H. Whiteman. 2005. Evolutionary ecology of facultative paedomorphosis in newts and salamanders. Biol. Rev. 80: 663–671.

Emerson, S.B. and M.A.R. Koehl. 1990. The interaction of behavioral and morphological change in the evolution of a novel locomotor type: 'Flying' frogs. Evolution 44: 1931–1946.

Jørgensen, C.B. 2000. Amphibian respiration and olfaction and their relationships: from Robert Townson (1794) to the present. Biol. Rev. 75: 297–345.

Maynard Smith, J. 1989. Evolutionary Genetics. Oxford University Press, Oxford, UK.

Rachowicz, L.J., R.A. Knapp, J.A.T. Morgan, M.J. Stice, V.T. Vredenburg, J.M. Parker et al. 2006. Emerging infectious disease as a proximate cause of amphibian mass mortality. Ecology 87: 1671–1683.

Schmid-Hempel, P. 2011. Evolutionary Parasitology: The Integrated Study of Infections, Immunology, Ecology, and Genetics. Oxford University Press, Oxford, UK.

Zamora-Camacho, F.J., L. Medina-Gálvez and S. Zambrano-Fernández. 2019. The roles of sex and morphology in burrowing depth of Iberian spadefoot toads in different biotic and abiotic environments. J. Zool. 309: 224–230.

CHAPTER 2

Origin, Evolution and Diversification of Extant Amphibians

Diego San Mauro,[1,*] *Ainhoa Agorreta*[1] and
Joan Garcia-Porta[2,3,*]

INTRODUCTION

Amphibians constitute one of the major branches of the vertebrate tree of life, and are key for understanding the colonization of terrestrial environments by tetrapod ancestors. Although amphibians encompass many extinct forms of early land vertebrates that date back to the Paleozoic Era (Milner 1993, Benton 2005, Carroll 2009), at present they are solely represented by three extant orders of markedly distinct morphologies and life styles (Fig. 1): Anura (frogs and toads), Caudata (salamanders and newts), and Gymnophiona (caecilians). Extant amphibians are often referred to as lissamphibians (Lissamphibia) due to their soft, moist skin, seriously loaded with mucous and poison glands. They are generally associated with humid environments, laying bare (amnion-less) eggs, and mainly becoming predators of small invertebrates in their adulthood (Duellman and Trueb 1994). Altogether, the three orders of extant amphibians sum up about 8,380 currently recognized species distributed across temperate and tropical landmasses worldwide (AmphibiaWeb 2021, Frost 2021).

Anurans are the most diverse and cosmopolitan group of extant amphibians, with 7,400 species grouped into about 55 families distributed in all continents excluding Antarctica. Except for a couple of exceptions, frogs and toads are generally tailless, short-bodied animals with relatively long and strong hind limbs adapted for jumping, burrowing, or (secondarily) swimming. Their specialized morphology fundamentally

[1] Department of Biodiversity, Ecology and Evolution. Complutense University of Madrid, 28040 Madrid, Spain.
  Email: ainhoaag@ucm.es
[2] Departament de Genètica, Microbiologia i Estadística, Facultat de Biologia, Universitat de Barcelona (UB), Av. Diagonal 645,Barcelona 08028, Spain.
[3] Institut de Recerca de la Biodiversitat (IRBio), Universitat de Barcelona (UB), Barcelona, Spain.
* Co-corresponding authors: dsanmaur@ucm.es; jgarciaporta@ub.edu

Figure 1. Diversity of the three orders of extant amphibians: Gymnophiona (a,b), Caudata (c–e), Anura (f–l). Species shown are: (a) *Rhinatrema bivittatum* from French Guiana, (b) *Gegeneophis primus* from India, (c) *Bolitoglossa orestes* from Venezuela, (d) *Salamandra salamandra* from Spain, (e) *Ambystoma mexicanum* captive bred, (f) *Pipa arrabali* from Venezuela, (g) *Ceratophrys calcarata* from Venezuela, (h) *Pristimantis pleurostriatus* from Venezuela, (i) *Bufo spinosus* from Morocco, (j) *Elachistocleis surinamensis* from Venezuela, (k) *Dendrobates tinctorius* from French Guiana, (l) *Hyalinobatrachium cappellei* from Venezuela. Photo credits: César Barrio-Amorós/Doc Frog Photography/CRWild (c, f, g, h, j, l), Ramachandran Kotharambath (b), Diego San Mauro (a, d, e, i, k).

related to the adaptation for jumping was likely the key for their greater biological success compared to the other two orders of extant amphibians.

In contrast to anurans, salamanders and newts have a relatively slender body with proportionate limbs and long tails. This body shape is likely resembling more closely the one that tetrapod ancestors displayed in the past. Although most salamanders are terrestrial walkers, there are many aquatic representatives as well. Extant caudatans are less diversified than anurans, with nearly 770 currently recognized species grouped into 10 families mostly distributed in landmasses derived from Laurasia.

The smallest and least known group of extant amphibians are the caecilians, with only slightly over 210 recognized species (10 families) of mainly Gondwanan distribution except for a couple of notable exceptions in Asia and Mesoamerica. Compared to the other two extant amphibian orders, caecilians are very distinct as they have an elongate, limbless body (often with conspicuous annulation) with sensory tentacles on the snout, all resulting from the adaptation to a primarily burrowing or dwelling life style.

Origin and Ancestry of Extant Amphibians

From a strict neontological point of view, all extant amphibians constitute a monophyletic group (=clade, see Text Box 1) whose sister taxon are the amniotes (mammals plus reptiles, including birds), altogether conforming the extant representatives of Tetrapoda (land vertebrates). According to the most recent investigations, the closest extant relatives of tetrapods are the lungfishes, followed more distantly by the coelacanths, and altogether conforming the extant Sarcopterygii (lobe-finned fishes branch of the vertebrate tree of life). These phylogenetic relationships (see Text Box 1) were subjected to debate in the past, but they are now robustly settled, mostly by the contribution of growing molecular phylogenetic and phylogenomic evidence that has accumulated in the last couple of decades (e.g., Zardoya and Meyer 2001, San Mauro 2010, Amemiya et al. 2013, Irisarri and Meyer 2016, Irisarri et al. 2017, Meyer et al. 2021).

Paleontologically speaking, things are a bit more complicated. Incidentally, the term amphibian often refers also to the many tetrapod forms that appeared early after the jump-to-land of vertebrates 385–360 million years ago (Ma), most of which are not directly related (i.e., sharing close ancestry) to the extant amphibian orders. Among these early amphibian tetrapods (traditionally classified as Labyrinthodontia or Stegocephalia; Carroll 2009, Laurin 2010), there are notorious Upper Devonian fossils like *Acanthostega* or *Ichthyostega*, which represent transitional steps in the colonization of terrestrial environments from sarcopterygian ancestors, along with older (more fish-like) forms like *Tiktaalik* or *Panderichthys* (Clack 2012). The main issue in terms of relating extinct and extant amphibian forms stems from the fact that the fossil record of early representatives of the major extant amphibian orders is relatively poor, as is that of putative common ancestors of the group (Carroll 2009, Ksepka et al. 2015). All this complicates their ancestry assessment and assignment relative to the major groups of amphibian tetrapods occurring during the Paleozoic Era.

There are three main hypotheses about the ancestry of extant amphibians (Fig. 2). The most generally accepted hypothesis considers them a monophyletic group that arose from a single lineage of Temnospondyli tetrapods, a group of large and robust amphibian animals that lived during the Paleozoic Era (e.g., Panchen and Smithson 1987, Milner 1988, Benton 1990, Bolt 1991, Trueb and Cloutier 1991, Ruta et al. 2003, Schoch 2019). In contrast, several authors have proposed the alternative hypothesis that the most likely ancestor of all extant amphibians arose within the Lepospondyli (e.g., Laurin and Reisz 1997, Laurin 1998, Vallin and Laurin 2004, Marjanović and Laurin 2013), a group of small-sized Paleozoic tetrapods that are also

Text Box 1: On Phylogenies and Phylogenetic Analysis

Generally speaking, a phylogeny (or evolutionary tree) is the graphic/diagrammatic representation of the historical relationships among elements that maintain lines of descent in time. In the biological context, such elements usually encompass organisms (e.g., species), genes, or DNA/protein sequences, although the use of phylogenies can be extended outside the biological context to, e.g., languages (comparative linguistics) or even other disciplines. Understanding the pattern of evolutionary relationships is essential in comparative studies of any kind (particularly in Biology) because there are statistical dependencies among elements sharing common ancestry.

In a phylogenetic tree, there are specific parts that can be identified. The tips of the tree (also known as external nodes, terminals or leaves) represent the end of an evolutionary line, such as a present form of an extant organism or a last record of a fossil. These tips are interconnected by lines (branches or lineages) to internal nodes (representing hypothetical ancestors) in a hierarchical, usually dichotomic, fashion. When the ultimate common ancestor is made explicit, the phylogenetic tree is rooted, meaning that there is a distinct ancestral line (origin) of the whole tree. Otherwise, an unrooted tree makes no assumption about the ancestral line or origin of the tree. Knowing the tree root enables setting a direction for the inferred evolutionary transformations. Two lineages sharing a common ancestor are said to be sister to each other and altogether constitute a clade or monophyletic group (i.e., contains a common ancestor and all its descendants). Alternatively, a group that contains a common ancestor but not all its descendants (only some of them) is called paraphyletic or non-monophyletic. As a particular case of these latter, a polyphyletic group contains organisms with different common ancestors and that are grouped together because of convergence of characters (see panel a in box figure). Only monophyletic groups provide unambiguous information about the evolutionary process and how lineages have emerged from each other.

Given that any historical episode is by definition unrecoverable, the reconstruction of a phylogeny is a process of inference (known as phylogenetic analysis) that yields a hypothetical depiction of ancestor-descendant relationships that reflect an indirect record of the evolutionary process. Phylogenies are inferred from observed characters that are assumed to be inherited, such as organisms' traits/features or molecular sequence data. There are different methods of phylogenetic analysis, from the earlier and simplest (e.g., parsimony and distances) to the more modern and sophisticated (e.g., probabilistic methods such as maximum likelihood and Bayesian inference). These latter have become the standard nowadays and they use complex models of molecular sequence evolution to approximate a statistical description of the process of change in the observed characters (e.g., the process of DNA or protein substitution). Apart from producing the relationships among the studied elements (tree topology), probabilistic methods can also inform about the rates of change (in the form of varying lengths of branches) or even time of divergence from shared common ancestors (dated tree or timetree). The phylogenies produced in each case are known as cladogram, phylogram, and chronogram, respectively (see panel b in box figure).

contd. ...

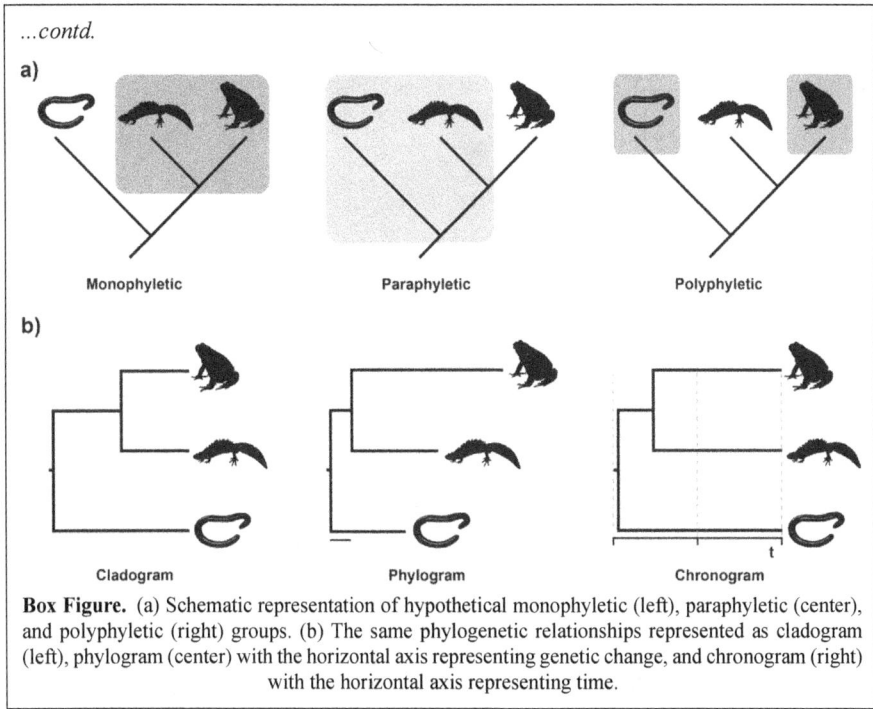

Box Figure. (a) Schematic representation of hypothetical monophyletic (left), paraphyletic (center), and polyphyletic (right) groups. (b) The same phylogenetic relationships represented as cladogram (left), phylogram (center) with the horizontal axis representing genetic change, and chronogram (right) with the horizontal axis representing time.

considered ancestral or related to extant amniotes. Some studies have also supported a diphiletic nature of extant amphibians, with Anura and Caudata (altogether conforming the clade Batrachia) arising from the Temnospondyli and Gymnophiona arising from the Lepospondyli (Carroll 2001, 2007, Carroll et al. 2004, Anderson 2008, Anderson et al. 2008). Neontologically speaking, this latter hypothesis (dubbed 'polyphyletic hypothesis') would imply a closer relationship of caecilians to amniotes than to batrachians (frogs and salamanders), but this is contradicted by overwhelming evidence (particularly from molecular data) supporting the monophyly of extant amphibians relative to amniotes (e.g., Zhang et al. 2005, Frost et al. 2006, San Mauro 2010, Irisarri et al. 2017, Siu-Ting et al. 2019).

The putative oldest fossil of an extant amphibian is *Gerobatrachus hottoni*, from the Lower Permian of Texas (Anderson et al. 2008). It is a nearly complete skeleton that shows a mixture of traits of Anura and Caudata and belongs to the Temnospondyli. Several studies (Anderson et al. 2008, Maddin et al. 2012, Schoch 2019, Schoch et al. 2020) have placed *Gerobatrachus* as either stem or sister of Batrachia, but with varying resolutions for the phylogenetic position of Gymnophiona (either within the Lepospondyli, or, more recently and widely-accepted, as sister to batrachians in a monophyletic Lissamphibia). However, this is not exempt of debate, and some authors have put into question that *Gerobatrachus* is indeed closely related to Lissamphibia at all (Sigurdsen and Green 2011, Marjanović and Laurin 2013).

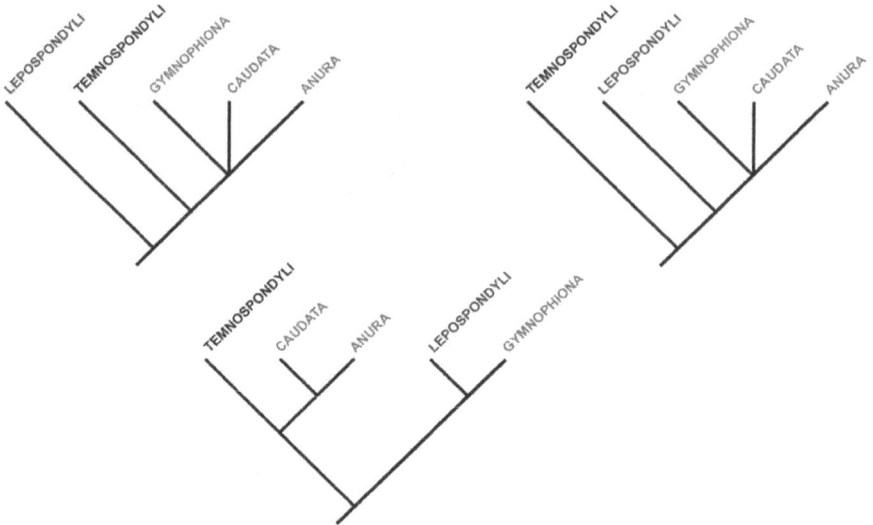

Figure 2. Phylogenetic relationships among extant and Paleozoic amphibians: Temnospondyls as ancestors of monophyletic lissamphibians (top left); Lepospondyls as ancestors of monophyletic lissamphibians (top right); Lissamphibians are diphyletic, frogs and salamanders related to temnospondyls, caecilians related to lepospondyls (bottom).

Not counting *Gerobatrachus*, which remains contentious, the oldest unambiguously lissamphibian fossil is *Triadobatrachus massinotii*, from the Lower Triassic of Madagascar (Rage and Rocek 1989, Ruta and Coates 2007, Ksepka et al. 2015). *Triadobatrachus* is clearly recognized as a stem or basal member of Anura, and possess many transitional traits that eventually conformed frog characteristics (like jumping locomotion), and that appeared better defined in the descendant fossil *Prosalirus bitis*, from the Lower Jurassic of Arizona.

The oldest fossil records of the salamander branch are from the Middle Jurassic: *Kokartus honorarius* from Kyrgyzstan (Skutschas and Martin 2011) and *Marmorerpeton* (two species) from England (Evans et al. 1988). Recently (Schoch et al. 2020), another fossil specimen, *Triassurus sixtelae*, from the Middle Triassic of Kyrgyzstan, has been proposed as the geologically oldest stem-group salamander. In a similar way, a recent fossil discovery, *Chinlestegophis jenkinsi*, from the Upper Triassic of Colorado has been proposed as the oldest stem representative of caecilians (Pardo et al. 2017). Nonetheless, the affinities of *Chinlestegophis* have been recently questioned (Santos et al. 2020), leaving *Eocaecilia micropodia*, from the Lower Jurassic of Arizona, as the unambiguously-accepted oldest stem caecilian (Jenkins and Walsh 1993, Jenkins et al. 2007, Maddin et al. 2012).

Taken together, the fossil record indicates that amphibians clearly have a long evolutionary history that probably extends back to the Permian, or even the Carboniferous, more than 300 Ma. Many molecular phylogenetic studies that have provided estimates for the origin of extant amphibians, including recent ones based on transcriptomic or genomic-scale data, have placed the time of divergence between caecilian and batrachian ancestors also in the Upper Paleozoic Era (Carboniferous-Permian), although often with notable variation in mean values or confidence interval width of age estimates (San Mauro 2010, Pyron 2011, Irisarri et al. 2017, Siu-Ting et al. 2019, Hime et al. 2021, for citing just a few).

Phylogeny and Systematics of Major Amphibian Lineages

The growing accumulation of morphological and (particularly) molecular evidence over the last couple of decades has boosted the resolution of important parts of the amphibian tree of life. Nevertheless, there are still many obscure or contentious relationships, especially at genus and species levels, still in need of further research and clarification. In this section, we provide an overview and discussion of the systematics and phylogenetic relationships of the main amphibian branches based on the large-scale study by Jetz and Pyron (2018) that is largely congruent with many other recent studies on particular clades or the entire Lissamphibia (San Mauro et al. 2005, 2014, Frost et al. 2006, Roelants et al. 2007, Zhang and Wake 2009, Pyron and Wiens 2011, Zhang et al. 2013, Feng et al. 2017, Hime et al. 2021, for citing just a few), as well as the information on dedicated databases on amphibian diversity (Frost et al. 2006, AmphibiaWeb 2021) (Fig. 3).

The monophyly of extant amphibians with respect to amniotes as well as respective monophyly of each of the three recognized orders (Anura, Caudata, Gymnophiona) are widely accepted nowadays, particularly with the overwhelming support of molecular phylogenetic evidence. Likewise, the vast majority of studies have found support for the sister group relationship between Anura and Caudata (clade Batrachia, Fig. 4), to the exclusion of Gymnophiona (Zardoya and Meyer 2001, Zhang et al. 2005, San Mauro et al. 2005, Frost et al. 2006, Roelants et al. 2007, San Mauro 2010, Pyron and Wiens 2011, Irisarri et al. 2017, Jetz and Pyron 2018, Siu-Ting et al. 2019, Hime et al. 2021, for citing just a few). Only a minority of studies have supported an alternative sister group relationship between Caudata and Gymnophiona (clade Procera, Fig. 4) (Feller and Hedges 1998, Zhang et al. 2003, Fong et al. 2012), but this is mainly attributed to incorrect selection of data or implementation of analysis (Siu-Ting et al. 2019).

In the case of Gymnophiona, South American Rhinatrematidae is sister to all other extant caecilian families, followed by Asian Ichthyophiidae as the sister group of the larger clade Teresomata (unranked taxonomically), which includes all other families of caecilians. Within this latter clade, African Scolecomorphidae are basal-most, followed by Herpelidae + Chikilidae, and then Caeciliidae + Typhlonectidae as the sister group of a clade containing Indotyphlidae as sister to Dermophiidae + Siphonopidae (Fig. 3). The extant caecilian families are distributed throughout tropical habitats in Africa, India, Seychelles, Southeast Asia, and Central and South America (Fig. 5), and their current distribution is consistent with an origin of the order in Gondwana, possibly prior to the fragmentation of Pangea, with later dispersals to Southeast Asia and Central America. In some ways, the phylogeny of caecilian families roughly depicts their complex process of morphological specialization to fossorial habits, such as reinforcement of the skull, reduction of eyes, and loss of tail (Duellman and Trueb 1994). Typhlonectidae is the only family that has representatives secondarily adapted to an aquatic or semi-aquatic lifestyle. As in the other two extant amphibian orders (Anura and Caudata), the caecilian phylogeny

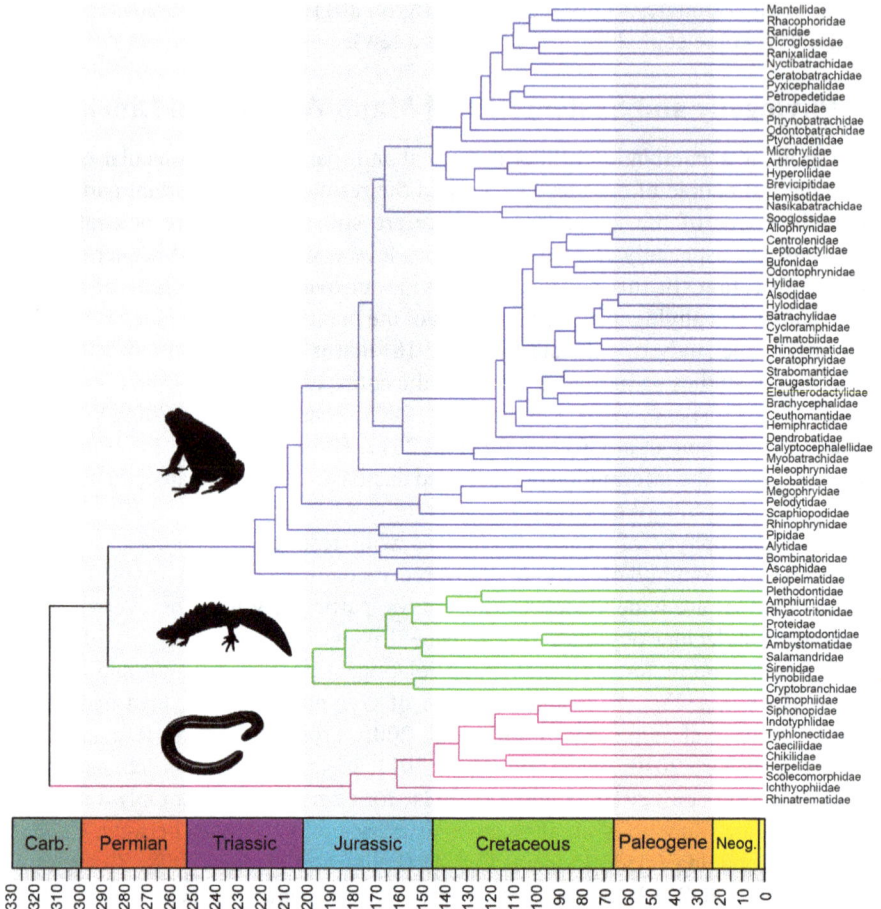

Figure 3. Phylogeny of all families of extant amphibians. The tree shown is a schematic summary chronogram based on the large-scale study by Jetz and Pyron (2018), and represents the best estimate of evolutionary relationships and divergence times among amphibians in agreement with other recent sources of phylogenetic data (see text).

also shows a trend towards greater complexity in reproductive biology modes, going from oviparity and aquatic larvae (in basal-most families like Rhinatrematidae and Ichthyophiidae) to direct development and viviparity in many teresomatan caecilians (San Mauro et al. 2014).

Regarding Caudata, Asian families Crytobranchidae and Hynobiidae are the sister clade of all other salamander families, followed by American family Sirenidae as sister to two major (species-richer) clades. One of these includes Salamandridae as sister to Ambystomatidae + Dicamptodontidae; the other one includes Proteidae as sister to Rhyacotritonidae and Amphiumidae + Plethodontidae (Fig. 3). Extant salamanders are currently distributed in Asia, Europe, and North America (with a lesser secondary dispersal of plethodontids in South America) (Fig. 5), which suggests an origin of the order in the Laurasian portion of supercontinent Pangaea.

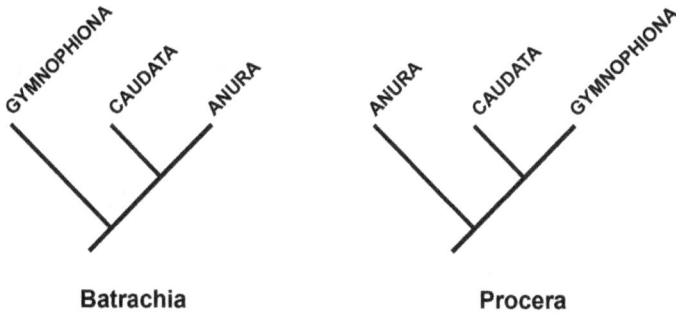

Batrachia   **Procera**

Figure 4. Competing hypotheses on the phylogenetic relationships among the three orders of extant amphibians: Batrachia (left) and Procera (right). Support is overwhelmingly higher for the Batrachia hypothesis.

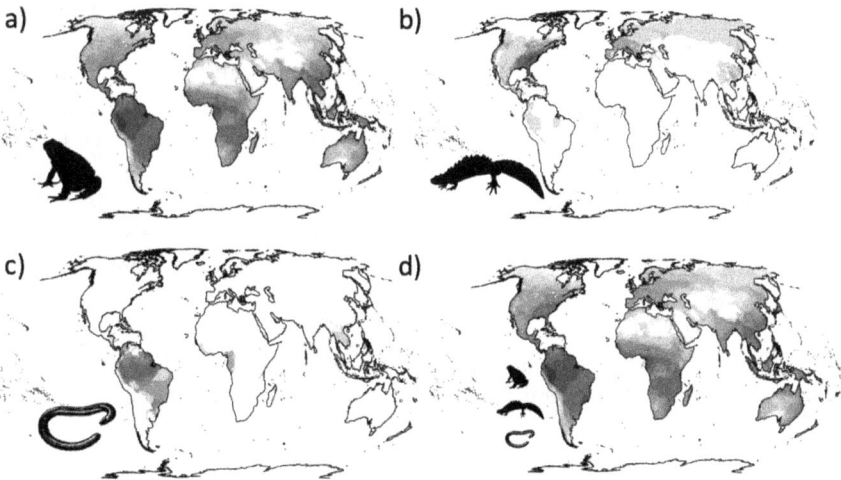

Figure 5. Diversity (in a log-scale) of each of the three orders of extant amphibians: (a) Anura, (b) Caudata, (c) Gymnophiona and (d) all amphibians. In all maps, the darker the shade, the higher the diversity (raw data obtained from IUCN, https://www.iucnredlist.org/resources/spatial-data-download).

There are both aquatic and terrestrial groups of salamanders, although the latter are intimately linked to the presence of water in the environment. The three basalmost families of salamanders (Cryptobranchidae, Hynobiidae, Sirenidae) present all external fertilization, but more derived forms have transitioned to internal fertilization in which the male produces a spermatophore that is collected by the female. Interestingly, and unlike in the other two extant amphibian orders, pedomorphosis (when juvenile characters of an ancestral form are retained in the adult individual of the descendant) is widespread among salamanders (Fig. 1e), having evolved independently in representatives of eight out of the ten families, including basal Cryptobranchidae and Sirenidae (Duellman and Trueb 1994).

In Anura, there are several superfamilies, collectively named Archaeobatrachia (=primitive/ancient frogs), that constitute a paraphyletic assemblage with respect to the most diverse and derived group of frogs, the Neobatrachia (=new/advanced frogs).

These (more basal) archaeobatrachian superfamilies include: Leiopelmatoidea [Ascaphidae + Leiopelmatidae], Discoglossoidea [Bombinatoridae + Alytidae], Pipoidea [Rhinophrynidae + Pipidae], and Pelobatoidea [Scaphiopodidae + Pelodytidae + Megophryidae + Pelobatidae]; this latter (also named Anomocoela) being the sister group of neobatrachians. Leiopelmatoidea (also named Amphicoela) is the basal-most clade of all archaeobatrachians (hence of all anurans), followed by Discoglossoidea and then Pipoidea (Fig. 3). In contrast to the archaeobatrachian assemblage, neobatrachians are clearly monophyletic, and contain two major (very species-rich) superfamilies that account for more than half of all extant amphibian species currently recognized: Hyloidea (=Nobleobatrachia) and Ranoidea (=Ranoides). Along these two larger clades, there are other smaller families that appear as more-basal early splits, such as Heleophrynidae (sister to all other neobatrachians), Nasikabatrachidae + Sooglossidae (sister to Ranoidea), and Calyptocephalellidae + Myobatrachidae (sister to Hyloidea) (Fig. 3). Hyloidea and Ranoidea are both very diverse morphologically and ecologically, containing each many additional families for which many phylogenetic relationships remain unclear.

Frogs and toads constitute the most diverse and cosmopolitan group of extant amphibians, and they are currently distributed worldwide (Fig. 5). The early fossil record of Anura has Lower Triassic representatives in both southern and northern areas of Pangaea (*Triadobatrachus* is from Madagascar and *Czatkobatrachus* is from Poland), which suggests that its ancestral lineage was present in this supercontinent before its Mesozoic fragmentation. Indeed, the fragmentation of the supercontinent may have triggered the divergence of some anuran lineages: Ascaphidae (North America) vs. Leiopelmatidae (New Zealand), Rhinophrynidae (Central and North America) vs. Pipidae (Africa and South America), and Pelobatoidea (Eurasia and North America) vs. Neobatrachia (originally Gondwanan) (Roelants and Bossuyt 2005). Fertilization is generally external in anurans, and conducted through a characteristic pseudocopulation posture called amplexus (Fig. 1j).

Drivers of Phenotypic and Species Diversification in Amphibians

The massive biodiversity of amphibians constitutes a great system to understand the processes that produce phenotypic and species diversification. These can be divided in two main types, adaptive and non-adaptive processes (Gittenberger 1991, 2004). Adaptive processes basically require the action of natural selection operating in different directions in different populations (Schluter 2000). The simplest way to understand how this process drives phenotypic (and species) diversification is through the classic Simpson's "adaptive landscape" (Simpson 1944). In this concept, inspired by Wright's fitness landscapes for gene frequencies (Wright 1932), fitness is visualized as the height of a surface that, in turn, varies as a function of the values of two or more phenotypic traits (Simpson 1944). In this way, the combination of trait values that allow higher fitness are visualized as 'peaks' in the surface while those combinations of trait values that determine low fitness are visualized as 'valleys' (Fig. 6).

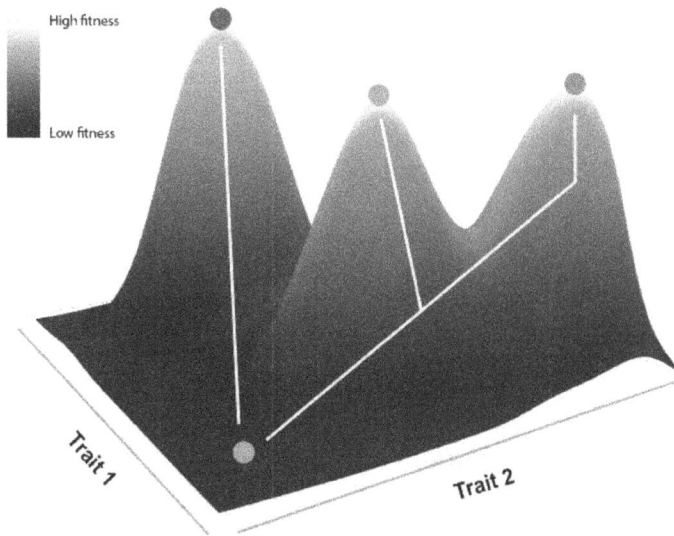

Figure 6. Conceptual depiction of an "adaptive landscape" where the height of the surface correlates with fitness and the two horizontal axes represent the variation existing along two traits. Groups can potentially experience phenotypic and species diversification as they "move" across this surface (circles).

In this 'landscape' formed by peaks and valleys, populations diverge because they are 'pulled' toward different adaptive peaks and away from the valleys of lower fitness. From an ecological perspective, each of the adaptive peaks can be seen as a particular combination of trait values that allow an efficient use of a particular niche (Schluter 2000). Organisms may suddenly be exposed to a great variety of empty niches, which can be visualized as adaptive landscapes with many unoccupied fitness peaks (Martin and Wainwright 2013). In a possible outcome derived from this situation, the divergent selective regimes that operate across the 'landscape' may rapidly pull different populations into the different adaptive peaks producing phenotypic (and genetic) divergence among populations. Under such scenario, speciation occurs as an incidental 'by product' of adaptive divergence (Nosil 2012). In this way, new niche availability or, as coined by Simpson, "ecological opportunity", can theoretically drive great levels of phenotypic and species diversification. Simpson proposed three different scenarios in which high levels of ecological opportunity could theoretically drive high amounts of evolutionary diversification (Simpson 1944, 1955). We introduce them and discuss their role in amphibian diversification as follows:

1. Extinction of Antagonists

This occurs when several niches become available after the extinction of the species that occupied these niches. This is typically the case after a massive extinction event. In these situations, surviving groups often rapidly fill the new-formed ecological vacancies producing outstanding levels of phenotypic and species diversity. In amphibians, we have clear signatures of sudden increases of species diversification after two major mass extinction events: the Permian mass extinction and the Cretaceous-Tertiary (K-T) mass extinction. The end of the Permian (around 250 Ma) is considered the most severe biotic crisis of the Phanerozoic, likely extinguishing more than 90% of all species (Benton and Twitchett 2003). This event had a massive toll on amphibian diversity, with a huge decrease in species numbers with the extinction of complete groups (Fortuny and Steyer 2019). Temnospondyls, however, responded to the end-Permian extinction by going from relatively low levels of phenotypic and species diversity to a sudden increase of phenotypic and species diversification (Tarailo 2018). Aside from the evidence provided by the fossil record, rates of diversification estimated from the phylogeny of extant amphibians also indicates a sudden increase of species diversification rates after the end-Permian extinction event (Roelants et al. 2007). The K-T mass extinction event (around 65 Ma) is famously known as the event that produced the extinction of non-avian dinosaurs and other groups (Sakamoto et al. 2016), releasing a huge number of ecological vacancies that facilitated the radiation of mammals (Meredith et al. 2011). Just like mammals, an increase in diversification rates after the K-T boundary event is also detected in anurans and salamanders (Roelants et al. 2007, Jetz and Pyron 2018). Interestingly, diversification dynamics in Gymnophiona appear to be largely unaffected by this mass extinction event, likely consequence of its specialized subterranean niche that may be highly resilient to substantial environmental changes (Jetz and Pyron 2018).

2. Exposure to New Environments

Organisms are exposed to new environments either when they disperse into a new area or when they are exposed to environmental changes. The most obvious examples of evolutionary diversification driven by the exposure to new environments is the great diversification that follows island colonization (Losos and Ricklefs 2009). The astonishing variety of ecologies, sizes and shapes found in many insular groups of anurans exemplify the role of island colonization at triggering diversification. Well studied examples include the Malagasy Mantellidae (Valero et al. 2017), the genus *Kaloula* in the Sulawesi (Blackburn et al. 2013) or the genus *Eleutherodactylus* in the Caribbean (Dugo-Cota et al. 2019). In all cases, insular groups radiate into a great variety of ecomorphs that increase levels of speciation and phenotypic disparity. Sometimes, closely related groups radiate into multiple islands in an archipelago producing small diversifications within each of the islands. These cases provide an excellent opportunity to test the association between morphological traits and particular ecologies and the extent to which the evolutionary process is predictable (Mahler et al. 2017). The Caribbean *Eleutherodactylus* offer a great case-study for that and shows that this genus generated replicated radiations in each of the main Caribbean islands, exhibiting a large degree of phenotypic convergence (Dugo-Cota et al. 2019). High levels of ecological opportunity are also expected when groups colonize new continental regions as it can be seen in the great regional radiations produced by the family Bufonidae as it expanded across the world (Van Bocxlaer et al. 2010).

3. *Acquisition of Key Innovations*

A key innovation is a trait (or combination of traits) that allows a new interaction with the environment. Very often, this grants the access to several new, previously inaccessible, niches that in turn may trigger high rates of phenotypic and species diversification (Losos 2009). Many putative key innovations have been proposed in amphibians. For example, the extraordinary adaptive success of the anuran diversification, not remotely paralleled by caecilians or salamanders, has been related to the early acquisition of two major innovations: powerful hind limbs and short stiffened vertebral column (the urostyle) for primary jumping locomotion, and a larval form that differs drastically from the adult in morphology and ecology (the tadpole, aquatic and primarily non-predator), that avoids niche competence between different life phases (Duellman and Trueb 1994). But other, more clade-restricted key innovations have also been linked to ecological opportunity and diversification. For example, the acquisition of new reproductive modes that facilitates ecological shifts between aquatic and terrestrial environments, such as direct development or the use of foam nests, can be seen as innovations that exposed groups to environments such as dry open areas (Pereira et al. 2007, Dugo-Cota et al. 2019).

Non-adaptive Diversification

All previous scenarios involve a significant action of natural selection at driving diversification. However, high rates of species diversification can also occur through purely non-adaptive processes, for example through allopatric speciation and neutral processes (Gittenberger 1991). Amphibians contain a flagship example of such scenario: the North American woodland salamanders (genus *Plethodon*). This group diversified at a great speed producing almost a new species every Ma during the last 8–11 Ma (Kozak et al. 2006). This group is also characterized by a great ecological and phenotypic conservatism. In this case, the high degree of niche specialization, coupled with habitat fragmentation, promoted the isolation of ecologically similar populations that slowly diverged and speciated through neutral processes (Kozak et al. 2006). A similar scenario likely occurs in the diversification of the highly phenotypically and ecologically similar species in the *Bufo bufo* species complex, which mainly diversified through climate-driven allopatric barriers (Garcia-Porta et al. 2012).

Dedication

We dedicate this chapter to the memory of late Professor David B. Wake for his invaluable contribution to the fields of herpetology and evolutionary biology.

Acknowledgements

JGP was supported by the postdoctoral fellowship program Beatriu de Pinós (2020 BP 00147, Government of Catalonia).

References

Amemiya, C.T., J. Alfoldi, A.P. Lee, S. Fan, H. Philippe, I. MacCallum et al. 2013. The African coelacanth genome provides insights into tetrapod evolution. Nature 496: 311–316.

AmphibiaWeb. 2021. AmphibiaWeb: Information on amphibian biology and conservation [web application]. Available at https://amphibiaweb.org/. University of California, Berkeley, CA, USA (Accessed: September 9, 2021).

Anderson, J.S. 2008. Focal review: The origin(s) of modern amphibians. Evol. Biol. 35: 231–247.

Anderson, J.S., R.R. Reisz, D. Scott, N.B. Fröbisch and S.S. Sumida 2008. A stem batrachian from the Early Permian of Texas and the origin of frogs and salamanders. Nature 453: 515–518.

Benton, M.J. 1990. Phylogeny of the major tetrapod groups: Morphological data and divergence dates. J. Mol. Evol. 30: 409–424.

Benton, M.J. and R.J. Twitchett. 2003. How to kill (almost) all life: the end-Permian extinction event. Trends Ecol. Evol. 18: 358–365.

Benton, M.J. 2005. Vertebrate Palaeontology. Blackwell Publishing Ltd, Malden, MA, USA.

Blackburn, D.C., C.D. Siler, A.C. Diesmos, J.A. McGuire, D.C. Cannatella and R.M. Brown. 2013. An adaptive radiation of frogs in a Southeast Asian island archipelago. Evolution 67: 2631–2646.

Bolt, J.R. 1991. Lissamphibian origins. pp. 194–222. *In*: Schultze, H.P. and L. Trueb [eds.]. Origins of the Major Groups of Tetrapods: Controversies and Consensus. Cornell University Press, Ithaca, NY, USA.

Carroll, R.L. 2001. The origin and early radiation of terrestrial vertebrates. J. Paleont. 75: 1202–1213.

Carroll, R.L., C. Boisvert, J. Bolt, D.M. Green, N. Philip, C. Rolian et al. 2004. Changing patterns of ontogeny from osteolepiform fish through Permian tetrapods as a guide to the early evolution of land vertebrates. pp. 321–343. *In*: Arratia, G., M.V.H. Wilson and R. Cloutier [eds.]. Recent Advances in the Origin and Early Radiation of Vertebrates. Pfeil, München, Germany.

Carroll, R.L. 2007. The Palaeozoic ancestry of salamanders, frogs and caecilians. Zool. J. Linn. Soc. 150: 1–140.

Carroll, R.L. 2009. The Rise of Amphibians: 365 Million Years of Evolution. Johns Hopkins University Press, Baltimore, MD, USA.

Clack, J.A. 2012. Gaining Ground: the Origin and Evolution of Tetrapods. Indiana University Press, Bloomington, IN, USA.

Duellman, W.E. and L. Trueb. 1994. Biology of Amphibians. Johns Hopkins University Press, Baltimore, MD, USA.

Dugo-Cota, Á., C. Vilà, A. Rodríguez and A. Gonzalez-Voyer. 2019. Ecomorphological convergence in *Eleutherodactylus* frogs: a case of replicate radiations in the Caribbean. Ecol. Lett. 22: 884–893.

Evans, S.E., A.R. Milner and F. Mussett. 1988. The earliest known salamanders (Amphibia: Caudata): a record from the middle Jurassic of England. Geobios 21: 539–552.

Feller, A.E. and S.B. Hedges. 1998. Molecular evidence for the early history of living amphibians. Mol. Phylogenet. Evol. 9: 509–516.

Feng, Y.J., D.C. Blackburn, D. Liang, D.M. Hillis, D.B. Wake, D.C. Cannatella et al. 2017. Phylogenomics reveals rapid, simultaneous diversification of three major clades of Gondwanan frogs at the Cretaceous–Paleogene boundary. Proc. Natl. Acad. Sci. U. S. A. 114: E5864–E5870.

Fong, J.J., J.M. Brown, M.K. Fujita and B. Boussau. 2012. A phylogenomic approach to vertebrate phylogeny supports a turtle-archosaur affinity and a possible paraphyletic Lissamphibia. PLoS One 7: e48990.

Fortuny, J. and J.-S. Steyer. 2019. New insights into the evolution of temnospondyls. J. Iber. Geol. 45: 247–250.

Frost, D.R., T. Grant, J. Faivovich, R.H. Bain, A. Haas, C.F.B. Haddad et al. 2006. The amphibian tree of life. Bull. Am. Mus. Nat. His. 297: 1–370.

Frost, D.R. 2021. Amphibian Species of the World: an Online Reference. Version 6.1 (9 September, 2021). Electronic Database accessible at https://amphibiansoftheworld.amnh.org/index.php. American Museum of Natural History, New York, USA.

Garcia-Porta, J., S.N. Litvinchuk, P.A. Crochet, A. Romano, P.H. Geniez, M. Lo-Valvo et al. 2012. Molecular phylogenetics and historical biogeography of the west-palearctic common toads (*Bufo bufo* species complex). Mol. Phylogenet. Evol. 63: 113–130.

Gittenberger, E. 1991. What about non-adaptive radiation? Biol. J. Linn. Soc. 43: 263–272.

Gittenberger, E. 2004. Radiation and adaptation, evolutionary biology and semantics. Org. Divers. Evol. 4: 135–136.

Hime, P.M., A.R. Lemmon, E.C. Moriarty Lemmon, E. Prendini, J.M. Brown, R.C. Thomson et al. 2021. Phylogenomics reveals ancient gene tree discordance in the amphibian tree of life. Syst. Biol. 70: 49–66.

Irisarri, I. and A. Meyer. 2016. The identification of the closest living relative(s) of Tetrapods: Phylogenomic lessons for resolving short ancient internodes. Syst. Biol. 65: 1057–1075.

Irisarri, I., D. Baurain, H. Brinkmann, F. Delsuc, J.Y. Sire, A. Kupfer et al. 2017. Phylotranscriptomic consolidation of the jawed vertebrate timetree. Nat. Ecol. Evol. 1: 1370–1378.

Jenkins, F.A. and D.M. Walsh. 1993. An early Jurassic caecilians with limbs. Nature 365: 246–249.

Jenkins, F.A., D.M. Walsh and R.L. Carroll. 2007. Anatomy of *Eocaecilia micropodia*, a limbed caecilian of the Early Jurassic. Bull. Mus. Comp. Zool. 158: 285–366.

Jetz, W. and R.A. Pyron. 2018. The interplay of past diversification and evolutionary isolation with present imperilment across the amphibian tree of life. Nat. Ecol. Evol. 2: 850–858.

Kozak, K.H., D.W. Weisrock and A. Larson. 2006. Rapid lineage accumulation in a non-adaptive radiation: phylogenetic analysis of diversification rates in eastern North American woodland salamanders (Plethodontidae: *Plethodon*). Proc. R. Soc. B Biol. Sci. 273: 539–546.

Ksepka, D.T., J.F. Parham, J.F. Allman, M.J. Benton, M.T. Carrano, K.A. Cranston et al. 2015. The fossil calibration database-A new resource for divergence dating. Syst. Biol. 64: 853–859.

Laurin, M. and R. Reisz. 1997. A new perspective on tetrapod phylogeny. pp. 9–59. *In*: Sumida, S.S. and K.L. Martin [eds.]. Amniote Origins—Completing the Transition to Land. Academic Press, New York, USA.

Laurin, M. 1998. The importance of global parsimony and historical bias in understanding tetrapod evolution. Part I. Systematics, middle ear evolution and jaw suspension. Ann. Sci. Nat. Zool. Biol. Anim. 13eme Série. 19: 1–42.

Laurin, M. 2010. How Vertebrates Left the Water. University of California Press, Berkeley and Los Angeles, CA, USA.

Losos, J.B. 2009. Lizards in an Evolutionary Tree: Ecology and Adaptive Radiation of Anoles. University of California Press, Berkeley and Los Angeles, CA, USA.

Losos, J.B. and R.E. Ricklefs. 2009. Adaptation and diversification on islands. Nature 457: 830–836.

Maddin, H.C., A.P. Russell and J.S. Anderson. 2012. Phylogenetic implications of the morphology of the braincase of caecilian amphibians (Gymnophiona). Zool. J. Linn. Soc. 166: 160–201.

Mahler, D.L., M.G. Weber, C.E. Wagner and T. Ingram. 2017. Pattern and process in the comparative study of convergent evolution. Am. Nat. 190: S13–S28.

Marjanović, D. and M. Laurin. 2013. The origin(s) of extant amphibians: A review with emphasis on the "lepospondyl hypothesis". Geodiversitas 35: 207–272.

Martin, C.H. and P.C. Wainwright. 2013. Multiple fitness peaks on the adaptive landscape drive adaptive radiation in the wild. Science 339: 208–211.

Meredith, R.W., J.E. Janečka, J. Gatesy, O.A. Ryder, C.A. Fisher, E.C. Teeling et al. 2011. Impacts of the Cretaceous Terrestrial Revolution and KPg extinction on mammal diversification. Science 334: 521–524.

Meyer, A., S. Schloissnig, P. Franchini, K. Du, J.M. Woltering, I. Irisarri et al. 2021. Giant lungfish genome elucidates the conquest of land by vertebrates. Nature 590: 284–289.

Milner, A.R. 1988. The relationships and origin of living amphibians. pp. 59–102. *In*: Benton, M.J. [ed.]. The Phylogeny and Classification of the Tetrapods. Clarendon Press, Oxford, UK.

Milner, A.R. 1993. The Paleozoic relatives of lissamphibians. Herpetol. Monogr. 7: 8–27.

Nosil, P. 2012. Ecological Speciation. Oxford University Press, Oxford, UK.

Panchen, A.L. and T.R. Smithson. 1987. Character diagnosis, fossils and the origin of tetrapods. Biol. Rev. 62: 341–438.

Pardo, J.D., B.J. Small and A.K. Huttenlocker. 2017. Stem caecilian from the Triassic of Colorado sheds light on the origins of Lissamphibia. Proc. Natl. Acad. Sci. U. S. A. 114: E5389–E5395.

Pereira, S.L., K.P. Johnson, D.H. Clayton and A.J. Baker. 2007. Mitochondrial and nuclear DNA sequences support a Cretaceous origin of Columbiformes and a dispersal-driven radiation in the Paleogene. Syst. Biol. 56: 656–672.

Pyron, R.A. 2011. Divergence time estimation using fossils as terminal taxa and the origins of lissamphibia. Syst. Biol. 60: 466–481.

Pyron, R.A. and J.J. Wiens. 2011. A large-scale phylogeny of Amphibia including over 2,800 species, and a revised classification of extant frogs, salamanders, and caecilians. Mol. Phylogenet. Evol. 61: 543–583.

Rage, J. and Z. Rocek. 1989. Redescription of *Triadobatrachus massinoti* (Piveteau, 1936) an anuran amphibian from the early Triassic. Paleontogr. Abt. A 206: 1–16.

Roelants, K. and F. Bossuyt. 2005. Archaeobatrachian paraphyly and pangaean diversification of crown-group frogs. Syst. Biol. 54: 111–126.

Roelants, K., D.J. Gower, M. Wilkinson, S.P. Loader, S.D. Biju, K. Guillaume et al. 2007. Global patterns of diversification in the history of modern amphibians. Proc. Natl. Acad. Sci. U. S. A. 104: 887–892.

Ruta, M., M.I. Coates and D.L.J. Quicke. 2003. Early tetrapod relationships revisited. Biol. Rev. 78: 251–345.

Ruta, M. and M.I. Coates. 2007. Dates, nodes and character conflict: addressing the lissamphibian origin problem. J. Syst. Palaeontol. 5: 69–122.

Sakamoto, M., M.J. Benton and C. Venditti. 2016. Dinosaurs in decline tens of millions of years before their final extinction. Proc. Natl. Acad. Sci. 113: 5036–5040.

San Mauro, D., M. Vences, M. Alcobendas, R. Zardoya and A. Meyer. 2005. Initial diversification of living amphibians predated the breakup of Pangaea. Am. Nat. 165: 590–599.

San Mauro, D. 2010. A multilocus timescale for the origin of extant amphibians. Mol. Phylogenet. Evol. 56: 554–561.

San Mauro, D., D.J. Gower, H. Müller, S.P. Loader, R. Zardoya, R.A. Nussbaum et al. 2014. Life-history evolution and mitogenomic phylogeny of caecilian amphibians. Mol. Phylogenet. Evol. 73: 177–189.

Santos, R.O., M. Laurin and H. Zaher. 2020. A review of the fossil record of caecilians (Lissamphibia: Gymnophionomorpha) with comments on its use to calibrate molecular timetrees. Biol. J. Linn. Soc. 131: 737–755.

Schluter, D. 2000. The Ecology of Adaptive Radiation. Oxford University Press, Oxford, UK.

Schoch, R.R. 2019. The putative lissamphibian stem-group: Phylogeny and evolution of the dissorophoid temnospondyls. J. Paleontol. 93: 137–156.

Schoch, R.R., R. Werneburg and S. Voigt. 2020. A Triassic stem-salamander from Kyrgyzstan and the origin of salamanders. Proc. Natl. Acad. Sci. U. S. A. 117: 11584–11588.

Sigurdsen, T. and D.M. Green. 2011. The origin of modern amphibians: a re-evaluation. Zool. J. Linn. Soc. 162: 457–469.

Simpson, G.G. 1944. Tempo and Mode in Evolution. Columbia University Press, New York, USA.

Simpson, G.G. 1955. Major Features of Evolution. Columbia University Press, New York, USA.

Siu-Ting, K., M. Torres-Sanchez, D. San Mauro, D. Wilcockson, M. Wilkinson, D. Pisani et al. 2019. Inadvertent paralog inclusion drives artifactual topologies and timetree estimates in phylogenomics. Mol. Biol. Evol. 36: 1344–1356.

Skutschas, P. and T. Martin. 2011. Cranial anatomy of the stem salamander *Kokartus honorarius* (Amphibia: Caudata) from the Middle Jurassic of Kyrgyzstan. Zool. J. Linn. Soc. 161: 816–838.

Tarailo, D.A. 2018. Taxonomic and ecomorphological diversity of temnospondyl amphibians across the Permian-Triassic boundary in the Karoo Basin (South Africa). J. Morphol. 279: 1840–1848.

Trueb, L. and R. Cloutier. 1991. A phylogenetic investigation of the inter- and intrarelationships of the Lissamphibia (Amphibia: Temnospondyli). pp. 223–313. *In*: Schultze, H.P. and L. Trueb [eds.]. Origins of the Major Groups of Tetrapods: Controversies and Consensus. Cornell University Press, Ithaca, NY, USA.

Valero, K.C.W., J. Garcia-Porta, A. Rodríguez, M. Arias, A. Shah, R.D. Randrianiaina et al. 2017. Transcriptomic and macroevolutionary evidence for phenotypic uncoupling between frog life history phases. Nat. Commun. 8: 15213.

Vallin, G. and M. Laurin. 2004. Cranial morphology and affinities of *Microbrachis*, and a reappraisal of the phylogeny and lifestyle of the first amphibians. J. Vert. Paleon. 24: 56–72.

Van Bocxlaer, I., S.P. Loader, K. Roelants, S.D. Biju, M. Menegon and F. Bossuyt. 2010. Gradual adaptation toward a range-expansion phenotype initiated the global radiation of toads. Science 327: 679–682.

Wright, S. 1932. The roles of mutation, inbreeding, crossbreeding, and selection in evolution. Proceedings of the Sixth International Congress on Genetics 1: 356–366.

Zardoya, R. and A. Meyer. 2001. On the origin of and phylogenetic relationships among living amphibians. Proc. Natl. Acad. Sci. U. S. A. 98: 7380–7383.

Zhang, P., Y.Q. Chen, H. Zhou, X.L. Wang and L.H. Qu. 2003. The complete mitochondrial genome of a relic salamander, *Ranodon sibiricus* (Amphibia: Caudata) and implications for amphibian phylogeny. Mol. Phylogenet. Evol. 28: 620–626.

Zhang, P., H. Zhou, Y.-Q. Chen, Y.-F. Liu and L.-H. Qu. 2005. Mitogenomic perspectives on the origin and phylogeny of living amphibians. Syst. Biol. 54: 391–400.

Zhang, P. and D.B. Wake. 2009. Higher-level salamander relationships and divergence dates inferred from complete mitochondrial genomes. Mol. Phylogenet. Evol. 53: 492–508.

Zhang, P., D. Liang, R.L. Mao, D.M. Hillis, D.B. Wake and D.C. Cannatella. 2013. Efficient sequencing of anuran mtDNAs and a mitogenomic exploration of the phylogeny and evolution of frogs. Mol. Biol. Evol. 30: 1899–1915.

CHAPTER 3

Amphibian Ecophysiology

Pablo Burraco[1,2,]* and *Caitlin Gabor*[3]

INTRODUCTION

The plastic ability to cope with environmental variation is ubiquitous across taxa and resides in the necessity of individuals to survive and reproduce under heterogenous conditions. Ectothermic organisms have a scarce capacity to metabolically maintain their body temperature and often show low vagility and high philopatry, thus plasticity can represent an essential strategy for them to face local environmental shifts (Seebacher et al. 2015). Plastic responses are often regulated by physiological pathways. Changes in the expression of these pathways can just include adaptive molecular alterations with no apparent phenotypic effects, but also can induce phenotypic variation (e.g., behavioural and morphological antipredator responses; Levis and Pfennig 2020). Unfortunately, the adaptiveness and evolvability of plastic responses in ectotherms can be greatly constrained by the *Anthropocene*, characterized by a faster pace of environmental change than previous Quaternary Epochs' conditions (Gunderson and Stillman 2015, Otto 2018). The study of the physiology underlying the maintenance, detection, and development of adaptive plasticity is therefore essential to understand eco-evolutionary dynamics, and for improving biodiversity conservation.

Among vertebrates, amphibians have the highest proportion of endangered species. Although the causes linked to the decline of amphibians are complex, they are mostly linked to the impact of human activities, such as habitat loss and climate change (Blaustein et al. 2011, Howard and Bickford 2014). The understanding of the physiological causes and consequences linked to amphibians global decline may provide key knowledge to develop effective conservation actions (Hayes et al. 2010). From the last fifty years, developmental biologists have studied amphibians to understand the physiological regulation of morphogenesis and tissue remodelling

[1] Institute of Biodiversity, Animal Health and Comparative Medicine, College of Medical, Veterinary and Life Sciences, University of Glasgow, Glasgow G12 8QQ, UK.
[2] Doñana Biological Station, Spanish National Research Council (EBD-CSIC), Seville 41092, Spain.
[3] Department of Biology, Texas State University, San Marcos, TX, 78666, USA.
* Corresponding author: burraco@ebd.csic.es

which they undergo from embryonic to juvenile stages through metamorphosis (Dickhoff et al. 1990, Kloas 2002, White and Nicoll 2013). This knowledge has also allowed both evolutionary and conservation ecologists to identify the mechanisms involved in developmental, growth, morphological, or behavioural responses to environmental variation (Hayes 1997, Moore et al. 2005, Wilczynski et al. 2005, Walls and Gabor 2019, Denver 2021). Likewise, the study of amphibian thermal physiology has allowed for better understanding the role of acclimation processes to temperature not only at individual levels but also from a comparative perspective, which have implications for the conservation of amphibians and other ectothermic species (Navas et al. 2008, Gerick et al. 2014, Novarro et al. 2018; also see Chapter 6 in this volume). However, further effort is definitely needed to fully understand the machinery underlying amphibian responses to environmental changes, a knowledge that will improve predictions on their resilience and evolvability under future scenarios.

In this chapter, we first discuss the main physiological pathways regulating developmental plasticity in amphibians, and the consequences for some life-history traits and health biomarkers of altering those pathways. Next, we comment on the effects on amphibian physiology of facing some environmental conditions. We also discuss the recent comparative approaches investigating the variation of some physiological mechanisms across amphibian species. Finally, we indicate some possible directions and approaches for future amphibian eco-evolutionary studies.

The Physiological Plasticity of Metamorphosis Timing: A Way to Cope with Environmental Stress

Istock (1967) hypothesized that "the evolutionary adaptations of the different phases of a complex life cycle are independent" and that such independence "is likely to make complex life cycles generally unstable over evolutionary time", thus complex-life cycles "will generate selective forces favouring a reduction or loss of one phase or the other". However, despite such expected evolutionary instability of life cycles with two or more ecologically distinct phases, aquatic eggs hatching as aquatic larva metamorphosing to terrestrial subadults/adults is the most frequent life history among anuran species (Gomez-Mestre et al. 2012). Also, complex-life cycles represent the ancestral state for anurans, whereas direct development has often evolved from that initial state (Gomez-Mestre et al. 2012). These findings highlight the ecological and evolutionary role of metamorphosis in anurans across their evolutionary history, indicated also by the fact that species with complex-life cycles have survived over millions of years. As discussed below, the success of this life history strategy is likely associated with the plastic capacity of most amphibians to shape the timing to and size at metamorphosis, despite possible carry over-effects on post-metamorphosis stages (Denver 2021).

Metamorphosis is a process dependent on the production, by the thyroid gland, of the precursor hormone thyroxine T4, a molecule that is transformed into T3, a more active derivate that acts on peripheral organs (Laudet 2011, Denver 2013). Thyroid hormones bind to high-affinity thyroid hormone receptors and activate a downstream signalling pathway (Laudet 2011). This cascade will finally involve the transcription of target genes that lead to the morphogenesis required to complete metamorphosis. In addition to the role of thyroid hormones, the glucocorticoid (GC) corticosterone, synergizes with thyroid hormones to promote tissue transformation (Denver 2021). The hypothalamic-pituitary-interrenal (HPI) axis releases the corticotropin-release factor that finally modulates the production of thyroid hormones and corticosterone. The activation of the HPI axis is shaped by environmental stimuli, and harsh conditions often upregulate the production of those hormones. If the activation of the HPI-axis takes place when the larva has achieved a minimum body size for transformation, development is promoted and metamorphosis is accelerated (Denver 2009, 2021). Amphibian larvae can often upregulate HPI-axis activity and then accelerate development in response to a broad number of environmental conditions such as starvation, pond drying or intra-specific competition (Glennemeier and Denver 2002, Crespi and Denver 2005, Gomez-Mestre et al. 2013, Burraco and Gomez-Mestre 2016, Charbonnier et al. 2018). Both embryo and larvae often show a great sensitiveness to environmental variation; for example, tadpoles can detect and respond to small variations in the level of a water body (Denver et al. 1998). However, adaptive developmental plasticity is often highly variable between genotypes, a pattern that might be linked to the existence of maintenance or production costs constraining or limiting such plastic ability (Relyea 2002, Denver 2009, Burraco et al. 2021, 2022a).

Developmental plasticity can involve trade-offs in amphibians. Individuals developing faster in response to stress tend to show reduced body size and fat reserves at metamorphosis (Denver 1997, Burraco et al. 2017), two traits that negatively correlate with survival at later stages (Scott et al. 2007, Cabrera-Guzmán et al. 2013; but see Earl and Whiteman 2015). Trade-offs between development and growth are commonly observed in amphibians in response to a broad number of stressors (e.g., Bekhet et al. 2014, Dananay and Benard 2018, Welch et al. 2019), suggesting a compromise between achieving an optimal size at metamorphosis and the time that larvae are exposed to risks (Werner 1986). Moreover, the impact of environmental conditions on developmental plasticity can be stage dependent. At early larval stages, thyroid receptors have not been developed yet, and this likely explains why detrimental conditions at those stages can delay or even inhibit metamorphosis (Denver 1997, Grimaldi et al. 2013). Overall, these findings give insights into the mechanisms underlying the impact of environmental conditions on amphibian life-history traits. However, the understanding of possible detrimental long-term consequences linked to the plastic activation of the amphibian endocrine cascade still need further exploration as, alternatively, compensatory responses might make up early poor conditions at metamorphosis (e.g., Charbonnier et al. 2018).

Because developmental plasticity is largely present across amphibian phylogeny, benefits of plastically altering larval period are expected to be bigger than the costs. Developmental acceleration does not only have implications for life-history traits such as body size at metamorphosis, but also for physiological pathways such as oxidative stress or the immune system, which are often markers of individual performance or life expectancy. To exemplify this, a well-studied response is the developmental acceleration commonly experienced by amphibian larvae facing pond desiccation, metamorphosing at smaller sizes and with lower fat storage

(Newman 1992, Denver 2009, Burraco et al. 2017). This developmental response involves the upregulation of the HPI-axis (Denver 2009), which often incurs a physiological cascade involving elevated metabolism, redox imbalance, or decreased immune function at metamorphosis (Gervasi and Foufopoulos 2008, Gomez-Mestre et al. 2013). Oxidative stress can be understood as a mediator of life-history trait trade-offs (Monaghan et al. 2009). Likewise, a reduction in the immune capacity can involve higher disease susceptibility (Warne et al. 2011, Kohli et al. 2019; Chapters 4 and 5 in this volume) and might be contributing to the decline of amphibian populations facing droughts (McMenamin et al. 2008, O'Regan et al. 2014; reviewed by Griffis-Kyle 2016). Furthermore, costs in response to desiccation can persist beyond larval stages, and they can even negatively impact fecundity (Cayuela et al. 2016). Other environmental conditions that involve developmental acceleration in amphibian larvae can also imbalance the redox machinery and the immune system of amphibians such as in response to high salinity, warm temperatures, toxic substances, or as consequence of compensatory growth responses to detrimental conditions at early life (Mann et al. 2009, Burraco and Gomez-Mestre 2016, Burraco et al. 2018, 2020a, Murillo-Rincón et al. 2017, Hidalgo et al. 2020). Therefore, developmental plasticity can involve a physiological toll, however we still need to understand the complex long-term implications. The use of some health biomarkers is just emerging for amphibians, where further research might improve our current knowledge on their decline and our predictions on future broad-scale patterns through the projection of mechanistic models into geographical space (i.e., species distribution models).

Role of Glucocorticoids in Responding to Human Altered/ Disturbed Habitats

What is becoming clear is that, in and of itself, the relationship of more environmental stress and higher constitutive corticosterone levels are not necessarily indicators of lower population health in amphibians (MacDougall-Shackleton et al. 2019, Romero and Beattie 2021). Additionally, it is not well understood whether stronger or dampened responses to stressors indicate a more adaptive response (Wingfield and Sapolsky 2003). In two literature reviews, there was no consensus as to why some species show higher stress response in urban populations compared to non-urban conspecifics and other species show the lower stress response or no difference between populations (Murray et al. 2019, Injaian et al. 2020). What is missing from these studies is data from the entire reactive scope (in addition to a focus on ectotherms). The reactive scope model of stress suggests that the longer the duration of an acute stressor and repeated prolonged stressors can induce phenotypic damage (Romero et al. 2009). Further, an efficient negative feedback along the HPI axis is important to return the GC levels to baseline (Romero et al. 2009), where shorter duration of the acute stress response could allow for a strong (and potentially adaptive) stress response while decreasing the damage (Vitousek et al. 2019).

Because GCs are an important component of life-history traits in amphibians, where high levels can be detrimental to early life growth and development (Denver 2009), amphibians could protect themselves from chronically elevated GCs by quickly shutting down the stress response (Wingfield 2013). Therefore, evaluating glucocorticoid flexibility in terms of how quickly the stress response is turned on or off or whether the species upregulates vs downregulates corticosterone in response to stressors could provide a greater understanding of how, or if, populations can cope with anthropogenic environmental change. Interestingly, Bókony et al. (2021) showed that tadpoles of the common toad (*Bufo bufo*) from anthropogenically modified habitats not only showed higher corticosterone release rates (upregulated) in response to stress, but they also had stronger negative feedback (quickly turned off), which may provide a mechanism for coping with anthropogenic change. Follow up studies exploring the relationship between aspects of the GC profile and fitness are still needed to fully explore the fitness benefits and or trade-offs involved in endocrine flexibility. Additionally, exploring whether the activation of endocrine coping mechanisms incurs in trans-generational effects (Donelan et al. 2020) would help understand the potential for amphibians to respond to rapid environmental change.

Whereas many studies have found that corticosterone levels are elevated in response to stressors, others are finding that the release of this hormone can be downregulated. For example, in response to the anthropogenic stressor of artificial light at night (ALAN), Forsburg et al. (2021) found that two species of anurans differentially modulated their corticosterone response. In one species, tadpoles of the Rio Grande leopard frog (*Rana berlandieri*) downregulated corticosterone release rates after exposure to pulsed lights at night over 14 days, whereas the Gulf coast toad (*Bufo valliceps*) upregulated corticosterone in these conditions. By downregulating corticosterone response to stressors instead of upregulating corticosterone (as found in many species) *R. berlandieri*, a species that is a slower developer, may acquire more time to grow before metamorphosing. Whereas, *B. valliceps* metamorphose quickly so may not benefit from downregulating corticosterone in response to the stressor. This study suggests that the hormonal regulation can depend on the species biology (i.e., developmental and growth rates) and that further studies across the amphibian phylogeny are needed. Further, follow-up studies are needed to evaluate if stronger negative feedback is adaptive in anthropogenic habitats as well as comparative studies are needed to examine if there are consistent patterns in the use of upregulation and downregulation of corticosterone in response to stressors that is associated with the life history of the amphibian species.

A particular example of the complexity of hormonal regulation in response to environmental conditions comes from research exploring the physiological anti-predatory responses in amphibian larvae. Middlemis Maher et al. (2013) described an increase in corticosterone levels in larvae exposed to predators, whereas other studies have found reductions in the levels of this hormone (Burraco et al. 2013, Burraco and Gomez-Mestre 2016) or even complex dynamics over exposure time or across populations (Dahl et al. 2012, Bennet et al. 2016). Variation in corticosterone levels in response to predators is likely explained by the role of this hormone in regulating behaviour and morphology (e.g., Hossie et al. 2010, Middlemis Maher et al. 2013), however, this association still needs further exploration as it might be species-dependent and/or shaped by other environmental conditions (Relyea 2004, Schoeppner and Relyea 2008, Cope et al. 2020).

Another area of research that has not been as well developed in amphibians is the link between physiology, the microbiome and environmental change. The microbial community can be shaped by environmental factors, disease, and the host. Microbial communities can in turn impact host health and ultimately their fitness (Warne et al. 2019). In amphibians, elevated corticosterone is also associated with disease (Warne et al. 2011, Gabor et al. 2013, 2015, 2018), and disease alters microbial communities which in turn also affects physiology (Warne et al. 2019). Indeed, ornate chorus frogs (*Pseudacris ornata*) tadpoles had elevated corticosterone release rates in lower quality environments (e.g., higher temperatures, more developed land). In turn, skin bacterial diversity varied with land cover characteristics (Goff et al. 2020). Such measures may aid in identification of at-risk populations and provide management implications for maintenance of high-quality environmental conditions as this may preserve bacterial diversity.

Insights from Divergent Physiological Evolution in Amphibians

Past and current habitat conditions can drive between-population responsiveness to environmental variation in amphibians. In this line, the proximity of amphibian populations to the sea or to agricultural systems correlates with their tolerance to salinity or pesticides (Cothran et al. 2013, Hua et al. 2015, Hopkins et al. 2016). Likewise, conditions such as temperature or salinity can involve local adaptation and explain geographic variation in amphibian populations (Räsänen et al. 2003, Merilä et al. 2004, Muir et al. 2014, Nowakowski et al. 2018). These evolutionary patterns are associated with divergence in the stress physiology machinery of amphibians and, in particular, to variation in corticosterone levels. In *Bufo bufo*, urban tadpoles have higher corticosterone levels, both constitutively and in response to induced stress, than tadpoles from agricultural or natural habitats (Bókony et al. 2021). In the same study, recovery rates of corticosterone levels were higher in tadpoles from agricultural and urban habitats, compared to natural habitats, which opens a new research avenue to understanding possible adaptation to stressors in amphibian populations inhabiting contrasting environments. In a different study, populations of the Jefferson's salamander (*Ambystoma jeffersoniaum*) inhabiting ponds within a gradient of pH and nitrate contamination, both constitutively and corticosterone levels in response to a stressor were associated to ponds with lower pH and higher nitrate levels (Chambers et al. 2013). Finally, in the rough-skinned newt (*Taricha granulosa*), populations living in close proximity to the ocean had lower corticosterone levels in response to salinity increase than populations inhabiting further from salt water, whereas baseline corticosterone levels did not differ between populations (Hopkins et al. 2016). These studies exemplify that both constitutive and plastic hormonal levels can evolve in response to divergent environmental conditions and give mechanistic insights to understand local adaptation and spatial distribution of populations.

The assessment of other mechanisms, including different health biomarkers mentioned above (microbiome, oxidative stress, or telomeres), can improve our understanding on the long-term consequences that divergent environmental conditions may have for individual and population resilience. The use of these biomarkers can be particularly useful in amphibian studies as long-term surveys are often not easily feasible. For example, populations exposed to divergent desiccation risk differ in their developmental plastic responses to pond drying, and populations under higher desiccation risk have higher developmental plasticity and shorter telomeres, suggesting physiological costs associated to the maintenance of developmental plasticity (Burraco et al. 2017). However, so far, most of the studies using these health biomarkers have been conducted on birds and mammals (Isaksson 2010, Olsson et al. 2018), thus we still need to validate the relevance of these parameters for amphibian health and fitness.

Comparative Studies to Explore the Role of Environmental Change on Amphibians

Recently, comparative studies are providing key knowledge to understand the role of environmental conditions on driving speciation and adaptation processes in amphibians. In this line, phylogenetic and trait-based predictions have allowed the estimation of species threatened with extinction, and to predict endangered or critically endangered species (González-del-Pliego et al. 2019). Also, large-scale phylogenetic analyses have revealed the role of climatic conditions on the evolution of niche specialization in amphibians (Bonetti and Wiens 2014), and have showed how higher speciation, lower extinction, and limited dispersal in the tropics than in temperate regions explain high species diversity in tropical amphibians (Pyron and Wiens 2013). Comparative approaches are also providing essential insights of physiological mechanisms underlying some evolutionary patterns in amphibians. The study of the variation in thermal physiological traits across amphibian species show that, despite the overall low evolvability of these traits, thermal physiology is associated to context-specific patterns that explain amphibian species divergence and distribution (Bonino et al. 2020, Bodensteiner et al. 2021). From a molecular perspective, climatic conditions can involve divergence in life histories but can also drive changes at the genome level, as denoted by an association between genome size and environmental temperature and humidity across anuran phylogeny, likely due to the effect of these environmental conditions on pre-metamorphic stages (Liedtke et al. 2018). At a narrower scale, comparative skin microbiome analyses from 89 Madagascar amphibian species showed that both host and environmental factors drive amphibian cutaneous communities, providing mechanistic information to understand divergence in disease resistance (Bletz et al. 2017). Finally, comparative transcriptomic analyses are helping to understand the genetic basis of speciation, particularly, understanding the pathways underlying fossorial adaptations of caecilians to their extreme environments (Torres-Sánchez et al. 2019).

Comparative experimental approaches testing for divergence in physiological mechanisms in response to specific environmental conditions can provide essential mechanistic information to understand speciation and adaptation in amphibians. The number of these studies is still limited, likely because they require a great sampling effort plus the conduction of experiments including physiological assessments across several species. Some studies have experimentally tested for among-species differences in the response to biotic and abiotic conditions such as temperature or the exposure to *Batrachochytrium dendrobatidis* (Searle et al. 2011, Duarte et al. 2012), representing an essential starting point to identify specific causes and consequences of amphibian speciation or decline. In studies investigating the physiology underlying a particular evolutionary process, the reconstruction of the ancestral state is essential. In larvae of spadefoot toad species, developmental plasticity is the ancestral state of reaction norms (Gomez-Mestre and Buchholz 2006). Among three species of spadefoot toads that show a marked divergence in their larval period, pond duration seems to be the main factor explaining such among-species differences in timing of metamorphosis (Kulkarni et al. 2011). Furthermore, the canalised development observed in one of these three species of spadefoot toads (*Scaphiopus couchii*) has evolved by genetic accommodation of endocrine pathways controlling metamorphosis (Kulkarni et al. 2017). The latter example provides evidence for a link between physiological mechanisms, adaptive plasticity and species divergence. Another comparative approach is to experimentally test differences in physiological response within a widespread species as this may provide insights into how and whether populations can adapt to environmental change and or disease. Novarro et al. (2018) explored how the red-backed salamander (*Plethodon cinereus*) copes with thermal heterogeneity associated with warming climates. In this study raising the ambient temperature elevated corticosterone release rates across populations, but the effects differed among sites with rates higher in the warmest most southern population. This example suggests that some populations may be more resilient to environmental change because the southernmost population altered their foraging rate which compensated for increased metabolic demands. Further experimental approaches are still needed to understand the mechanisms underlying adaptation and speciation processes across the amphibian phylogeny. These experiments should, ideally, include realistic scenarios in which several factors interact simultaneously along with measuring multiple physiological metrics.

Future Directions for Amphibian Ecophysiology

Ultimately, studies are needed to determine if natural selection favours, for example, a single optimal hormonal response to the environment or if corticosterone stress responses and recovery rates are flexible, meaning not only individuals respond by changing their physiology in different environments but individuals differ in their response consistently indicating a genetic basis to their response. Additionally, we lack long term studies in amphibians that ultimately link altered fitness to physiological responses associated with anthropogenic change. Such questions may work well in Head Start programs in zoos because they are raising multiple generations of amphibians (Harding et al. 2016). These programs allow for studies both in the lab and follow-up field approaches using sufficient sample size to obtain the power needed to understand the implications of multiple environmental stressors on physiology of declining amphibian species (Fig. 1). Additionally, it is possible to compare the physiological state/health of individuals that were captive raised vs wild animal brought into the lab to establish captive populations. Rearing wild caught eggs to tadpoles in the lab would essentially be a common garden experiment that

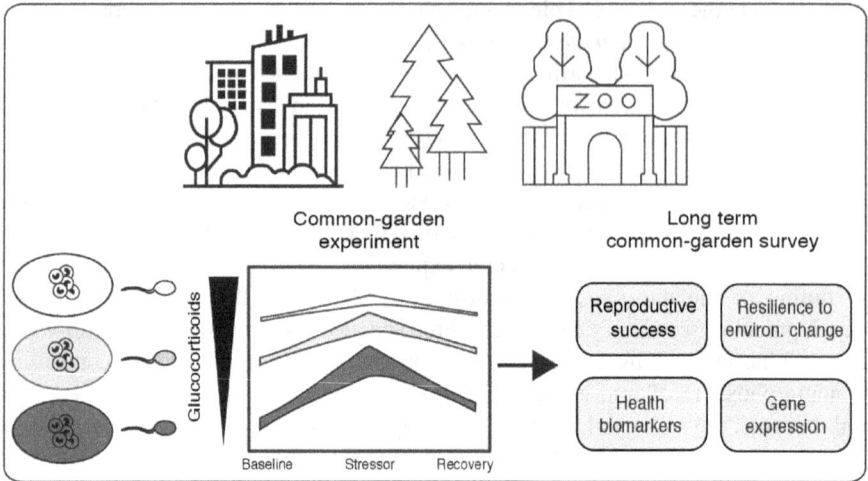

Figure 1. Common garden experiments can allow understanding the effect of genetic and environmental variation on amphibian physiology. In particular, long-term experiments are needed to understand the consequences for fitness of the onset of stress physiology mechanisms in response to environmental change. For example, these experiments would allow investigating whether differences in glucocorticoids release and recovery rates in response to environmental change have later consequences for individual performance. The use of captive animal facilities, and, specifically, Head Start programs in zoos, can contribute to the study of possible long-lasting effects (i.e., during a lifetime or across generations) of such endocrine activation on lifetime reproductive outcome, physiological biomarkers, gene expression, or resilience of individuals from populations facing contrasting environments.

can provide insights into fixed vs plastic traits. Wild caught, captive raised eggs could provide a point of comparisons with eggs that come from captive animals and raised in the lab to compare reproductive output and overall health. To balance out the work with endangered species, it could be helpful to tie this work in with studies using "tolerant" species to evaluate how the physiological responses of endangered species defer in response to rapid anthropogenic changes compared to closely related more tolerant species. However, we acknowledge that these approaches might be challenging, especially for amphibians with aquatic and terrestrial phases that often require conditions not easily achievable in animal facilities or laboratories. The latter can clearly limit long-term experiments, and likely explain the limited number of studies addressing fitness consequences of conditions experienced at early life.

Another gap in amphibian studies are the links between environmental change, physiology and gene expression. Studying gene expression in relation to exposure to environmental change will provide insights into mechanisms underlying organismal response to change (e.g., Liedtke et al. 2021, Levis et al. 2021), and these studies could provide insights into sublethal thresholds that eventually could have negative consequences to fitness (Connon et al. 2018). Transcriptome screening has been used to examine early signs of disease and other stressors in fish and can thus provide early warning biomarker (Jeffries et al. 2016, Bass et al. 2017). Putting these different methods together will provide the best ways of exploring the effects and mitigating the damage from environmental change. Similarly, ageing markers such as telomere length estimates provide useful insights because they are an integrative marker of the conditions experienced by an organism during its lifetime, and telomere length/shortening often correlate with important life-history traits like life expectancy or reproductive outcome (but further info is needed, especially for ectotherms; Olsson et al. 2018, Burraco et al. 2022b). Finally, the development of sequencing technologies, together with the improved knowledge on amphibian husbandry and physiology, will greatly improve our understanding on environmental conditions and amphibian performance, both at the individual and transgenerational level.

References

Bass, A.L., S.G. Hinch, A.K. Teffer, D.A. Patterson and K.M. Miller. 2017. A survey of microparasites present in adult migrating Chinook salmon (*Oncorhynchus tshawytscha*) in south-western British Columbia determined by high-throughput quantitative polymerase chain reaction. J. Fish Dis. 40: 453–477.

Bekhet, G.A., H.A. Abdou, S.A. Dekinesh, H.A. Hussein and S.S. Sebiae. 2014. Biological factors controlling developmental duration, growth and metamorphosis of the larval green toad, *Bufo viridis viridis*. J. Basic Appl. Zool. 67: 67–82.

Bennet, A.M., J.N. Longhi, E.H. Chin, G.P. Burness, L.R. Kerr and D.L. Murray. 2016. Acute changes in whole body corticosterone in response to perceived predation risk: A mechanism for anti-predator behavior in anurans? Gen. Comp. Endocrinol. 229: 62–66.

Blaustein, A.R., B.A. Han, R.A. Relyea, P.T. Johnson, J.C. Buck, S.S. Gervasi et al. 2011. The complexity of amphibian population declines: understanding the role of cofactors in driving amphibian losses. Annals New York Acad. Sci. 1223: 108–119.

Bletz, M.C., H. Archer, R.N. Harris, V.J. McKenzie, F.C. Rabemananjara, A. Rakotoarison et al. 2017. Host ecology rather than host phylogeny drives amphibian skin microbial community structure in the biodiversity hotspot of Madagascar. Front. Microbiol. 8: 1530.

Bodensteiner, B.L., G.A. Agudelo-Cantero, A.Z.A. Arietta, A.R. Gunderson, M.M. Muñoz, J.M. Refsnider et al. 2021. Thermal adaptation revisited: How conserved are thermal traits of reptiles and amphibians? J. Exp. Zool. A 335: 173–194.

Bókony, V., N. Ujhegyi, K.Á. Hamow, J. Bosch, B. Thumsová, J. Vörös et al. 2021. Stressed tadpoles mount more efficient glucocorticoid negative feedback in anthropogenic habitats due to phenotypic plasticity. Sci. Total Environ. 753: 141896.

Bonetti, M.F. and J.J. Wiens. 2014. Evolution of climatic niche specialization: a phylogenetic analysis in amphibians. Proc. R. Soc. B 281(1795): 20133229.

Bonino, M.F., F.B. Cruz and M.G. Perotti. 2020. Does temperature at local scale explain thermal biology patterns of temperate tadpoles? J. Thermal Biol. 94: 102744.

Burraco, P. and I. Gomez-Mestre. 2016. Physiological stress responses in amphibian larvae to multiple stressors reveal marked anthropogenic effects even below lethal levels. Physiol. Biochem. Zool. 89: 462–472.

Burraco, P., A. Laurila and G. Orizaola. 2021. Limits to compensatory responses to altered phenology in amphibian larvae. Oikos 130: 231–239.

Burraco, P., A.E. Valdés and G. Orizaola. 2020a. Metabolic costs of altered growth trajectories across life transitions in amphibians. J. Anim. Ecol. 89: 855–866.

Burraco, P., C. Díaz-Paniagua and I. Gomez-Mestre. 2017. Different effects of accelerated development and enhanced growth on oxidative stress and telomere shortening in amphibian larvae. Sci. Rep. 7: 7494.

Burraco, P., G. Orizaola, P. Monaghan and N.B. Metcalfe. 2020b. Climate change and ageing in ectotherms. Global Change Biol. 26: 5371–5381.

Burraco, P., L.J. Duarte and I. Gomez-Mestre. 2013. Predator-induced physiological responses in tadpoles challenged with herbicide pollution. Curr. Zool. 59: 475–484.

Burraco, P., M. Iglesias-Carrasco, C. Cabido and I. Gomez-Mestre. 2018. Eucalypt leaf litter impairs growth and development of amphibian larvae, inhibits their antipredator responses and alters their physiology. Conserv. Physiol. 6: coy066.

Burraco, P., M.A. Rendón, C. Díaz-Paniagua and I. Gomez-Mestre. 2022a. Maintenance of phenotypic plasticity is linked to oxidative stress in spadefoot toad larvae. Oikos 2022(5): e09078.

Burraco, P., P.M. Lucas and P. Salmón. 2022b. Telomeres in a spatial context: A tool for understanding ageing pattern variation in wild populations. Ecography 2022(6): e05565.

Cabrera-Guzmán, E., M.R. Crossland, G.P. Brown and R. Shine. 2013. Larger body size at metamorphosis enhances survival, growth and performance of young cane toads (*Rhinella marina*). PLoS ONE 8: e70121.

Cayuela, H., D. Arsovski, E. Bonnaire, R. Duguet, P. Joly and A. Besnard. 2016. The impact of severe drought on survival, fecundity, and population persistence in an endangered amphibian. Ecosphere 7: e01246.

Chambers, D.L., J.M. Wojdak, P. Du and L.K. Belden. 2013. Pond acidification may explain differences in corticosterone among salamander populations. Physiol. Biochem. Zool. 86: 224–232.

Charbonnier, J.F., J. Pearlmutter, J.R. Vonesh, C.R. Gabor, Z.R. Forsburg and K.L. Grayson. 2018. Cross-life stage effects of aquatic larval density and terrestrial moisture on growth and corticosterone in the spotted salamander. Diversity 10(3): 68.

Connon, R.E., K.M. Jeffries, L.M. Komoroske, A.E. Todgham and N.A. Fangue. 2018. The utility of transcriptomics in fish conservation. J. Exp. Biol. 221: jeb148833.

Cope, K.L., M.W. Schook and M.F. Benard. 2020. Exposure to artificial light at night during the larval stage has delayed effects on juvenile corticosterone concentration in American toads, *Anaxyrus americanus*. Gen. Comp. Endocrinol. 295: 113508.

Cothran, R.D., J.M. Brown and R.A. Relyea. 2013. Proximity to agriculture is correlated with pesticide tolerance: evidence for the evolution of amphibian resistance to modern pesticides. Evol. Appl. 6: 832–841.

Crespi, E.J. and R.J. Denver. 2005. Roles of stress hormones in food intake regulation in anuran amphibians throughout the life cycle. Comp. Biochem. Physiol. A 141: 381–390.

Dahl, E., G. Orizaola, S. Winberg and A. Laurila. 2012. Geographic variation in corticosterone response to chronic predator stress in tadpoles. J. Evol. Biol. 25: 1066–1076.

Dananay, K.L. and M.F. Benard. 2018. Artificial light at night decreases metamorphic duration and juvenile growth in a widespread amphibian. Proc. R. Soc. B 285: 20180367.

Denver, R.J., N. Mirhadi and M. Phillips. 1998. Adaptive plasticity in amphibian metamorphosis: Response of *Scaphiopus Hammondii* tadpoles to habitat desiccation. Ecology 79: 1859–1872.

Denver, R.J. 1997. Proximate mechanisms of phenotypic plasticity in amphibian metamorphosis. American Zoologist 37(2): 172–184.

Denver, R.J. 2009. Stress hormones mediate environment-genotype interactions during amphibian development. Gen. Comp. Endocrinol. 164: 20–31.

Denver, R.J. 2013. Neuroendocrinology of amphibian metamorphosis. Curr. Top. Develop. Biol. 103: 195–227.

Denver, R.J. 2021. Stress hormones mediate developmental plasticity in vertebrates with complex life cycles. Neurobiol. Stress 14: 100301.

Dickhoff, W.W., C.L. Brown, C.V. Sullivan and H.A. Bern. 1990. Fish and amphibian models for developmental endocrinology. J. Exp. Zool. 256(S4): 90–97.

Donelan, S.C., J.K. Hellmann, A.M. Bell, E. Luttbeg, J.L. Orrock, M.J. Sheriff et al. 2020. Transgenerational plasticity in human-altered environments. Trends Ecol. Evol. 35: 115–124.

Duarte, H., M. Tejedo, M. Katzenberger, F. Marangoni, D. Baldo, J.F. Beltrán et al. 2012. Can amphibians take the heat? Vulnerability to climate warming in subtropical and temperate larval amphibian communities. Global Change Biol. 18: 412–421.

Earl, J.E. and H.H. Whiteman. 2015. Are commonly used fitness predictors accurate? A meta-analysis of amphibian size and age at metamorphosis. Copeia 103: 297–309.

Forsburg, Z.R., A. Guzman and C.R. Gabor. 2021. Artificial light at night (ALAN) affects the stress physiology but not the behavior or growth of *Rana berlandieri* and *Bufo valliceps*. Environ. Pollut. 277: 116775.

Gabor, C.R., M.C. Fisher and J. Bosch. 2013. A non-invasive stress assay shows that tadpole populations infected with *Batrachochytrium dendrobatidis* have elevated corticosterone levels. PloS One, 8(2): e56054.

Gabor, C.R., M.C. Fisher and J. Bosch. 2015. Elevated corticosterone levels and changes in amphibian behavior are associated with *Batrachochytrium dendrobatidis* (Bd) infection and Bd lineage. PLoS One 10(4): e0122685.

Gabor, C.R., S.A. Knutie, E.A. Roznik and J.R. Rohr. 2018. Are the adverse effects of stressors on amphibians mediated by their effects on stress hormones? Oecologia 186(2): 393–404.

Gerick, A.A., R.G. Munshaw, W.J. Palen, S.A. Combes and S.M. O'Regan. 2014. Thermal physiology and species distribution models reveal climate vulnerability of temperate amphibians. J. Biogeogr. 41: 713–723.

Gervasi, S.S. and J. Foufopoulos. 2008. Costs of plasticity: responses to desiccation decrease post-metamorphic immune function in a pond-breeding amphibian. Funct. Ecol. 22: 100–108.

Glennemeier, K.A. and R.J. Denver. 2002. Small changes in whole-body corticosterone content affect larval Rana pipiens fitness components. General and Comparative Endocrinology 127(1): 16–25.

Goff, C.B., S.C. Walls, D. Rodriguez and C.R. Gabor. 2020. Changes in physiology and microbial diversity in larval ornate chorus frogs are associated with habitat quality. Conserv. Physiol. 8: coaa047.

Gomez-Mestre, I. and D.R. Buchholz. 2006. Developmental plasticity mirrors differences among taxa in spadefoot toads linking plasticity and diversity. Proc. Natl. Acad. Sci. USA 103: 19021–19026.

Gomez-Mestre, I., R.A. Pyron and J.J. Wiens. 2012. Phylogenetic analyses reveal unexpected patterns in the evolution of reproductive modes in frogs. Evolution 66: 3687–3700.

Gomez-Mestre, I., S. Kulkarni and D.R. Buchholz. 2013. Mechanisms and consequences of developmental acceleration in tadpoles responding to pond drying. PloS One 8(12): e84266.

González-del-Pliego, P., R.P. Freckleton, D.P. Edwards, M.S. Koo, B.R. Scheffers, R.A. Pyron and W. Jetz. 2019. Phylogenetic and trait-based prediction of extinction risk for data-deficient amphibians. Current Biology 29(9): 1557–1563.

Griffis-Kyle, K.L. 2016. Physiology and ecology to inform climate adaptation strategies for desert amphibians. Herpetol. Conserv. Biol. 11: 563–582.

Grimaldi, A., N. Buisine, T. Miller, Y.B. Shi and L.M. Sachs. 2013. Mechanisms of thyroid hormone receptor action during development: lessons from amphibian studies. Biochim. Biophys. Acta 1830: 3882–3892.

Gunderson, A.R. and J.H. Stillman. 2015. Plasticity in thermal tolerance has limited potential to buffer ectotherms from global warming. Proc. R. Soc. B 282: 20150401.

Harding, G., R.A. Griffiths and L. Pavajeau. 2016. Developments in amphibian captive breeding and reintroduction programs. Conserv. Biol. 30: 340–349.

Hayes, T.B. 1997. Amphibian metamorphosis: an integrative approach. Am. Zool. 37: 121–123.

Hayes, T.B., P. Falso, S. Gallipeau and M. Stice. 2010. The cause of global amphibian declines: a developmental endocrinologist's perspective. J. Exp. Biol. 213: 921–933.

Hidalgo, J., F. Álvarez-Vergara, I. Peña-Villalobos, C. Contreras-Ramos, J.C. Sanchez-Hernandez and P. Sabat. 2020. Effect of salinity acclimation on osmoregulation, oxidative stress, and metabolic enzymes in the invasive *Xenopus laevis*. J. Exp. Zool. A 333: 333–340.

Hopkins, G.R., E.D. Brodie, Jr., L.A. Neuman-Lee, S. Mohammadi, G.A. Brusch IV, Z.M. Hopkins et al. 2016. Physiological responses to salinity vary with proximity to the ocean in a coastal amphibian. Physiol. Biochem. Zool. 89: 322–330.

Hossie, T.J., B. Ferland-Raymond, G. Burness and D.L. Murray. 2010. Morphological and behavioural responses of frog tadpoles to perceived predation risk: a possible role for corticosterone mediation? Ecoscience 17(1): 100–108.

Howard, S.D. and D.P. Bickford. 2014. Amphibians over the edge: silent extinction risk of Data Deficient species. Divers. Distrib. 20: 837–846.

Hua, J., D.K. Jones, B.M. Mattes, R.D. Cothran, R.A. Relyea and J.T. Hoverman. 2015. The contribution of phenotypic plasticity to the evolution of insecticide tolerance in amphibian populations. Evol. Appl. 8: 586–596.

Injaian, A.S., C.D. Francis, J.Q. Ouyang, D.M. Dominoni, J.W. Donald, M.J. Fuxjager et al. 2020. Baseline and stress-induced corticosterone levels across birds and reptiles do not reflect urbanization levels. Conserv. Physiol. 8.doi: 10.1093/conphys/coz110.

Isaksson, C. 2010. Pollution and its impact on wild animals: a meta-analysis on oxidative stress. EcoHealth 7: 342–350.

Istock, C.A. 1967. The evolution of complex life cycle phenomena: an ecological perspective. Evolution 21: 592–605.

Jeffries, K.M., R.E. Connon, B.E. Davis, L.M. Komoroske, M.T. Britton, T. Sommer, T. et al. 2016. Effects of high temperatures on threatened estuarine fishes during periods of extreme drought. J. Exp. Biol. 219: 1705–1716.

Kloas, W. 2002. Amphibians as a model for the study of endocrine disruptors. Int. Rev. Cytol. 216: 1–57.

Kohli, A.K., A.L. Lindauer, L.A. Brannelly, M.E.B. Ohmer, C. Richards-Zawacki, L. Rollins-Smith et al. 2019. Disease and the drying pond: examining possible links among drought, immune function, and disease development in amphibians. Physiol. Biochem. Zool. 92: 339–348.

Kulkarni, S.S., I. Gomez-Mestre, C.L. Moskalik, B.L. Storz and D.R. Buchholz. 2011. Evolutionary reduction of developmental plasticity in desert spadefoot toads. J. Evol. Biol. 24: 2445–2455.

Kulkarni, S.S., R.J. Denver, I. Gomez-Mestre and D.R. Buchholz. 2017. Genetic accommodation via modified endocrine signalling explains phenotypic divergence among spadefoot toad species. Nature Comm. 8: 993.

Laudet, V. 2011. The origins and evolution of vertebrate metamorphosis. Curr. Biol. 21: R726–R737.

Levis, N.A. and D.W. Pfennig. 2020. Plasticity-led evolution: A survey of developmental mechanisms and empirical tests. Evol. Develop. 22: 71–87.

Levis, N.A., P.W. Kelly, E.A. Harmon, I.M. Ehrenreich, D.J. McKay and D.W. Pfennig. 2021. Transcriptomic bases of a polyphenism. J. Exp. Zool. B: Mol. Dev. Evol. In press.

Liedtke, H.C., D.J. Gower, M. Wilkinson and I. Gomez-Mestre. 2018. Macroevolutionary shift in the size of amphibian genomes and the role of life history and climate. Nature Ecol. Evol. 2: 1792–1799.

Liedtke, H.C., E. Harney and I. Gomez-Mestre. 2021. Cross-species transcriptomics uncovers genes underlying genetic accommodation of developmental plasticity in spadefoot toads. Mol. Ecol. 30: 2220–2234.

Mann, R.M., R.V. Hyne, C.B. Choung and S.P. Wilson. 2009. Amphibians and agricultural chemicals: review of the risks in a complex environment. Environ. Pollut. 157: 2903–2927.

MacDougall-Shackleton, S.A., F. Bonier, L.M. Romero and I.T. Moore. 2019. Glucocorticoids and "stress" are not synonymous. Integr. Org. Biol. 1(1): obz017.

McMenamin, S.K., E.A. Hadly and C.K. Wright. 2008. Climatic change and wetland desiccation cause amphibian decline in Yellowstone National Park. Proc. Natl. Acad. Sci. USA 105: 16988–16993.

Merilä, J., F. Söderman, R. O'hara, K. Räsänen and A. Laurila. 2004. Local adaptation and genetics of acid-stress tolerance in the moor frog, *Rana arvalis*. Conserv. Genet. 5: 513–527.

Middlemis Maher, J., E.E. Werner and R.J. Denver. 2013. Stress hormones mediate predator-induced phenotypic plasticity in amphibian tadpoles. Proc. R. Soc. B 280: 20123075.

Monaghan, P., N.B. Metcalfe and R. Torres. 2009. Oxidative stress as a mediator of life history trade-offs: mechanisms, measurements and interpretation. Ecology Letters 12(1): 75–92.

Moore, F.L., S.K. Boyd and D.B. Kelley. 2005. Historical perspective: hormonal regulation of behaviors in amphibians. Horm. Behav. 48: 373–383.

Muir, A.P., R. Biek and B.K. Mable. 2014. Local adaptation with high gene flow: temperature parameters drive adpatation to altitude in the common frog (*Rana temporaria*). Mol. Ecol. 23: 561–574.

Murillo-Rincón, A.P., A. Laurila and G. Orizaola. 2017. Compensating for delayed hatching reduces offspring immune response and increases life-history costs. Oikos 126: 565–571.

Murray, M.H., C.A. Sánchez, D.J. Becker, K.A. Byers, K.E. Worsley-Tonks and M.E. Craft. 2019. City sicker? A meta-analysis of wildlife health and urbanization. Front. Ecol. Environ. 17: 575–583.

Navas, C.A., F.R. Gomes and J.E. Carvalho. 2008. Thermal relationships and exercise physiology in anuran amphibians: integration and evolutionary implications. Comp. Biochem. Physiol. A 151: 344–362.

Newman, R.A. 1992. Adaptive plasticity in amphibian metamorphosis. BioScience 42: 671–678.

Novarro, A.J., C.R. Gabor, C.B. Goff, T.D. Mezebish, L.M. Thompson and K.L. Grayson. 2018. Physiological responses to elevated temperature across the geographic range of a terrestrial salamander. J. Exp. Biol. 221: jeb178236.

Nowakowski, A.J., J.I. Watling, M.E. Thompson, G.A. Brusch, A. Catenazzi, S.M. Whitfield et al. 2018. Thermal biology mediates responses of amphibians and reptiles to habitat modification. Ecol. Lett. 21: 345–355.

O'Regan, S.M., W.J. Palen and S.C. Anderson. 2014. Climate warming mediates negative impacts of rapid pond drying for three amphibian species. Ecology 95: 845–855.

Olsson, M., E. Wapstra and C. Friesen. 2018. Ectothermic telomeres: it's time they came in from the cold. Phil. Trans. R. Soc. B 373: 20160449.

Otto, S.P. 2018. Adaptation, speciation and extinction in the Anthropocene. Proc. R. Soc. B 285: 20182047.

Pyron, R.A. and J.J. Wiens. 2013. Large-scale phylogenetic analyses reveal the causes of high tropical amphibian diversity. Proceedings of the Royal Society B: Biological Sciences 280(1770): 20131622.

Räsänen, K.R., A. Laurila and J. Merilä. 2003. Geographic variation in acid stress tolerance of the moor frog, *Rana arvalis*. I. Local adaptation. Evolution 57: 352–362.

Relyea, R.A. 2002. Costs of phenotypic plasticity. Am. Nat. 159: 272–282.

Relyea, R.A. 2004. Fine-tuned phenotypes: tadpole plasticity under 16 combinations of predators and competitors. Ecology 85: 172–179.

Romero, L.M., M.J. Dickens and N.E. Cyr. 2009. The reactive scope model—a new model integrating homeostasis, allostasis, and stress. Horm. Behav. 55: 375–389.

Romero, L.M. and U.K. Beattie. 2021. Common myths of glucocorticoid function in ecology and conservation. J. Exp. Zool. A: Ecol. Integr. Physiol. In press.

Seebacher, F., C.R. White and C.E. Franklin. 2015. Physiological plasticity increases resilience of ectothermic animals to climate change. Nature Clim. Change 5: 61–66.

Schoeppner, N.M. and R.A. Relyea. 2008. Detecting small environmental differences: risk-response curves for predator-induced behavior and morphology. Oecologia 154: 743–754.

Scott, D.E., E.D. Casey, M.F. Donovan and T.K. Lynch. 2007. Amphibian lipid levels at metamorphosis correlate to post-metamorphic terrestrial survival. Oecologia 153: 521–532.

Searle, C.L., S.S. Gervasi, J. Hua, J.I. Hammond, R.A. Relyea, D.H. Olson et al. 2011. Differential host susceptibility to *Batrachochytrium dendrobatidis*, an emerging amphibian pathogen. Conserv. Biol. 25: 965–974.

Torres-Sánchez, M., D.J. Gower, D. Alvarez-Ponce, C.J. Creevey, M. Wilkinson and D. San Mauro. 2019. What lies beneath? Molecular evolution during the radiation of caecilian amphibians. BMC Genomics 20: 354.

Vitousek, M.N., C.C. Taff, T.A. Ryan and C. Zimmer. 2019. Stress resilience and the dynamic regulation of glucocorticoids. Integr. Comp. Biol. 59: 251–263.

Walls, S.C. and C.R. Gabor. 2019. Integrating behavior and physiology into strategies for amphibian conservation. Front. Ecol. Evol. 7: 234.

Warne, R.W., E.J. Crespi and J.L. Brunner. 2011. Escape from the pond: stress and developmental responses to ranavirus infection in wood frog tadpoles. Funct. Ecol. 25: 139–146.

Warne, R.W., L. Kirschman and L. Zeglin. 2019. Manipulation of gut microbiota during critical developmental windows affects host physiological performance and disease susceptibility across ontogeny. Journal of Animal Ecology 88(6): 845–856.

Welch, A.M., J.P. Bralley, A.Q. Reining and A.M. Infante. 2019. Developmental stage affects the consequences of transient salinity exposure in toad tadpoles. Integr. Comp. Biol. 59: 1114–1127.

Werner, E.E. 1986. Amphibian metamorphosis: Growth rate, predation risk, and the optimal size at transformation. The American Naturalist 128(3): 319–341.

White, B.A. and C.S. Nicoll. 2013. Hormonal control of amphibian metamorphosis. pp. 363–396. *In:* Gilbert, L.I. and E. Frieden [eds.]. Metamorphosis: A Problem in Developmental Biology. Springer, Boston, MA, USA.

Wilczynski, W., K.S. Lynch and E.L. O'Bryant. 2005. Current research in amphibians: studies integrating endocrinology, behavior, and neurobiology. Horm. Behav. 48: 440–450.

Wingfield, J.C. and R.M. Sapolsky. 2003. Reproduction and resistance to stress: when and how. J. Neuroendocrinol. 15: 711–724.

Wingfield, J.C. 2013. The comparative biology of environmental stress: Behavioural endocrinology and variation in ability to cope with novel, changing environments. Anim. Behav. 85: 1127–1133.

CHAPTER 4

Eco-Immunology of Amphibians

Mar Comas[1,2,]* and *Gregorio Moreno-Rueda*[2]

INTRODUCTION

All organisms are exposed to pathogens and parasites, which, by taking resources from them, reduce their fitness and increase their mortality (Schmid-Hempel 2011). Parasites are supposed to be about 40% of overall species on the Earth, but they have a major relevance in ecology, as intervene in about 75% of ecological interactions (Dobson et al. 2008). Both pathogens and parasites consume host resources and hence often reduce host energy stores, harming body condition (Sánchez et al. 2018) and, consequently, decreasing their reproductive success (Rätti et al. 1993, Marzal et al. 2005, Møller et al. 2009). Hence, pathogens and parasites constitute a strong selective pressure, affecting host population structure and ecosystem functioning (Chapter 5). These ecological impacts translate to evolutionary responses by hosts trying to avoid the harmful effects of parasites and pathogens. The consequence frequently is an evolutionary arms race between pathogens/parasites and their hosts (Gómez et al. 2010). Moreover, parasites often prove decisive in sexual selection, with females usually choosing less parasitized males (Hamilton and Zuk 1982, Able 1996, Møller et al. 1999). Therefore, both pathogens and parasites have consequences in most aspects of the host's life history (Clayton and Moore 1997, Combes 2001, Schmid-Hempel 2011).

Given that pathogens and parasites reduce host fitness, hosts have evolved an array of mechanisms in order to avoid parasites or reduce their impact on fitness. These mechanisms include behavioral adaptations, as for example avoiding places where parasites are more probable (Hart 1990) or mating preferably with unparasitized individuals (Able 1996); anatomical adaptations like the skin (Elias 2007) or parts of the body designed to physically remove parasites (Clayton et al. 2005); chemical secretions (Moreno-Rueda 2017); or physiological adaptations

---
[1] Department of Biological Sciences, Dartmouth College, Hanover, NH, USA.
[2] Dpto. de Zoología, Facultad de Ciencias, Universidad de Granada, E-18071, Granada, Spain.
  Email: gmr@ugr.es
* Corresponding author: marcomas@ugr.es

such as fever (Kluger et al. 1996) and, of course, the immune system properly said (Schmid-Hempel 2011, Demas and Nelson 2012, Malagoli and Ottaviani 2014). The immune system is composed of a complex set of cells and molecules, which interact among themselves and with the organism in a number of processes whose goal is to maintain the homeostasis when challenged by external (pathogens, parasites, or accidents) or internal (cancer) attacks. In brief, when a cell is damaged or an alien element is detected, a cascade of signals is produced in the organism which activates the leucocytes, which, in different ways, destroy the alien elements or the malfunctioning cells.

Proved the importance of pathogens and parasites for host fitness, an effective immune system may be crucial for maximizing host survival (Møller and Saino 2004). One could think that the best immune system is the one providing the maximal immune response to pathogens. However, this is not the optimal immune system. First, an excessively reactive immune system may attack the own healthy cells, provoking immunopathology (Råberg et al. 1998). Moreover, the immune system typically generates oxidant reagents, with the goal of destroying invasive pathogens (Sorci and Faivre 2009). But these oxidant substances also may affect host tissues. In the second place, an immune response requests a number of resources such as amino acids and other substances, as well as energy. The immune system requests a large amount of amino acids to build antibodies, cytokines and other molecules (Lochmiller and Deerenberg 2000). These amino acids are detracted from other physiological functions, such as skeletal muscles, so mounting an immune response may be negative for locomotor capacity (Zamora-Camacho et al. 2015). Other substances, such as carotenoids, have an important function as immunomodulators (Møller et al. 2000), but they also are frequently used as pigments, which have a role as signals in several communication contexts (Hill 2004). Hence, when an animal is combating a pathogen by mounting an immune response, carotenoids are frequently mobilized from the tissues where they act as signals to intervene in the immune response (Blount et al. 2003, McGraw and Ardia 2003, Alonso-Alvarez et al. 2004). Finally, the immune response is typically energetically costly (Demas 2004). Mounting an immune response requests a rise in metabolism for the synthesis of a set of proteins, cell division and fever generation. Energy invested in the immune response, obviously, cannot be allocated to other functions such as growth, reproduction or other self-maintenance processes (Ardia and Schat 2008). In fact, ironically, the immune response may negatively affect survival (Moret and Schmid-Hempel 2000, Hanssen et al. 2004, Eraud et al. 2009). Therefore, the immune response must be optimal according to the ecological context and the payoff between costs and benefits.

From the aforementioned arguments, one may deduce that the optimal immune response may vary according to different scenarios. First, not all immune responses are equally costly, some responses being more costly (in energetic terms) than others (Buehler et al. 2008). Second, not all pathogens or parasites are the same. Some pathogens are more virulent than others, so a strong and quick response is more profitable. Other pathogens have so low virulence that the best option may be a response of tolerance -that is, not cleaning the pathogen (Råberg et al. 2009). And lastly, hosts must evaluate how many resources to allocate to combat the infection and how many resources to invest in other fitness-related functions, which probably is context-dependent. For example, older blue-footed boobies, *Sula nebouxii*, invest more in reproduction when sham-infected than younger boobies, probably because the host "interprets" that it will not survive for another breeding opportunity (Velando et al. 2006). The fact that the optimal immune response is not always the same conducts to a context-dependent variation in the optimal immune response. This variation is expected to be interspecific and intraspecific, and occur across geography, ontogeny, seasonally, and according to the genre (e.g., Forbes et al. 2006, Brown et al. 2015, Assis et al. 2020). This variation is what Immunological Ecology (also called Ecological Immunology or Eco-immunology) studies (Viney and Riley 2014).

Eco-immunology studies the ecological-dependent variation of the immune response and tries to understand how organisms respond optimally to their parasites. This discipline has traditionally been frequently applied to birds (Ardia and Schat 2008). However, Eco-immunology has received little attention in amphibians. The lack of eco-immunological studies in amphibians is astonishing, as amphibians have a series of characteristics that make them unique among vertebrates and special cases for the study of Eco-immunology. They are ectotherms (like reptiles), but also have a dependence on water that does not exist in other tetrapods, hydrothermal regulation being very important to them (Chapter 6). Furthermore, they typically have a complex cycle with a metamorphosis in which the anatomy and internal organs are completely remodeled, to the point that the animal goes from being aquatic (with gills) to being terrestrial (with lungs) (Chapter 9). Their complex life cycle does them to deal with two very different environments (water and earth), even when they are adults (as amphibians typically need water to breed). Their skin is very permeable in general, which makes them more sensitive to pathogens such as *Batrachochytrium* (Chapter 5), especially considering that the skin is one of the main barriers against pathogens in tetrapods. These characteristics mean that the conclusions reached from studies on other tetrapods do not necessarily apply to amphibians. But amphibian Eco-immunology is not only important because amphibians are very different to other tetrapods, it is also important due to the fact that amphibians are experiencing a strong worldwide decline (Stuart et al. 2004) and part of that decline is a consequence of diseases (Skerratt et al. 2007), such as the chytridiomycosis and other diseases (Chapter 5). For these reasons, the study of immunology in amphibians is important for amphibian conservation (Carey et al. 1999, Rollins-Smith 2001, Blaustein et al. 2012; also see Acevedo-Whitehouse and Duffus 2009, Sadd and Schmid-Hempel 2009, Hawley and Altizer 2011). In this chapter, we will review the state of the art of Eco-immunology applied to amphibians, indicating the main gaps in our knowledge.

Amphibian Immune System

Vertebrate immune system is conserved across evolution, so the amphibian immune system is similar to that of other vertebrates (Du Pasquier et al. 1989, Robert and Ohta 2009, Rollins-Smith and Woodhams 2012). The amphibian immune system is composed of a network of physiological processes in which we can find a number of cells (white cells) and proteins involved. White cells or leucocytes are mainly developed in the thymus, spleen, liver, kidney, and bone marrow. Nevertheless, hematopoiesis occurs in different organs according to the amphibian group (Grayfer and Robert 2016). The main organs containing lymphoid tissues may vary among amphibians, and even between larvae and adult states in the same species (Rollins-Smith and Woodhams 2012). For example, the main lymphoid organs in caecilians are the thymus, spleen and liver, in caudate are the thymus, spleen, liver and gut, while in anurans are the kidney and, in adults but not in tadpoles, the bone marrow.

As in other vertebrates, the immune system in amphibians is complex. Traditionally, the immune system is divided into innate and adaptive immune responses. The innate immune response is a quick and non-specific response to pathogens detected in the organism. Meanwhile, the adaptive immune response unfolds a specific response against a determinate pathogen (Flajnik 2018). The adaptive immune response is specific and more effective but takes a time to be prepared.

The immune system can also be divided into humoral and cellular immune systems. The humoral immune system is composed of a number of proteins released into the plasma, such as antibodies (immunoglobulins), lysozymes, and the complement system. The cellular immune system is composed of white cells not activated by antibodies, which intervene in phagocytosis. Humoral and cellular immune systems intervene during both the innate and adaptive responses and are interrelated.

Innate Immune System

The innate immune system is composed by an array of leucocytes, such as macrophages, neutrophils, basophils, eosinophils, and lymphocytes, among other components. Leucocytes patrol the tissues searching for unhealthy cells and alien cells. These white cells, as well as other cells of the organism, have toll-like receptors (TLR), which intervene in the innate recognition of pathogens. When a white cell contacts a pathogen, these receptors bind to determinate molecules of the surface of the pathogen called pathogen-associated molecular pattern (PAMP) (Roach et al. 2005). In the case of phagocytes, such as macrophages, neutrophils or dendritic cells, the bound pathogen is engulfed in phagosomes and destroyed by lysis in the phagolysosomes (the union between the phagosome and a lysosome). Phagocytosis implies not only digestive enzymes, but also free radicals produced in the phagolysosomes. When in contact with a pathogen, eosinophils secrete proteins and free radicals to kill parasites. Natural killer cells are specialized to destroy compromised host cells (for example cells infested with viruses). Infested cells typically have low levels of membrane proteins of type I major histocompatibility complex (MCH), allowing natural killers to identify the infected cells. The major histocompatibility complex (MHC) is very important in the immune system, as this complex is composed of membrane proteins determining self-recognition as well as non-self-recognition.

Some leucocytes as macrophages and lymphocytes, as well as infected and damaged cells, release cytokines to the surrounding environment. Cytokines are a type of protein that act as intercellular signals, attracting more leucocytes by chemotaxis. Cytokines also control the division of white cells, in this way regulating the relative importance of different arms of the immune system. Inflammatory cytokines are fundamental in the inflammatory process, which involves a change in the local vascular system, increasing the plasma flow to the affected region. This favors the access of leucocytes to the attacked zone, as well as several proteins such as immunoglobulins, proteins of the complement system, and lysozymes.

Other cytokines also intervene in the generation of fever. Fever is a response consisting of increasing body temperature, which is phylogenetically very well conserved and also occurs in ectotherms such as amphibians (Sherman et al. 1991, Sherman and Stephens 1998, Sherman 2008). While endotherms increase their body temperature by rising metabolism, ectotherms such as amphibians do so with thermoregulatory behaviors (behavioral fever). Fever increases the immune response, favoring the proliferation, migration and activity of various white cells, in addition to a direct adverse effect on the pathogens themselves.

At the same time, pathogens are attacked by the humoral immune system. In the humoral system, generalist antibodies produced by a number of leucocytes intervene. Antibodies bind to antigens on the surface of pathogens. Antibodies may neutralize the pathogen, avoiding its binding to healthy cells. Also provoke the agglutination of pathogens, facilitating their elimination by phagocytes. At this moment, the proteins of the complement system are important. They attach to antibodies and trigger a cascade of reactions facilitating the phagocytosis of marked cells, the inflammation and the attraction of leucocytes, and activating the membrane attack complex, a range of proteins that generate pores in the membrane of pathogenic cells conducting to its death.

The main function of the innate immune system is to terminate the infection or, at least, to retard the growth of the infectious pathogens, providing time for the adaptive response to intervene.

Adaptive Immune System

During the innate immune response, dendritic cells (and other cells such as macrophages) take the rest of the engulfed pathogens to present them to lymphocytes. Dendritic cells express MHC proteins of class II, which are useful to present these fragments of destroyed pathogens to lymphocytes (Castell-Rodríguez et al. 1999). For this reason, MHC is important for disease resistance, and determinate alleles of the MHC have been related to resistance against determinate pathogens (Barribeau et al. 2008, Teacher et al. 2009, Savage and Zamudio 2011, Bataille et al. 2015). Fragments of the lysed pathogens include antigens, which contain epitopes, the part of the antigen recognized by the immune system. Epitopes bind to MHC proteins and are translated to the membrane of the dendritic cell. Dendritic cells are frequently called at this moment antigen-presenting cells. Antigen-presenting cells search for immature T-cells in order to present the epitope. The epitope binds to a T cell receptor (TCR), activating the T-cell. Mature T-cells may act as cytotoxic T-cells, which search for infected cells and destroy them with cytotoxins. Cytotoxins generate pores in target cells, finally provoking their apoptosis. Other T-cells differentiate into helper T-cells, which interact with B-cells to activate those B-cells producing specific antibodies against the pathogen.

B lymphocytes are specialized in producing antibodies. B-cells recognize antigens in their native form (i.e., they do not need the intervention of MHC complex, as T-cells do). B-cells produce an enormous diversity of antibodies through a number of processes in which intervene hypermutation, with random genetic mutations of the genes coding the different parts of the immunoglobulins. This produces numerous lineages of B-cells, each with a specific antigen receptor (immunoglobulin) in its membrane. In this way, B-cells rearrange the expression of immunoglobulins to find one immunoglobulin matching the pathogen antigen. When they catch a fragment of the pathogen (antigen) by binding to the B cell receptor (a transmembrane immunoglobulin), the antigen is presented to helper T-cells. Then, helper T-cells secrete cytokines that activate the replication of B-cells. Finally, the lymphocyte B that has created the specific immunoglobulin against the pathogen replicates generating a line of lymphocytes producing antibodies. Moreover, some lymphocytes are differentiated in memory cells which will be important in future contacts with the same pathogen. Most of B-cells (90%) are dedicated to secreting specific antibodies, which bind to antigens in the pathogen provoking the agglutination, activating the complement, opsonization (which favors phagocytosis), the binding to cytotoxic cells, or neutralization (avoiding the pathogen can bind to organism cells). The rest of B-cells (10%) become long-live memory B-cells. A percentage of T-cells also become memory T-cells.

Once B lymphocytes are producing the specific antibodies against the pathogen, we can speak about acquired immunity. Specific immunoglobulins will attack specific pathogens, neutralizing the pathogen. This first response is slow and depends on aleatory processes (such as hypermutation) and the genetic diversity of alleles synthesizing immunoglobulins. Memory cells, however, have additional importance. In the case of an ulterior infection by the same pathogen, memory cells allow a fast and strong response (secondary response) that efficiently and quickly will eliminate the pathogen.

The Amphibian Skin and Antimicrobial Peptides

The skin is the first anatomical barrier against pathogens and parasites (see a deep review on amphibian skin as a defense against pathogens in Varga et al. 2019). Most vertebrates harbor rigid structures in the skin that help to prevent infections, such as scales, feathers or hair. Amphibians, however, have nude skin which proportionates a faint barrier front the environment, with the exception of caecilians which have dermal scales. This environment, moreover, frequently is humid, favoring the presence of pathogens. Amphibian skin is very permeable, hence being sensible to pathogen attacks. For this reason, amphibians are sensitive to pathogens such as *Batrachochytrium* (but see Grogan et al. 2018). Consequently, amphibians typically cover their skin with mucous secretions that usually include antimicrobial peptides (review in Rollins-Smith 2009; also see König et al. 2015). Antimicrobial peptides are secreted in granular glands and are actives against bacteria, fungi, viruses and protozoa (Zasloff 2002). These antimicrobial peptides seem to be produced by symbiont bacteria (Mangoni et al. 2001). The role of antimicrobial peptides in the amphibian immune system is so important that amphibian species expressing skin peptides less active against *Batrachochytrium* are the more threatened (Woodhams et al. 2006). Skin peptides can vary geographically (Tennessen et al. 2009) and this variation may have consequences on susceptibility to pathogens. For example, *Litoria genimaculata* frogs living at high elevations have less effective peptides against *Batrachochytrium* than populations at low elevations, and so high-altitude populations show a greater intensity of infection (Woodhams et al. 2010).

In addition to these antimicrobial peptides, in the mucous layer on the skin, we can also find elements of the immune system, such as phagocytes and lysozymes. In addition, the amphibian skin is inhabited by a number of symbiont bacteria, which seem to have an important role in limiting the establishment of pathogenic organisms, by secreting antibiotics, competitive exclusion, or by activating the immune system (Woodhams et al. 2007, Holden et al. 2015, Knutie et al. 2017; review in Colombo et al. 2015).

Ontogeny and Senescence of the Immune System in Amphibians

A particular feature of amphibians as vertebrates is that they have complex life cycles. Consequently, a typical amphibian has an aquatic larval phase and a terrestrial or semi-aquatic adult phase (although amphibians present a lot of variations in this scheme, Chapter 9). This implies that the environment, and therefore the pathogens, that an amphibian must face are different depending on the phase of its ontogeny. Hence, along the ontogeny, the amphibian immune system not only must maturate but must also become adapted to new pathogens different depending on the amphibians' phase and environment (Rollins-Smith 1998). In addition, adults often live both on land and in water, so they must deal with the pathogens of both environments. How this lifestyle affects the amphibian immune system is much unknown until now. The antibodies expressed by larvae and adults differ (Flajnik et al. 1987), probably because the larvae and adults face a different array of pathogens. Lymphocytes, for example, are less abundant in larvae, which seem to be less immunocompetent than adults. In fact, amphibian larvae tolerate allografts from other individuals better than adults. Adult amphibians, indeed, present more MHC molecules in cell membranes than larvae, probably for this reason being more reluctant to accept allografts from other individuals (Flajnik et al. 1987). Consequently, amphibians show different susceptibility to disease according to the ontogenetic state in which they are (de Jesús Andino et al. 2012, Abu Bakar et al. 2016).

The complex life cycle also involves a deep metamorphosis in which the internal organs and tissues of the larvae are strongly modified. These processes are very delicate for the immune system since the lysis of old tissues can be mistaken by the immune system as an attack. Furthermore, the immune system can identify the new tissues that appear as foreign, giving rise to an autoimmune response and therefore to an immunopathology. Indeed, MHC molecules in cells are different in larvae and adult tissues (Flajnik et al. 1986), hence larval and adult cells are histoimcompatible. Probably, this is a reason why the immune system is suppressed during metamorphosis (Izutsu 2009). In fact, during metamorphosis, the expression of interleukin 2 is inhibited (Ruben et al. 1992) and lymphocyte populations are reduced (Rollins-Smith et al. 1997), conducing to immunosuppression. Another possible reason is that metamorphosis is energetically very costly, immune system being suppressed in order to save energy. In any case, metamorphosis is a very sensitive phase to pathogen attacks.

Independently of ontogeny, the immune system also varies seasonally in an individual. Concretely, hibernation inhibits immunocompetence and hematopoietic organs shrink (Cooper et al. 1992) and the activity of antimicrobial skin peptides (Matutte et al. 2000). Immunosuppression during hibernation may occur in order to save energy, but also pathogens are less active in cold environments and, given that amphibians are quiescent, their probability to contact with pathogens is reduced (Rollins-Smith and Woodhams 2012).

A recent study across different species has shown a negligible aging in ectotherms (Reinke et al. 2022). For example, in the long-lived turtle *Trachemys scripta* the humoral immune responses are maintained with age (Zimmerman et al. 2013). Immune senescence is relatively well-known in vertebrates, but information for amphibians is very limited (Torroba and Zapata 2003). Still, the amphibian immune system has been found to show senescence. Saad et al. (1994) showed deteriorated immune organs (thymus, spleen and bone marrow) and lower antibody titers in old versus young *Bufotes viridis* toads. A similar finding was reported by Plytycz et al. (1995) for the thymus of *Rana temporaria*. Zamora-Camacho and Comas (2018), moreover, showed in *Epidalea calamita* toads that cutaneous immune response decreased linearly with age in a cross-sectional study.

Tradeoffs

Mounting an immune response implies an elevation of metabolism (Sherman and Stephens 1998, Llewellyn et al. 2012), proliferation of cells, synthesis of numerous proteins (antibodies, proteins acting as signals, enzymes and others such as proteins of the complement; see Klasing 2004), and generation of free radicals (directly to combat pathogens and indirectly due to the rise of cellular activity) that may impair the oxidative status of tissues (Sorci and Faivre 2009). Therefore, an immune response requests energy and limiting resources such as amino acids and vitamins (Klasing 1998). Given that energy and resources are not unlimited, when the organism is mounting an immune response, it needs to sequester these resources from other physiological compartments. In this way, the organism facing a pathogen should employ the best option to maximize fitness, which is not always to maximize the immune response. Hence, when an immune response arises, trade-offs between the immune response and other fitness-related physiological processes frequently emerge. The more relevant trade-off of immune response may be found with growth (van der Most et al. 2011), reproduction (Bonneaud et al. 2003), mating effort (Faivre et al. 2003) and locomotion (Zamora-Camacho et al. 2015).

The immune response is typically traded with growth, as the two immune response and growth require the same resources (structural elements such as proteins and energy) and generate oxidative stress (Alonso-Alvarez et al. 2007). Several studies support the existence of a trade-off between growth and immune capacity in amphibians, although these studies focus on larval growth. Little is known about the trade-off of adult growth (amphibians are indeterminate growers) and tissue regeneration with the immune capacity. Tadpoles in temporal ponds typically grow faster (in order to escape the pond before the complete desiccation). Experimental studies have clearly shown that a cost associated to this fast growth is a lower immunocompetence (*Lithobates sylvaticus* frogs, Gervasi and Foufopoulos 2008; *Spea multiplicata* toads, Bagwill et al. 2016; *L. pipiens* frogs, Brannelly et al. 2019). Similarly, in an experimental test of the trade-off between growth rate and immune capacity, Murillo-Rincón et al. (2017) delayed hatching in a group of *Rana arvalis* tadpoles, and this provoked accelerated growth and consequently reduced immunocompetence. But moreover, they found that immune challenged tadpoles grown slower than control and sham-control tadpoles, demonstrating that the trade-off between immune response and growth goes in the two directions. Additionally supporting the existence of a trade-off between growth and immunity in amphibians, Johnson et al. (2012) found that amphibian species with quicker development were more susceptible to a parasitic trematode.

Corticosterone has been frequently suggested as a mediator in the trade-off between growth rate and immune capacity. Corticosterone is released in stressful situations, such as pond desiccation, and its function is to regulate the allocation of energy to those processes more important for immediate survival (Sapolsky et al. 2000), frequently conducing to immunomodulation (Falso et al. 2015). In this way, it is expected that, in the stressing situation of being a tadpole in a desiccating pond, corticosterone would deviate the resources towards growth, accelerating metamorphosis (Gomez-Mestre et al. 2013; also see Chapter 3), hence reducing the energy available for the immune system. This idea was tested by Bagwill et al. (2016) in the toad *Spea multiplicata*. Their findings supported the trade-off between growth rate and immune capacity (estimated as leucocyte concentration in the spleen), but corticosterone titers were unrelated to the reduction in immune capacity. By contrast, experimental supplementation with corticosterone reduced immune capacity in *Lithobates sylvaticus* tadpoles while accelerating the growth rate in individuals not infected with ranavirus (Kirschman et al. 2018).

Locomotion is implied in different tasks of animals, such as searching for food or mates, patrolling the territory, chasing prey, and escaping from predators. Every of these activities are important for maximizing fitness and imply intense muscle activity, with the concomitant consumption of energy and production of free radicals. Not surprisingly, physical activity is associated to immunosuppression in animals (van Dijk and Matson 2016). The trade-off between different types of locomotion and immunity in amphibians is still poorly studied. Kärvemo et al. (2020) found that *Pelophylax lessonae* frogs infected with *Batrachochytrium* dispersed less distance that healthy frogs, but it is unclear if the reduced movement was a consequence of a trade-off between the immune system and the locomotion capacity, or a consequence of the morbidity associated with the illness. Nevertheless, Brown and Shine (2014) did show that adult *Rhinella marina* toads moving long distances suffer a down-regulation of some components of the immune system. However, an immune challenge did not affect short-distance swimming capacity of the newt *Pleurodeles waltl* (Zamora-Camacho et al. 2020). Therefore, still is poorly known how the immune system affects locomotion and vice versa. Moreover, amphibians display numerous types of locomotion which act in different contexts (Chapter 8), so we only have started to scratch the surface of the trade-off between the immune system and the plethora of locomotion modes in amphibians.

Given that predators may provoke the dead of the animals, when exposed to predators, one may expect an immunosuppression in order to improve escape capacity (Martin et al. 2012). Indeed, tadpoles exposed to predators show reduced immune capacity (Seiter 2011; but see Rumschlag and Boone 2018). However, Ramsey et al. (2010) reported an increase of skin antimicrobial peptides in *Xenopus laevis* when chased, which is explained as an anticipation to possible skin damaged produced by the predator. On the other hand, immune challenged amphibians, given the possible trade-off between locomotion capacity and immune response, could use less energy-demanding strategies such as hiding. However, in the frog *Pelophylax perezi*, the immersion time (a way to be hidden underwater) in response to a potential predator was unaltered by an immune challenge (Iglesias-Carrasco et al. 2018).

Investment in an immune response also may affect the investment in reproduction. For example, McCallum and Trauth (2007) found that male *Acris crepitans* frogs whose immune system was challenged by the inoculation of sheep blood cells showed a lower development of their gonads and spermatic cells. Hence, males combating a pathogen have less resources available for reproduction, which may be indicated by their sexual signals (the immunocompetence handicap hypothesis, Folstad and Karter 1992). In other words, it is expected that sexually selected traits that work as signals indicating individual quality during the mating process are sensible to pathogens, only high-quality males being able to combat pathogens and show elaborated signals (Hamilton and Zuk 1982). This idea was tested in amphibians by Cornuau et al. (2014). These authors inoculated males of *Lissotriton helveticus* newts with an antigen and examined the effect on the development of several sexually selected traits involved in mating (e.g., the crest) and on the courtship behavior. However, they did not find any effect of the stimulation of the immune system on sexual signals in this species, providing no support for the hypothesis. Iglesias-Carrasco et al. (2016), by contrast, reported in the same species a positive correlation between immune capacity and the size of several sexually selected traits. Desprat et al. (2015) enhanced sexual signals (both acoustic and visual) in males of the frog *Hyla arborea* by supplying them with testosterone. According to the immunocompetence handicap hypothesis, an immunosuppression was expected, but the authors failed to find a negative effect of testosterone on the immune system. On the other hand, in another study, *H. arborea* frogs whose immune system was challenged with a mitogen, showed reduced carotenoid-dependent coloration, suggesting that coloration may be a good indicator of health and immunocompetence in this species (Desprat et al. 2017). More studies are undoubtedly needed. Moreover, similar negative effects of immune challenge on female reproduction are also expected but barely studied.

In general, studies on the trade-offs between the immune system and other traits physiologically relevant for fitness in amphibians are lacking, especially in comparison with birds (see Ardia and Schat 2008). More studies on the eco-immunological trade-offs in amphibians are needed, particularly if we consider the peculiarities of amphibians in relation to the other tetrapods, which cast doubts about whether conclusions reached in other groups are applicable to amphibians. In fact, several trade-offs well-established in other taxonomic groups have not been found in amphibians. For example, the trade-off between sexual signals and immunocompetence (Desprat et al. 2015). Besides demonstrating the existence of trade-offs, the subjacent causes of such trade-offs are still poorly studied, in any group in general, and amphibians in particular. It is to say, a question arises if trade-offs are caused by the limitation of energy, limited structural resources such as amino acids, limited availability of determined vitamins (carotenoids, tocopherol...), or the trade-off emerges in order to maintain the oxidative status in a situation with two competing physiological processes in which free radicals are produced.

Implications for Amphibian Conservation

Amphibians are suffering an important global decline. Although the causes of this decline seem to be multifactorial, emerging diseases are one of the main ones (see Chapter 5). Amphibian declines are frequently associated to mortality in mass during metamorphosis and are more frequent in cold environments (review in Carey et al. 1999). This pattern has been associated to diseases, given that, first, metamorphic amphibians suffer immunosuppression (Rollins-Smith 1998) and the immune system is less active when the temperature is low (Raffel et al. 2006). Therefore, in the framework of a global change, any perturbation that negatively affects amphibian immunocompetence might increase their susceptibility to emerging diseases, contributing to their decline (Acevedo-Whitehouse and Duffus 2009, Blaustein et al. 2012, Rollins-Smith 2017). For example, several pollutants may reduce the immune function so doing amphibians more susceptible to diseases (e.g., Rohr et al. 2008, de Jesús Andino et al. 2017, Milotic et al. 2017, Hall et al. 2020). Moreover, pond-desiccation has been linked to depression in immune response (Gervasi and Foufopoulos 2008). The current climatic warming probably will reduce the hydroperiod of ponds, especially in Mediterranean and arid environments, conducing to reduced immune capacity and so contributing to the decline of amphibian populations (Kohli et al. 2019). Therefore, understanding the ecological immunology of amphibians may be fundamental in order to provide successfully conservation measurements.

References

Able, D.J. 1996. The contagion indicator hypothesis for parasite-mediated sexual selection. Proc. Natl. Acad. Sci. USA 93: 2229–2233.

Abu Bakar, A., D.S. Bower, M.P. Stockwell, S. Clulow, J. Clulow and M.J. Mahony. 2016. Susceptibility to disease varies with ontogeny and immunocompetence in a threatened amphibian. Oecologia 181: 997–1009.

Acevedo-Whitehouse, K. and A.L.J. Duffus. 2009. Effects of environmental change on wildlife health. Philos. Trans. R. Soc. B 364: 3429–3438.

Alonso-Alvarez, C., S. Bertrand, G. Devevey, M. Gaillard, J. Prost, B. Faivre et al. 2004. An experimental test of the dose-dependent effect of carotenoids and immune activation on sexual signals and antioxidant activity. Am. Nat. 164: 651–659.

Alonso-Alvarez, C., S. Bertrand, B. Faivre and G. Sorci. 2007. Increased susceptibility to oxidative damage as a cost of accelerated somatic growth in zebra finches. Funct. Ecol. 21: 873–879.

Ardia, D.R. and K.A. Schat. 2008. Ecoimmunology. pp. 421–441. *In:* Davison, F., B. Kaspers and K. Schat [eds.]. Avian Immunology. Academic Press, London, UK.

Assis, V.R., S.T. Gardner, K.M. Smith, F.R. Gomes and M.T. Mendonça. 2020. Stress and immunity: Field comparisons among populations of invasive cane toads in Florida. J. Exp. Zool. A 333: 779–791.

Bagwill, A.L., M.B. Lovern, T.A. Worthington, L.M. Smith and S.T. McMurry. 2016. Effects of water loss on New Mexico spadefoot toad (*Spea multiplicata*) development, spleen cellularity, and corticosterone levels. J. Exp. Zool. A 325: 548–561.

Barribeau, S.M., J. Villinger and B. Waldman. 2008. Major histocompatibility complex based resistance to a common bacterial pathogen of amphibians. PLoS ONE 3: e2692.

Bataille, A., S.D. Cashins, L. Grogan, L.F. Skerratt, D. Hunter, M. McFadden et al. 2015. Susceptibility of amphibians to chytridiomycosis is associated with MHC class II conformation. Proc. R. Soc. B 282: 20143127.

Blaustein, A.R., S.S. Gervasi, P.T.J. Johnson, J.T. Hoverman, L.K. Belden, P.W. Bradley et al. 2012. Ecophysiology meets conservation: understanding the role of disease in amphibian population declines. Philos. Trans. R. Soc. B 367: 1688–1707.

Blount, J.D., N.B. Metcalfe, T.R. Birkhead and P.F. Surai. 2003. Carotenoid modulation of immune function and sexual attractiveness in zebra finches. Science 300: 125–127.

Bonneaud, C., J. Mazuc, G. Gonzalez, C. Haussy, O. Chastel, B. Faivre et al. 2003. Assessing the cost of mounting an immune response. Am. Nat. 161: 367–379.

Brannelly, L.A., M.E.B. Ohmer, V. Saenz, C.L. Richards-Zawacki and Y. Vindenes. 2019. Effects of hydroperiod on growth, development, survival and immune defences in a temperate amphibian. Funct. Ecol. 33: 1952–1961.

Brown, G.P. and R. Shine. 2014. Immune response varies with rate of dispersal in invasive cane toads (*Rhinella marina*). PLoS ONE 9: e99734.

Brown, G.P., C. Kelehear, C.M. Shilton, B.L. Phillips and R. Shine. 2015. Stress and immunity at the invasion front: a comparison across cane toad (*Rhinella marina*) populations. Biol. J. Linn. Soc. 116: 748–760.

Buehler, D.M., T. Piersma, K.D. Matson and B.I. Tieleman. 2008. Seasonal redistribution of immune function in a migrant shorebird. Annual-cycle effects override adjustments to thermal regime. Am. Nat. 172: 783–796.

Carey, C., N. Cohen and L.A. Rollins-Smith. 1999. Amphibian declines: an immunological perspective. Dev. Comp. Immunol. 23: 459–472.

Castell-Rodríguez, A.E., A. Hernández-Peñaloza, E.A. Sampedro-Carrillo, M.A. Herrera-Enriquez, S.J. Alvarez-Pérez and A. Rondán-Zarate. 1999. ATPase and MHC class II molecules co-expression in *Rana pipiens* dendritic cells. Dev. Comp. Immunol. 23: 473–485.

Clayton, D.H. and J. Moore [eds.]. 1997. Host-parasite evolution: general principles and avian models. Oxford University Press, Oxford, UK.

Clayton, D.H., B.R. Moyer, S.E. Bush, T.G. Jones, D.W. Gardiner, B.B. Rhodes et al. 2005. Adaptive significance of avian beak morphology for ectoparasite control. Proc. R. Soc. B 272: 811–817.

Colombo, B.M., T. Scalvenzi, S. Benlamara and N. Pollet. 2015. Microbiota and mucosal immunity in amphibians. Front. Immunol. 6: 111.

Combes, C. 2001. Parasitism: the ecology and evolution of intimate interactions. University of Chicago Press, Chicago, USA.

Cooper, E.L., R.K. Wright, A.E. Klempau and C.T. Smith. 1992. Hibernation alters the frog's immune system. Cryobiology 29: 616–631.

Cornuau, J.H., D.S. Schmeller, R. Pigeault and A. Loyau. 2014. Resistance of morphological and behavioral sexual traits of the palmate newt (*Lissotriton helveticus*) to bacterial lipopolysaccharide treatment. Amphibia-Reptilia 35: 63–71.

de Jesús Andino, F., B.P. Lawrence and J. Robert. 2017. Long term effects of carbaryl exposure on antiviral immune responses in *Xenopus laevis*. Chemosphere 170: 169–175.

de Jesús Andino, F., G. Chen, Z. Li, L. Grayfer and J. Robert. 2012. Susceptibility of *Xenopus laevis* tadpoles to infection by the ranavirus Frog-Virus 3 correlates with a reduced and delayed innate immune response in comparison with adult frogs. Virology 432: 435–443.

Demas, G.E. 2004. The energetics of immunity. A neuroendocrine link between energy balance and immune function. Horm. Behav. 45: 173–180.

Demas, G.E. and A.C. Nelson [eds.]. 2012. Ecoimmunology. Oxford University Press, New York, USA.

Desprat, J.L., T. Lengagne, A. Dumet, E. Desouhant and N. Mondy. 2015. Immunocompetence handicap hypothesis in tree frog: trade-off between sexual signals and immunity? Behav. Ecol. 26: 1138–1146.

Desprat, J.L., T. Lengagne and N. Mondy. 2017. Immune challenges and visual signalling in tree frogs. Sci. Nat. 104: 101.

Dobson, A., K.D. Lafferty, A.M. Kuris, R.F. Hechinger and W. Jetz. 2008. Homage to Linnaeus: how many parasites? How many hosts? Proc. Natl. Acad. Sci. USA 105(Supplement 1): 11482–11489.

Du Pasquier, L., J. Schwager and M.F. Flajnik. 1989. The immune system of *Xenopus*. Annu. Rev. Immunol. 7: 251–275.

Elias, P.M. 2007. The skin barrier as an innate immune element. Semin. Immunopathol. 29: 3–14.

Eraud, C., A. Jacquet and B. Faivre. 2009. Survival cost of an early immune soliciting in nature. Evolution 63: 1036–1043.

Faivre, B., A. Grégoire, M. Préault, F. Cézilly and G. Sorci. 2003. Immune activation rapidly mirrored in a secondary sexual trait. Science 300: 103.

Falso, P.G., C.A. Noble, J.M. Diaz and T.B. Hayes. 2015. The effect of long-term corticosterone treatment on blood cell differentials and function in laboratory and wild-caught amphibian models. Gen. Comp. Endocrinol. 212: 73–83.

Flajnik, M.F., J.F. Kaufman, E. Hsu, M. Manes, R. Parisot and L. Du Pasquier. 1986. Major histocompatibility complex-encoded class I molecules are absent in immunologically competent *Xenopus* before metamorphosis. J. Immunol. 137: 3891–3899.

Flajnik, M.F., E. Hsu, J.F. Kaufman and L. Du Pasquier. 1987. Changes in the immune system during metamorphosis of *Xenopus*. Immunol. Today 8: 58–64.

Flajnik, M.F. 2018. A cold-blooded view of adaptive immunity. Nat. Rev. Immunol. 18: 438–453.

Folstad, I. and A.J. Karter. 1992. Parasites, bright males, and the immunocompetence handicap. Am. Nat. 139: 603–622.

Forbes, M.R., D.L. McRuer and D. Shutler. 2006. White blood cell profiles of breeding American toads (*Bufo americanus*) relative to sex and body size. Comp. Clin. Pathol. 15: 155–159.

Gervasi, S.S. and J. Foufopoulos. 2008. Costs of plasticity: responses to desiccation decrease post-metamorphic immune function in a pond-breeding amphibian. Funct. Ecol. 22: 100–108.

Gómez, J.M., M. Verdú and F. Perfectti. 2010. Ecological interactions are evolutionarily conserved across the entire tree of life. Nature 465: 918–921.

Gomez-Mestre, I., S. Kulkarni and D.R. Buchholz. 2013. Mechanisms and consequences of developmental acceleration in tadpoles responding to pond drying. PLoS ONE 8: e84266.

Grayfer, L. and J. Robert. 2016. Amphibian macrophage development and antiviral defenses. Dev. Comp. Immunol. 58: 60–67.

Grogan, L.F., J. Robert, L. Berger, L.F. Skerratt, B.C. Scheele, J.G. Castley et al. 2018. Review of the amphibian immune response to Chytridiomycosis, and future directions. Front. Immunol. 9: 2536.

Hall, E.M., J.L. Brunner, B. Hutzenbiler and E.J. Crespi. 2020. Salinity stress increases the severity of ranavirus epidemics in amphibian populations. Proc. R. Soc. B 287: 20200062.

Hamilton, W.D. and M. Zuk. 1982. Heritable true fitness and bright birds: A role for parasites? Science 218: 384–387.

Hanssen, S.A., D. Hasselquist, I. Folstad and K.E. Erikstad. 2004. Costs of immunity. Immune responsiveness reduces survival in a vertebrate. Proc. R. Soc. B 271: 925–930.

Hart, B.L. 1990. Behavioral adaptations to pathogens and parasites: five strategies. Neurosci. Biobehav. Rev. 14: 273–294.

Hawley, D.M. and S.M. Altizer. 2011. Disease ecology meets ecological immunology: understanding the links between organismal immunity and infection dynamics in natural populations. Funct. Ecol. 25: 48–60.

Hill, G.E. 2004. A red bird in a brown bag: The function and evolution of colorful plumage in the house finch. Oxford University Press, New York, USA.

Holden, W.M., S.M. Hanlon, D.C. Woodhams, T.M. Chappell, H.L. Wells, S.M. Glisson et al. 2015. Skin bacteria provide early protection for newly metamorphosed southern leopard frogs (*Rana sphenocephala*) against the frog-killing fungus, *Batrachochytrium dendrobatidis*. Biol. Conserv. 187: 91–102.

Iglesias-Carrasco, M., M.L. Head, M.D. Jennions and C. Cabido. 2016. Condition-dependent trade-offs between sexual traits, body condition and immunity: the effect of novel habitats. BMC Evol. Biol. 16: 135.

Iglesias-Carrasco, M., M.L. Head and C. Cabido. 2018. Effect of an immune challenge on the anti-predator response of the green Iberian frog (*Pelophylax perezi*): the influence of urban habitats. Biol. J. Linn. Soc. 124: 447–455.

Izutsu, Y. 2009. The immune system is involved in *Xenopus* metamorphosis. Front. Biosci. 14: 141–149.

Johnson, P.T.J., J.R. Rohr, J.T. Hoverman, E. Kellermanns, J. Bowerman and K.B. Lunde. 2012. Living fast and dying of infection: host life history drives interspecific variation in infection and disease risk. Ecol. Lett. 15: 235–242.

Kärvemo, S., G. Wikström, L.A. Widenfalk, J. Höglund and A. Laurila. 2020. Chytrid fungus dynamics and infections associated with movement distances in a red-listed amphibian. J. Zool. 311: 164–174.

Kirschman, L.J., E.J. Crespi and R.W. Warne. 2018. Critical disease windows shaped by stress exposure alter allocation trade-offs between development and immunity. J. Anim. Ecol. 87: 235–246.

Klasing, K.C. 1998. Nutritional modulation of resistance to infectious diseases. Poult. Sci. 77: 1119–1125.

Klasing, K.C. 2004. The cost of immunity. Acta Zool. Sinica 50: 961–969.

Kluger, M.J., W. Kozak, C. Conn, L. Leon and D. Soszynski. 1996. Role of fever in disease. Ann. N.Y. Acad. Sci. 856: 224–233.

Knutie, S.A., C.L. Wilkinson, K.D. Kohl and J.R. Rohr. 2017. Early-life disruption of amphibian microbiota decreases later-life resistance to parasites. Nat. Commun. 8: 86.

Kohli, A.K., A.L. Lindauer, L.A. Brannelly, M.E.B. Ohmer, C. Richards-Zawacki, L.A. Rollins-Smith et al. 2019. Disease and the drying pond: examining possible links among drought, immune function, and disease development in amphibians. Physiol. Biochem. Zool. 92: 339–348.

König, E., O.R. Bininda-Emonds and C. Shaw. 2015. The diversity and evolution of anuran skin peptides. Peptides 63: 96–117.

Llewellyn, D., M.B. Thompson, G.P. Brown, B.L. Phillips and R. Shine. 2012. Reduced investment in immune function in invasion-front populations of the cane toad (*Rhinella marina*) in Australia. Biol. Invasions 14: 999–1008.

Lochmiller, R.L. and C. Deerenberg. 2000. Trade-offs in evolutionary immunology: just what is the cost of immunity? Oikos 88: 87–98.

Malagoli, D. and E. Ottaviani [eds.]. 2014. Eco-immunology: Evolutive aspects and future perspectives. Springer, Dordrecht, The Netherlands.

Mangoni, M.L., R. Miele, T.G. Renda, D. Barra and M. Simmaco. 2001. The synthesis of antimicrobial peptides in the skin of *Rana esculenta* is stimulated by microorganisms. FASEB J. 15: 1431–1432.

Martin, L.B., A.J. Brace, A. Urban, C.A.C. Coon and A.L. Liebl. 2012. Does immune suppression during stress occur to promote physical performance? J. Exp. Biol. 215: 4097–4103.

Marzal, A., F. de Lope, C. Navarro and A.P. Møller. 2005. Malarial parasites decrease reproductive success: an experimental study in a passerine bird. Oecologia 142: 541–545.

Matutte, B., K.B. Storey, F.C. Knoop and J. Conlon. 2000. Induction of synthesis of an antimicrobial peptide in the skin of the freeze-tolerant frog, *Rana sylvatica*, in response to environmental stimuli. FEBS Lett. 483: 135–138.

McCallum, M.L. and S.E. Trauth. 2007. Physiological trade-offs between immunity and reproduction in the northern cricket frog (*Acris crepitans*). Herpetologica 63: 269–274.

McGraw, K.J. and D.R. Ardia. 2003. Carotenoids, immunocompetence, and the information content of sexual colors: An experimental test. Am. Nat. 162: 704–712.

Milotic, D., M. Milotic and J. Koprivnikar. 2017. Effects of road salt on larval amphibian susceptibility to parasitism through behavior and immunocompetence. Aquat. Toxicol. 189: 42–49.

Møller, A.P., P. Christe and E. Lux. 1999. Parasitism, host immune function, and sexual selection. Q. Rev. Biol. 74: 3–20.

Møller, A.P., C. Biard, J.D. Blount, D.C. Houston, P. Ninni, N. Saino et al. 2000. Carotenoid-dependent signals: indicators of foraging efficiency, immunocompetence or detoxification ability? Avian Poult. Biol. Rev. 11: 137–159.

Møller, A.P. and N. Saino. 2004. Immune response and survival. Oikos 104: 299–304.

Møller, A.P., E. Arriero, E. Lobato and S. Merino. 2009. A meta-analysis of parasite virulence in nestling birds. Biol. Rev. 84: 567–588.

Moreno-Rueda, G. 2017. Preen oil and bird fitness: a critical review of the evidence. Biol. Rev. 92: 2131–2143.

Moret, Y. and P. Schmid-Hempel. 2000. Survival for immunity: The price of immune system activation for bumblebee workers. Science 290: 1166–1168.

Murillo-Rincón, A.P., A. Laurila and G. Orizaola. 2017. Compensating for delayed hatching reduces offspring immune response and increases life-history costs. Oikos 126: 565–571.

Plytycz, B., J. Mika and J. Bigaj. 1995. Age-dependent changes in thymuses in the European common frog, *Rana temporaria*. J. Exp. Zool. 273: 451–460.

Råberg, L., M. Grahn, D. Hasselquist and E. Svensson. 1998. On the adaptive significance of stress-induced immunosuppression. Proc. R. Soc. B 265: 1637–1641.

Råberg, L., A.L. Graham and A.F. Read. 2009. Decomposing health: tolerance and resistance to parasites in animals. Philos. Trans. R. Soc. B 364: 37–49.

Raffel, T.R., J.R. Rohr, J.M. Kiesecker and P.J. Hudson. 2006. Negative effects of changing temperature on amphibian immunity under field conditions. Funct. Ecol. 20: 819–828.

Ramsey, J.P., L.K. Reinert, L.K. Harper, D.C. Woodhams and L.A. Rollins-Smith. 2010. Immune defenses against *Batrachochytrium dendrobatidis*, a fungus linked to global amphibian declines, in the South African clawed frog, *Xenopus laevis*. Infect. Immun. 78: 3981–3992.

Rätti, O., R. Dufva and R.V. Alatalo. 1993. Blood parasites and male fitness in the pied flycatcher. Oecologia 96: 410–414.

Reinke, B.A., H. Cayuela, F.J. Janzen, J.-F. Lemaître, J.-M. Gaillard, A.M. Lawing et al. 2022. Diverse aging rates in ectothermic tetrapods provide insights for the evolution of aging and longevity. Science 376: 1459–1466.

Roach, J.C., G. Glusman, L. Rowen, A. Kaur, M.K. Purcell, K.D. Smith et al. 2005. The evolution of vertebrate Toll-like receptors. Proc. Natl. Acad. Sci. USA 102: 9577–9582.

Robert, J. and Y. Ohta. 2009. Comparative and developmental study of the immune system in *Xenopus*. Dev. Dyn. 238: 1249–1270.

Rohr, J.R., A.M. Schotthoefer, T.R. Raffel, H.J. Carrick, N. Halstead, J.T. Hoverman et al. 2008. Agrochemicals increase trematode infections in a declining amphibian species. Nature 455: 1235–1239.

Rollins-Smith, L.A., K.S. Barker and A.T. Davis. 1997. Involvement of glucocorticoids in the reorganization of the amphibian immune system at metamorphosis. Dev. Immunol. 5: 145–152.

Rollins-Smith, L.A. 1998. Metamorphosis and the amphibian immune system. Immunol. Rev. 166: 221–230.

Rollins-Smith, L.A. 2001. Neuroendocrine-immune system interactions in amphibians. Immunol. Res. 23: 273–280.

Rollins-Smith, L.A. 2009. The role of amphibian antimicrobial peptides in protection of amphibians from pathogens linked to global amphibian declines. Biochim. Biophys. Acta 1788: 1593–1599.

Rollins-Smith, L.A. and D.C. Woodhams. 2012. Amphibian immunity. pp. 92–143. *In:* Demas, G.E. and A.C. Nelson [eds.]. Ecoimmunology. Oxford University Press, New York, USA.

Rollins-Smith, L.A. 2017. Amphibian immunity-stress, disease, and climate change. Dev. Comp. Immunol. 66: 111–119.

Ruben, L.N., M.A. Scheinman, R.O. Johnson, S. Shiigi, R.H. Clothier and M. Balls. 1992. Impaired T cell functions during amphibian metamorphosis: IL-2 receptor expression and endogenous ligand production. Mech. Dev. 37: 167–172.

Rumschlag, S.L. and M.D. Boone. 2018. High juvenile mortality in amphibians during overwintering related to fungal pathogen exposure. Dis. Aquat. Organ. 131: 13–28.

Saad, A.H., M.H. Mansour, V. Dorgham and N. Badir. 1994. Age-related changes in the immune response of *Bufo viridis*. Dev. Comp. Immunol. 18: 162–165.

Sadd, B.M. and P. Schmid-Hempel. 2009. Principles of ecological immunology. Evol. Appl. 2: 113–121.

Sánchez, C.A., D.J. Becker, C.S. Teitelbaum, P. Barriga, L.M. Brown, A.A. Majewska et al. 2018. On the relationship between body condition and parasite infection in wildlife: a review and meta-analysis. Ecol. Lett. 21: 1869–1884.

Sapolsky, R.M., L.M. Romero and A.U. Munck. 2000. How do glucocorticoids influence stress responses? Integrating permissive, suppressive, stimulatory, and preparative actions. Endocr. Rev. 21: 55–89.

Savage, A.E. and K.R. Zamudio. 2011. MHC genotypes associate with resistance to a frog-killing fungus. Proc. Natl. Acad. Sci. USA 108: 16705–16710.

Schmid-Hempel, P. 2011. Evolutionary Parasitology: The integrated study of infections, immunology, ecology, and genetics. Oxford University Press, Oxford, UK.

Seiter, S.A. 2011. Predator presence suppresses immune function in a larval amphibian. Evol. Ecol. Res. 13: 283–293.

Sherman, E., L. Baldwin, G. Fernandez and E. Deurell. 1991. Fever and thermal tolerance in the toad *Bufo marinus*. J. Thermal Biol. 16: 297–301.

Sherman, E. 2008. Thermal biology of newts (*Notophthalmus viridescens*) chronically infected with a naturally occurring pathogen. J. Thermal Biol. 33: 27–31.

Sherman, E. and A. Stephens. 1998. Fever and metabolic rate in the toad *Bufo marinus*. J. Thermal Biol. 23: 49–52.

Skerratt, L.F., L. Berger, R. Speare, S. Cashins, K.R. McDonald, A.D. Phillott et al. 2007. Spread of chytridiomycosis has caused the rapid global decline and extinction of frogs. EcoHealth 4: 125–134.

Sorci, G. and B. Faivre. 2009. Inflammation and oxidative stress in vertebrate host-parasite systems. Philos. Trans. R. Soc. B 364: 71–83.

Stuart, S.N., J.S. Chanson, N.A. Cox, B.E. Young, A.S.L. Rodrigues, D.L. Fischman et al. 2004. Status and trends of amphibian declines and extinctions worldwide. Science 306: 1783–1786.

Teacher, A.G.F., T.W.J. Garner and R.A. Nichols. 2009. Evidence for directional selection at a novel major histocompatibility class I marker in wild common frogs (*Rana temporaria*) exposed to a viral pathogen (*Ranavirus*). PLoS ONE 4: e4616.

Tennessen, J.A., D.C. Woodhams, P. Chaurand, L.K. Reinert, D. Billheimer, Y. Shyr et al. 2009. Variations in the expressed antimicrobial peptide repertoire of northern leopard frog (*Rana pipiens*) populations suggest intraspecies differences in resistance to pathogens. Dev. Comp. Immunol. 33: 1247–1257.

Torroba, M. and A.G. Zapata. 2003. Aging of the vertebrate immune system. Microsc. Res. Tech. 62: 477–481.

van der Most, P.J., B. de Jong, H.K. Parmentier and S. Verhulst. 2011. Trade-off between growth and immune function: a meta-analysis of selection experiments. Funct. Ecol. 25: 74–80.

van Dijk, J.G.B. and K.D. Matson. 2016. Ecological immunology through the lens of exercise immunology: New perspective on the links between physical activity and immune function and disease susceptibility in wild animals. Integr. Comp. Biol. 56: 290–303.

Varga, J.F.A., M.P. Bui-Marinos and B.A. Katzenback. 2019. Frog skin innate immune defences: sensing and surviving pathogens. Front. Immunol. 9: 3128.

Velando, A., H. Drummond and R. Torres. 2006. Senescent birds redouble reproductive effort when ill. Confirmation of the terminal investment hypothesis. Proc. R. Soc. B 273: 1443–1448.

Viney, M.E. and E.M. Riley. 2014. From immunology to eco-immunology: more than a new name. pp. 1–19. *In:* Malagoli, D. and E. Ottaviani [eds.]. Eco-immunology: Evolutive aspects and future perspectives. Springer, Dordrecht, The Netherlands.

Woodhams, D.C., N. Kenyon, S.C. Bell, R.A. Alford, S. Chen, D. Billheimer et al. 2010. Adaptations of skin peptide defences and possible response to the amphibian chytrid fungus in populations of Australian green-eyed treefrogs, *Litoria genimaculata*. Divers. Distrib. 16: 703–712.

Woodhams, D.C., L.A. Rollins-Smith, C. Carey, L. Reinert, M.J. Tyler and R.A. Alford. 2006. Population trends associated with skin peptide defenses against chytridiomycosis in Australian frogs. Oecologia 146: 531–540.

Woodhams, D.C., V.T. Vredenburg, M.-A. Simon, D. Billheimer, B. Shakhtour, Y. Shyr et al. 2007. Symbiotic bacteria contribute to innate immune defenses of the threatened mountain yellow-legged frog, *Rana muscosa*. Biol. Conserv. 138: 390–398.

Zamora-Camacho, F.J., S. Reguera, M.V. Rubiño-Hispán and G. Moreno-Rueda. 2015. Eliciting an immune response reduces sprint speed in a lizard. Behav. Ecol. 26: 115–120.

Zamora-Camacho, F.J. and M. Comas. 2018. Early swelling response to phytohemagglutinin is lower in older toads. PeerJ 6: e6104.

Zamora-Camacho, F.J., M. Comas and G. Moreno-Rueda. 2020. Immune challenge does not impair short-distance escape speed in a newt. Anim. Behav. 167: 101–109.

Zasloff, M. 2002. Antimicrobial peptides of multicellular organisms. Nature 415: 389–395.

Zimmerman, L.M., S.G. Clairardin, R.T. Paitz, J.W. Hicke, K.A. LaMagdeleine, L.A. Vogel et al. 2013. Humoral immune responses are maintained with age in a long-lived ectotherm, the red-eared slider turtle. J. Exp. Biol. 216: 633–640.

CHAPTER 5

Amphibian Crisis and the Impact of Emerging Pathogens

Barbora Thumsová,[1] *Jaime Bosch*[2,]* and *Gonçalo M. Rosa*[3]

INTRODUCTION

Infectious diseases are prominent drivers of the most dramatic events in human and wildlife history, being responsible for population declines and species extinctions across the whole planet (Daszak et al. 2000, Cunningham et al. 2017). Emerging infectious diseases are defined as those that are newly recognized, newly appeared in the population, or have rapidly increased in incidence or geographic range (Daszak et al. 2000). Their emergence is largely associated with a wide range of environmental and ecological factors but is also tightly linked to anthropogenic activities (Daszak et al. 2000, Jones et al. 2008, Santini et al. 2018). Recent examples include the devil facial tumor disease in Australia (Patton et al. 2020), white-nose syndrome impacting bats in North America (Hoyt et al. 2021), and the Ebola virus disease in West African primates, including humans (WHO Ebola Response Team 2014). Yet, the emergence of infectious diseases has also increased globally, as is the case with the Coronavirus disease 2019 pandemic (caused by SARS-CoV-2; WHO 2020) or chytridiomycosis epizootic in amphibians, leading to catastrophic impacts over entire ecosystems (Scheele et al. 2019a). Biodiversity loss has implications beyond wildlife, threatening food security and risking human health (Pongsiri et al. 2009, Cardinale et al. 2012), which creates an urgent need to improve our understanding of pathogens and the processes which underpin their emergence.

[1] Museo Nacional de Ciencias Naturales-CSIC, José Gutiérrez Abascal 2, 28006 Madrid, Spain; Asociación Herpetológica Española, José Gutiérrez Abascal 2, 28006 Madrid, Spain.
Email: barbora.thums@gmail.com
[2] IMIB-Biodiversity Research Institute, University of Oviedo-CSIC-Principality of Asturias.Gonzalo Gutiérrez Quirós s/n, 33600 Mieres, Spain.
[3] Institute of Zoology, Zoological Society of London, Regent's Park, NW1 4RY London, UK; Centre for Ecology, Evolution and Environmental Changes (cE3c), Faculdade de Ciências, Universidade de Lisboa, 1749-016 Lisboa, Portugal.
Email: goncalo.m.rosa@gmail.com
* Corresponding author: jaime.bosch@csic.es

Despite the many contributing factors to amphibian declines, such as habitat loss, overexploitation, climate change or contamination, emerging infectious diseases have been described among the top threats to populations (Collins and Storfer 2003, Rachowicz et al. 2006, Skerratt et al. 2007). Several pathogens have been linked to outbreaks and mass mortality events in amphibian assemblages such as *Amphibiocystidium* (the causative agent for dermocystidiosis; González-Hernández et al. 2010), some bacterial agents of the family Chlamydiaceae (Martel et al. 2012a,b) as well as recently discovered parasitic protists belonging to the phylum Perkinsea (linked to mortality events of larval and juvenile ranid frogs; Davis et al. 2007, Landsberg et al. 2013, Isidro-Ayza et al. 2017). Yet, there is little overall understanding of these pathogens and their dynamics, which heavily contrasts with the advances of the past three decades in chytridiomycosis (Berger et al. 1998) and ranavirosis research (Cunningham et al. 1996, Bollinger et al. 1999).

Two chytrid fungi of the genus *Batrachochytrium* are the causative agents of chytridiomycosis and have been linked to the decline or extinction of hundreds of amphibian species worldwide (Lips 2016, Scheele et al. 2019a, Fisher and Garner 2020). Indeed, chytridiomycosis has been described as the worst infectious disease ever recorded among vertebrates (Gascon et al. 2007, Skerratt et al. 2007, Scheele et al. 2019a). Since its discovery (Berger et al. 1998, Longcore et al. 1999, Pessier et al. 1999), *Batrachochytrium dendrobatidis* (*Bd*) has been recorded across all continents where amphibians occur (Fisher et al. 2009, Lips 2016). The recent discovery of the salamander chytrid, *B. salamandrivorans* (*Bsal*) increased the concerns about a new wave of declines after local extinctions of European fire salamanders (*Salamandra salamandra*) in the Netherlands (Martel et al. 2013, Spitzen-van der Sluijs et al. 2013). Efforts for active and passive surveillance have documented that *Bsal* is continuing to spread across Europe, and its presence in the wild was already confirmed in Belgium, Germany, the Netherlands, and Spain, linked to drastic die-offs in all cases (Martel et al. 2014, 2020, Spitzen-van der Sluijs et al. 2016, Lötters et al. 2020, Schulz et al. 2020). Asia seems to be the biodiversity hotspot for both chytrid species and the original source of other parasitic lineages emerging elsewhere (Martel et al. 2014, O'Hanlon et al. 2018). Human-mediated transmission has been pointed to as the main cause of the intercontinental chytrid spread, with the pet trade more recently implicated in the emergence of *Bsal* in Europe (Schloegel et al. 2012, Nguyen et al. 2017, Fitzpatrick et al. 2018).

While *Bd* infects a large range of Anura, urodele and caecilian species (Olson and Ronnenberg 2014), *Bsal* is predominantly a urodele parasite (Martel et al. 2014), yet able to infect anurans at low intensities (Stegen et al. 2017). Both pathogens infect epidermal cells of amphibian skin, but *Bsal* invades deeper layers, resulting in a more erosive and ulcerative disease (Berger et al. 2005, Martel et al. 2013). These disruptions impair the essential osmoregulatory function of the amphibian skin, which leads to severe osmotic imbalance, cardiac arrest and finally death of the affected individual (Voyles et al. 2007, 2009). The symptoms of chytridiomycosis can additionally include lethargy, lack of appetite, or increased skin sloughing (Voyles et al. 2007).

Human activities are considered to play a significant role in the emergence of the large double-stranded DNA viruses of the genus *Ranavirus* (Schloegel et al. 2009). These Iridoviridae viruses can lead to fatal ranavirosis in a broad range of ectothermic vertebrates, from fish to amphibians and reptiles (Duffus et al. 2015, Price et al. 2017). Amphibian-associated ranaviruses belong to three main groups known as *Frog virus 3* (FV3)-like, *Common midwife toad virus* (CMTV)-like, and *Ambystoma tigrinum virus* (ATV)-like viruses (Jancovich et al. 2015, Price et al. 2017). Strains from all three major groups have been described from across the world, sometimes linked to episodes of mass mortality and morbidity (Granoff et al. 1965, Mazzoni et al. 2009, Duffus et al. 2015, Stöhr et al. 2015, Price et al. 2017). In Europe, the first mortalities associated with the presence of ranaviruses were described in the former Yugoslavia (Fijan et al. 1991). These were then followed by die-offs in Spain (misdiagnosed as red-leg syndrome; Márquez et al. 1995) and in the UK (Cunningham et al. 1996). Disease outbreaks and rapid population declines in Northern Spain led to the characterization of CMTV-like viruses as a new ranavirus lineage (Price et al. 2014). Additional mortality events in Denmark, the Netherland and Portugal were also associated with CMTV-like viruses, these becoming the most common ranaviruses in continental Europe (Ariel et al. 2009a, Duffus et al. 2015, Rijks et al. 2016, Price et al. 2017, Rosa et al. 2017).

While amphibian chytrids colonize and infect the host skin, ranaviruses tend to first infect the oral cavity, later targeting internal organs (liver, spleen, and kidneys) leading to their failure in fatal cases (Miller et al. 2011, Saucedo et al. 2019). Ranavirus-infected individuals may develop overt signs of disease including systemic hemorrhage, skin ulceration or extensive tissue necrosis (Cunningham et al. 1996, Gray et al. 2009), yet subclinical infections have also been reported without any evidence of demographic declines (Brenes et al. 2014, Rosa et al. 2017).

It is clear that pathogens, even when acting alone, may seriously impact wildlife diversity. However, while some species or populations experience extensive mortality events, others may be asymptomatic and disease-induced population declines may not be observed. Several factors contribute to these different outcomes from interspecific susceptibility (Conlon et al. 2004, Woodhams et al. 2006, Harrison et al. 2019), including genetic differences (Luquet et al. 2012), to abiotic factors and ecological interactions (Lips et al. 2003, Berger et al. 2004, Doddington et al. 2013). Determining how hosts, pathogens and the environment interact and influence the final disease outcome has been the focus of many research efforts (reviewed by Brunner et al. 2015, Bienentreu and Lesbarrères 2020). However, despite hundreds of publications, we are only scratching the surface, particularly regarding effective ways to mitigate the impacts of emerging diseases on wild amphibian populations.

In this chapter, we perform a comprehensive compilation of current knowledge and describe how the wide range of biotic and abiotic factors may mediate the impact of pathogens on amphibian assemblages, focusing on infection risk and the range of host responses. Finally, we report how the understanding of a synergistic manifestation of these factors can contribute to the development of effective management to reduce the damaging impacts of emerging diseases.

Abiotic Factors

The environment has been identified as a key driver underpinning the infection dynamics and its outcome on the amphibian host-pathogen systems (Lips 2016). Environmental (co)variates modify these interactions either by changing the relative virulence of the pathogen or by compromising amphibian immunity resulting in increased susceptibility (Rollins-Smith 2020). The final host response therefore greatly varies through geographical areas and environmental conditions (Table 1).

Climate

Temperature, Seasonality and Climate Change

Temperature is an important factor in disease ecology as it can affect not only the pathogen distribution, severity and seasonality of infections, but also the final host response (e.g., Berger et al. 2004, Murray et al. 2013, Cohen et al. 2017, Rohr and Cohen 2020). It is largely known that the physiological function of amphibian ranaviruses and chytrids is highly temperature-dependent. In laboratory conditions, ranaviruses thrive at 28°C but stop replicating when temperatures surpass 31°C (Ariel et al. 2009b, La Fauce et al. 2012). In fact, ranavirosis incidents are strongly seasonal being more frequent and severe at higher temperatures, while pathogen loads drop at temperatures lower than 12°C leading to a reduction in mortality (Brand et al. 2016). This was clearly shown by combining data on ranavirus epidemics in Common frogs (*Rana temporaria*) in the UK where higher temperatures have historically driven the severity and frequency of ranavirosis incidents in wild populations (Price et al. 2019).

Conversely, a negative correlation has been observed between temperature and chytrid infections, resulting in higher pathogen impact at lower temperatures (around 10°C in temperate regions, and around 20°C in tropical areas; e.g., Phillott et al. 2013, Fernández-Beaskoetxea et al. 2015, Hudson et al. 2019). Another important factor determining the severity of *Bd* infection described mainly in tropical geographical areas is rainfall. While *Bd* prevalence generally increases during the wet season (~ 250 mm/month), higher infection burdens and disease susceptibility are recorded predominantly during the dry season (~ 100 mm/month; e.g., Ruggeri et al. 2018a, Moura-Campos et al. 2021; but see Nava-González et al. 2019). These findings have also been supported experimentally, with individuals of Coquí frogs (*Eleutherodactylus coqui* and *E. portoricensis*) being more susceptible to chytridiomycosis when subjected to limited water treatments resembling drought (Longo et al. 2010). Yet, as temperature variation is greatly affected by altitude, *Bd*-induced mortality is usually restricted to higher elevations, where water temperatures are cooler (e.g., Walker et al. 2010, Doddington et al. 2013). Indeed, *Bd* epizootics have been mostly documented for montane regions of both temperate and tropic regions (e.g., Berger et al. 1998, Bosch et al. 2001, Catenazzi et al. 2011, Rosa et al. 2013). On the other hand, although lowlands are expected to be too warm to harbour *Bd* at high infection intensities, particular microhabitats can provide a thermal optimum to elicit severe infections followed by mortalities (e.g., Doddington et al. 2013).

Table 1. Main abiotic and biotic factors influencing *Bd* and ranavirus infections. Abbreviations key: FO, field observation; E, experimentally proven; L, larvae; J, juveniles (including recently metamorphosed); A, adult; Pos, positive relationship; Neg, negative relationship; mixed, mixed results; -, not known

Factor	FO/E	Life history stage	Load	Prevalence	Proven mortality effect	Proven population effect	Examples
Bd							
Temperature	FO+E	L, J, A	Neg	Neg	✓	✓	Geiger et al. 2011, Phillott et al. 2013, Fernández-Beaskoetxea et al. 2015, Hudson et al. 2019
Temperature variability	E	L, J, A	Pos	Pos	✓	-	Raffel et al. 2012, 2015
Seasonality	FO+E	L, A	Pos	mixed	✓	-	Kriger and Hero 2007, Longo et al. 2010, Longo et al. 2013, Valencia-Aguilar et al. 2016, Ruggeri et al. 2018a, Hudson et al. 2019, Nava-González et al. 2019
Climate warming	FO	L, J	Pos	Pos	✓	✓	Bosch et al. 2007, Clare et al. 2016
Elevation	FO	L, J, A	mixed	mixed	✓	✓	Kriger and Hero 2008, Walker et al. 2010, Rosa et al. 2013
Waterbody hydroperiod	FO	L, J, A	Pos	mixed	✓	✓	Terrell et al. 2014, Medina et al. 2015, Bosch et al. 2018, Ruggeri et al. 2018a
Water flow	FO	L, J, A	Neg	Pos	✓	-	Medina et al. 2015, Ruggeri et al. 2018b
Canopy cover	FO	J, A	Pos	Pos	-	-	Becker et al. 2012, Roznik et al. 2015, Scheele et al. 2015
pH	FO	L, J, A	-	mixed	-	-	Battaglin et al. 2016, Blooi et al. 2017, Kärvemo et al. 2018
Salinity	FO+E	L, J, A	Neg	Neg	✓	-	Heard et al. 2014, Stockwell et al. 2012, 2015
UVR	FO+E	L	Neg	Neg	-	-	Ortiz-Santaliestra et al. 2011, Hite et al. 2016
Agrochemicals	FO+E	L, J, A	mixed	mixed	✓	-	Davidson et al. 2007, Hanlon et al. 2012, Battaglin et al. 2016, Rohr et al. 2017

O3 pollution	FO + E	L, J	Pos	-	✓	-	Bosch et al. 2021a
Microplastics	E	L, J	Pos	-	✓	-	Bosch et al. 2021b
Reproductive effort	FO	J, A	-	Pos	✓	✓	An and Waldman 2016, Valenzuela-Sánchez et al. 2022
Interspecific body size	FO + E	J, A	Pos	Pos	✓	✓	Bradley et al. 2019, Scheele et al. 2019a
Migration	FO	A	Neg	Neg	-	-	Daversa et al. 2018b
Behavioral fever	FO + E	J, A	Neg	Neg	-	-	Richards-Zawacki 2010, Karavlan and Venesky 2016
Species richness	FO + E	L, A	mixed	mixed	-	-	Searle et al. 2011, Venesky et al. 2014, Becker et al. 2014, Lambertini et al. 2021
Presence of amplification hosts	FO + E	L, A	Pos	Pos	✓	-	Fernández-Beaskoetxea et al. 2016, Lambertini et al. 2021
Host density	FO + E	L, J, A	mixed	mixed	✓	-	Rachowicz and Briggs 2007, Briggs et al. 2010, Fernández-Beaskoetxea et al. 2016, Bosch et al. 2020c
Non-amphibian hosts	FO + E	L, J, A	-	Pos	-	-	McMahon et al. 2013, Oficialdegui et al. 2019
Coinfection	FO + E	L, J, A	mixed	mixed	✓	-	Whitfield et al. 2013, Rosa et al. 2017, Longo et al. 2019, Bosch et al. 2020a
Microbiome diversity	FO	L, J, A	Neg	Neg	✓	✓	Bates et al. 2018, Jervis et al. 2021
Zooplankton abundance	FO + E	L, J, A	Neg	Neg	-	-	Schmeller et al. 2014, Blooi et al. 2017

Table 1 contd. ...

...Table 1 contd.

Factor	FO/E	Life history stage	Load	Prevalence	Proven mortality effect	Proven population effect	Examples
ranavirus							
Temperature	FO + E	L, J, A	Pos	Pos	✓	-	Brand et al. 2016, Price et al. 2019, Bosch et al. 2020a
Seasonality	FO	A	Pos	Pos	✓	✓	Rosa et al. 2019
Climate warming	FO	A	-	-	✓	-	Price et al. 2019
Canopy cover	FO	L	-	-	✓	-	Gahl and Calhoun 2010
Waterbody hydroperiod	E	L	-	Neg	-	-	Hoverman et al. 2011
Salinity	FO + E	L, J	Pos	-	✓	-	Hall et al. 2020
Agrochemicals	FO + E	L, J	Pos	-	✓	-	Hua et al. 2017, Pochini and Hoverman 2017, Cusaac et al. 2021
Larval development duration	E	L	-	Pos	-	-	Hoverman et al. 2011
Behavioral fever	E	J, A	Neg	-	✓	-	Sauer et al. 2019
Amplification hosts	FO	A	-	-	✓	-	North et al. 2015
Coinfection	FO	L, J, A	mixed	mixed	-	-	Whitfield et al. 2013, Rosa et al. 2017, Bosch et al. 2020a
Microbiome diversity	E	J	Neg	-	✓	-	Harrison et al. 2019

Whereas *Bd* grows best in culture between 17°C and 25°C and dies at above 29°C (Piotrowski et al. 2004), *Bsal* has its temperature optimum slightly lower, thriving between 10°C and 15°C, with temperatures of $\geq$ 25°C killing the fungus (Martel et al. 2013). However, other studies suggest that there is a considerable variation among *Bd* strains in thermal tolerance, and new thermal profiles for *Bd* have emerged (e.g., Voyles et al. 2017). Besides, optimal *in vitro* conditions may not necessarily be determinant of its survival, reproduction, and pathogenicity (Sonn et al. 2017). As ectotherms, amphibian immune functions perform better at warmer temperatures and slower at cooler or extreme temperatures (Wright and Cooper 1981, Raffel et al. 2006), which facilitates *Bd* proliferation out of its optimal temperature range (reviewed by Rollins-Smith et al. 2011, Rollins-Smith 2020). As a result, minimum, and not warmer temperatures in temperate regions are better predictors of *Bd* infection (Fernández-Beaskoetxea et al. 2015).

Climate change is predicted to drive catastrophic effects on human, animal, and environmental ecosystems (Bellard et al. 2012, Knight and Harrison 2013). The impacts go beyond producing shifts in population distributions, with documented cases of species extinctions (Thomas et al. 2004). Likewise, it is widely expected that the nuclear changes in climate will result in the suppression of amphibian immune function leading to an increased risk of disease outbreaks (Harvell et al. 2002, Altizer et al. 2013; but see Lafferty and Mordecai 2016). These synergistic interactions are complex and adversely affect amphibian populations in a context-dependent manner. For instance, Pounds et al. (2006) proposed the *chytrid-thermal optimum hypothesis* which derived from the chytridiomycosis outbreak that occurred in the highlands of Central and South America; the authors assumed that the warming of these areas led to increased cloud cover which has altered the daytime radiant heating of microenvironments resulting in daytime cooling. This shift was suggested to be responsible for the creation of new local environment conditions that meet the ecological requirements for optimal growth and increased pathogenicity of *Bd*. Another study used long-term monitoring of amphibian population dynamics in a temperate high mountain system in central Spain, showing that *Bd*-mediated outbreaks were associated with rising temperatures in recent years (Bosch et al. 2007).

Warming tolerance is often used to evaluate species' extinction risk posed by climate change. This metric is expressed by the difference between environmental maximum temperature and species' critical thermal maximum (CT_{max}), which represents the upper limit of the ability of animals to counterbalance temperature increase and marks the loss of homeostasis (Luttershmidt and Hutchison 1997, Deutsch et al. 2008; see Chapter 6). This means that a species becomes more susceptible to increased environmental temperature when its CT_{max} value is small because even a slight environmental temperature increase makes the body temperature of individuals reach its lethal limit (Welbergen et al. 2008, Greenspan et al. 2017a). Nevertheless, pathogens may alter the thermal thresholds of their hosts with infected and diseased amphibians showing a decrease in maximum thermal tolerance (Sherman 2008, Greenspan et al. 2017a). This effect may be life stage and/ or species-specific (Fernández-Loras et al. 2019a) which add-on to the increased uncertainty in estimates of species' vulnerability to climate change.

The relationship between CT_{max} and infection has been translated into a conceptual model of the *tolerance mismatch hypothesis* (Nowakowski et al. 2016). This suggests that infection risk will decrease as the difference in thermal tolerance of host and pathogen (tolerance mismatch) increases. It means that lethal susceptibility to infection is supposed to increase with lower values of the individuals' CT_{max} (Rollins-Smith 2020). Contrarily, individuals which can reach higher CT_{max} values increase their chances to clear the infection by creating conditions incompatible with the thermal limits of the pathogen. However, other authors have failed to prove this hypothesis right; Fernández-Loras et al. (2019a) found that Common midwife toad (*Alytes obstetricans*) tadpoles and toadlets infected with *Bd* were not able to clear infections when exposed to their CT_{max} (around 37.5°C, which is almost 10°C higher than the CT_{max} of *Bd*).

A refinement of the *tolerance mismatch hypothesis* led to the *thermal mismatch hypothesis*, which has emerged to predict how infection risk is affected by temperature across climate zones (Cohen et al. 2017). The authors suggest that infectious disease outbreaks are more likely to occur at temperatures where the performance gap between pathogen and host is greatest in favor of the pathogen. In other words, the hypothesis predicts that cold-adapted hosts should be especially susceptible to disease under unusually warm conditions, whereas warm-adapted hosts are more prone to infection under cooler conditions. Assuming that smaller-bodied parasites/pathogens can adapt more quickly to the environmental temperature shifts (yet limited by extremes) than their larger-bodied hosts, this allows parasites to outperform their hosts under unusual, but not extreme, conditions (Cohen et al. 2017, 2019). This hypothesis has been supported by global meta-analysis and multiple laboratory experiments, where cooler climate host species experience overall greater *Bd*-induced mortality at relatively warm temperatures and vice versa (Cohen et al. 2018, 2019, Sauer et al. 2018, 2020).

Lastly, in addition to temperature shifts, climate change impacts evaporation and precipitation which is highly problematic to amphibians, whose life history and physiology are dependent on water availability (Corn 2005, Ryan et al. 2014, Chapter 6). Moreover, changes to precipitation regimes and hydroperiods may also lead to increased susceptibility to pathogens (Kohli et al. 2019); for example, pond-desiccation has been linked to depression in immune response (Gervasi and Foufopoulos 2008). Yet, given, for example, *Bd* thermal optimum, some studies have suggested that increased drought conditions may reduce the risk of chytridiomycosis (Kriger 2009). An experimental approach using susceptible Yosemite toads (*Anaxyrus canorus*) did not detect an effect of reduced water availability on disease risk post metamorphosis (Lindauer and Voyles 2019).

Habitat

Water Bodies and Disease Dynamics

Aquatic habitat morphology is known to play a role in host-parasite/pathogen interactions (e.g., Anlauf and Moffitt 2008), with amphibian pathogens having different dynamics depending on the type of aquatic system. It has been documented that species strongly associated with permanent aquatic habitats (lentic or lotic) can be at higher risk of infection (e.g., Lips et al. 2003, Skerratt et al. 2010, Murray et al. 2011, Medina et al. 2015, Ruggeri et al. 2018b). Contrarily, amphibians breeding in temporary water bodies, where desiccation prevents the survival of the pathogen, are generally less affected (Scheele et al. 2015). Moreover, increased pathogen occurrence in ephemeral ponds has been associated with short distances to perennial water sources (Scheele et al. 2015); this suggests that perennial water may act as a source, with amphibian movements facilitating pathogen spread into temporary ponds.

For a better understanding of how *Bd* interacts with amphibians in different habitats, Ruggeri et al. (2018b) compared *Bd* prevalence and infection loads of tadpoles from lentic and lotic systems in the Brazilian Atlantic Forest. They found that, although higher infection intensities corresponded to the tadpoles of lentic waters, *Bd* was more prevalent in lotic water. The same pattern was also found in Ecuador where *Bd* was more associated with stream breeding than pond breeding amphibians (Jervis et al. 2021). This finding could explain the fact that flowing water facilitates the spread of the zoospores and increases the probability that more individuals get infected. The authors hypothesized that as *Bd* zoospores are not able to swim long distances, the standing water could allow the creation of zoospore pools, where the same tadpoles could re-infect and so increase their infection load.

Canopy Cover and Infection Risk

As an important driver of microclimate within an ecosystem, canopy cover is a predictor of infection prevalence and intensity of temperature-dependent pathogens (e.g., Beyer et al. 2015, Scheele et al. 2015, Stoler et al. 2016). Trees slow air movement and regulate the amount of solar radiation, resulting in a significant reduction of water temperature and preventing its large fluctuation (Whitmore 1998, Werner and Glennemeier 1999). This leads to the creation of suitable conditions for the host or, conversely, for its pathogen.

A previous study showed a correlation between lower canopy cover and ranavirus mortality in amphibians (Gahl and Calhoun 2010), which is consistent with the general ranavirus dynamics where severity increases with higher temperatures. The opposite effect was detected for *Bd* where, due to lower water temperature, a higher *Bd* prevalence was detected in populations from forested wetlands than in populations from open wetlands (Becker et al. 2012). The same pattern emerged after the reduction of the rainforest canopy cover caused by a severe tropical cyclone in Australia. The habitat disturbance led to increasing temperature and rates of evaporative water loss in the microhabitat of the endangered rainforest frogs (*Litoria rheocola*) (Roznik et al. 2015). These changes reduced *Bd* infection risk in frogs by an average of 11–28% in damaged areas. These studies corroborate that microclimatic variation arising from changes in vegetation cover can drive the spatial distribution of pathogens and influence the probability of epidemic disease outbreaks.

Moisture, Substrate and Pathogen Environmental Persistence

Several studies have shown how moisture is critical to ensure the environmental persistence of aquatic pathogens and consequently affect the disease dynamic. It has been experimentally demonstrated that *Bd* needs, at least, a thin film of water for zoospore survival, movement, and colonization of a new host (Piotrowski et al. 2004, Kriger 2009). Field studies have also been consistent in detecting correlations between *Bd* infection parameters (or *Bd*-associated mortalities) and moisture levels, either on the substrate or through water connectivity (Sapsford et al. 2013, Terrell et al. 2014, Raffel et al. 2015).

When favorable environmental conditions are met, the long-term persistence of a pathogen may be ensured. Laboratory studies have documented zoospore persistence in sterilized lake water and moist sand for up to 7–12 weeks (Johnson and Speare 2003, 2005). In the field, Kolby et al. (2015) suggested the possible spread of *Bd* to terrestrial substrates and the potential deposition of *Bd*-positive residue on riparian vegetation. As *Bsal* produces two types of spores, the transmission and persistence strategy of the fungus slightly differs. In addition to motile spores, which are able to active swim, it also produces infectious encysted spores that float at the water-air interface and can remain infective for at least 31 days in filtered pond water. Moreover, transmission from forest soil was demonstrated up to 48 hours after the soil had been in contact with an infected animal, suggesting that soil may be a potential vector for *Bsal* (Stegen et al. 2017).

Ranaviruses have also been found to persist under different conditions in the soil, in addition to water and carcasses of animals that have died of infection (Brunner et al. 2015, Brunner and Yarber 2018). Even though they seem to cope with adverse conditions such as drying or freezing (Nazir et al. 2012, Johnson and Brunner 2014), viral persistence does better in both sterile and unsterile pond water compared to soil (Nazir et al. 2012). A similar pattern was also shown when testing viral persistence at different temperatures where *Ranavirus* had significantly lower survival in sediment compared to freshwater (Munro et al. 2016). While many studies experimentally indicate the possibility of both chytrid fungi and ranaviruses persisting in the abiotic environment, serious evidence from the field is lacking, if not completely missing. Therefore, more research to better understand the environmental persistence of these amphibian pathogens is urgently needed.

Water Parameters

Water parameters such as pH, conductivity, dissolved oxygen and salinity have been poorly evaluated, showing that these variables may be highly context-dependent on disease dynamics (Bienentreu and Lesbarrères 2020). For example, high pH water has been associated with both low (Battaglin et al. 2016) and high (Blooi et al. 2017, Kärvemo et al. 2018) *Bd* infection prevalence. Contrarily, no significant relationship between pH and pathogen incidence has been found for both *Bd* (Strauss and Smith 2013) and ranavirus (Hall et al. 2018).

The negative effect of salinity on both the probability and intensity of chytrid infection in the threatened growling grass frog (*Litoria raniformis*) has been explored assuming that warm, saline wetlands may be refuges from chytridiomycosis (Heard et al. 2014). These findings have been supported by experimental studies, demonstrating that sodium chloride (NaCl) has fungicidal properties that reduce the mortality rates of infected individuals in captivity, and suggesting that saline solution may be an effective treatment tool (Stockwell et al. 2012, 2015, Clulow et al. 2017). However, the negative impact of salt on amphibians has to be considered before the implementation of any mitigation measure, especially when increased salinity may lead to local population extinction (Karraker et al. 2008).

The opposite effect has been observed for ranavirus (Hall et al. 2020). Populations exposed to salt runoff are slightly more susceptible to ranavirosis outbreaks. Experimental chronic exposure to high salinity levels reduced larval wood frog (*Lithobates sylvaticus*) tolerance to infection leading to greater mortality at lower viral loads. Therefore, the suppressed immune function of reared larvae has supported several components of the stress-induced susceptibility hypothesis (Hall et al. 2020). These findings demonstrated that only a small change in habitat quality can significantly increase transmission efficiency and the severity of ranavirus epidemics (Hall et al. 2020).

Ultraviolet Radiation

The global changes in ozone, surface reflectivity, aerosols and cloudiness in the recent past have altered the amount of ultraviolet radiation (UVR) reaching the Earth's surface (e.g., Kerr and McElroy 1993, McKenzie et al. 2011). Several empirical and experimental studies have shown that the increased levels of UVR can be detrimental to amphibian development (Häkkinen et al. 2001), hatching success (Blaustein et al. 1997), and survival (Bancroft et al. 2008). It has been suggested that UVR can also impair immune response (Tevini 1993), which would influence amphibian susceptibility to infections. Indeed, the fact that this phenomenon may increase the likelihood of *Bd* outbreaks has been reported by several authors. This is the case for some regions in Central America where exposure to increased levels of UVR appeared to coincide with outbreaks of chytridiomycosis (Lips 1998). This hypothesis has also been tested experimentally in several North American species where synergistic interactions between *Bd* and UVR have led to reduced survival; yet, radiation did not necessarily affect the probability of infection (Garcia et al. 2006, Searle et al. 2010).

The association between UVR and *Bd* infection has first found by Ortiz-Santaliestra et al. (2011). Particularly the larvae of the spiny toad (*Bufo spinosus*) showed a significantly lower prevalence of *Bd* infection in tadpoles exposed to natural UVR, suggesting that *Bd* is more likely to die from UVR exposure than non-exposed individuals. Another explication could be that potential sublethal effects such as erratic swimming, observed to be caused by UVR (Nagl and Hofer 1997) could reduce the number of encounters between conspecifics, leading to reduce transmission (Ortiz-Santaliestra et al. 2011). The result of Ortiz-Santaliestra et al. (2011) was consistent with a previous study where the prevalence of *Bd* infection in *A. obstetricans* was inversely proportional to the amount of radiation received (Walker et al. 2010).

A more complex study illustrated how habitat structure could explain considerable variation in *Bd* infection. Hite et al. (2016) showed that despite UVR killing the free-living infectious stage of *Bd*, permanent ponds with more UVR exposure fostered multi-season host larvae fueling *Bd* production, which eventually led to higher infection prevalence. However, these results did not suggest that UVR exposure increases host susceptibility to *Bd*, but that the net effect of UVR on disease depends on multiple factors which are directly and indirectly mediated through community ecology (Penczykowski et al. 2014, Hite et al. 2016).

Agrochemicals

Intentional and nonintentional pollution from agricultural fertilizers and pesticides (e.g., herbicides, insecticides, and fungicides) is known to threaten the functionality of aquatic ecosystems worldwide (e.g., Matson et al. 1997, Bennett et al. 2001). Nonetheless, despite the general recognition as a cause of the amphibian population declines, the causal relationship between those factors is still little understood (Schmidt 2004, Hayes et al. 2010). Pesticide exposure can ultimately lead to death by intoxication (e.g., Bridges and Semlitsch 2000, Brühl et al. 2013), yet the threat from agrochemicals may indirectly impair amphibian survival in several other complex ways, including through interaction with both host and pathogen (Rohr et al. 2017). Chemical products can affect the physiological functions of pathogens, but also have an immunosuppressive effect on the host, rendering mixed outcomes (e.g., Davidson et al. 2007, Gahl et al. 2011, Pochini and Hoverman 2017, Rohr et al. 2017, Cusaac et al. 2021). In fact, antimicrobial skin peptides and the skin microbiome are amphibians' primary barriers against dermal pathogens (Rollins-Smith et al. 2005), and chemical products compromise their effectiveness making amphibians more susceptible to infection (Brunner et al. 2015, Rohr et al. 2017, McCoy and Peralta 2018).

Generally, disease-driven mortalities related to elevated concentrations of pesticides seem to be more common for individuals infected by ranavirus than by *Bd*. Particularly fungicides and herbicides may have more negative impacts on *Bd* than on the amphibian individual (but see Battaglin et al. 2016, Rohr et al. 2017, Goodman et al. 2021). For example, the fungicide thiiohanate-methyl can simultaneously clear *Bd* and facilitate the growth and development of tadpoles (Hanlon et al. 2012). Another study has shown no increase in chytridiomycosis-induced mortality after exposure of wood frogs (*L. sylvaticus*) to the glyphosate-based herbicide, which finally appeared to affect the pathogen more than the host (Gahl et al. 2011). Yet, when previously infecting the same species with ranavirus, Pochini and Hoverman (2017) observed increased insecticide (carbaryl and thiamethoxam) toxicity. Moreover, prior insecticide exposure exacerbated disease-induced mortality, but with minimal effects on infection prevalence and transmission of the virus. Lastly, when attempting to disentangle the contributing factors to hellbender declines in North America, Cusaac et al. (2021) tested the interaction of both *Bd* and/or ranavirus, following exposure to glyphosate herbicide. Whereas no animals resulted infected by *Bd*, the combination of ranavirus and herbicide resulted in 100% mortality.

There is evidence to suggest that populations can also evolve increased constitutive or inducible tolerance to pesticides (Crispo 2007, Hua et al. 2013, 2015). While the evolution of constitutive tolerance is more common in populations that are close to agriculture fields and, therefore constantly exposed to a pesticide, inducible tolerance is more likely when populations are far and infrequently exposed to pesticides (e.g., Crispo 2007). However, as a consequence of nonadaptive trade-offs, populations that express pesticide tolerance may suffer increased susceptibility to pathogens (Hua et al. 2015, 2017). Hua et al. (2017) have found that populations of *L. sylvaticus* living closer to agriculture and having higher constitutive tolerance experienced higher viral loads than populations far from agriculture.

Other pollutants

Tropospheric ozone (O_3) is a major air pollutant, widely affecting rural and forested areas of the Northern hemisphere (Sutton et al. 2011, CLRTAP 2017). Despite the scarcity of knowledge on O_3 impacts on amphibian health, its effects have been documented at physiological and immunological levels (Dohm et al. 2001, 2005, Mautz and Dohm 2004). Therefore, a potential role of oxidant air pollutants in regional disease-related declines of amphibian populations has been suggested. A recent study found that high O_3 levels significantly delayed the rate of development of tadpoles and increased *Bd* infection, providing the first evidence that O_3 may have an unfavorable effect on amphibian disease risk (Bosch et al. 2021a).

Microplastics are a new class of pollutants that are increasing global concerns due to their negative effect on aquatic biodiversity (e.g., Lambert and Wagner 2018). Compelling experimental evidence showed an interaction between chytridiomycosis and microplastic pollution (Bosch et al. 2021b). Microplastics were accumulated to a greater extent in non-*Bd*-exposed tadpoles, likely due to *Bd*-induced damage of the mouthparts in infected animals. However, the authors found that microplastics ingestion increased *Bd* load intensities in a dose-dependent manner.

Biotic Factors

Understanding the host range and the traits influencing susceptibility to pathogens is fundamental to predicting the dynamics of infectious diseases within the communities (Craft et al. 2008, Daszak et al. 2000, Table 1). Both chytrid fungi and ranaviruses are able to infect multiple species, which exhibit a great variation in their susceptibility to infections (e.g., Martel et al. 2014, Duffus et al. 2015, Price et al. 2017). Some hosts can develop signs of the disease and experience high mortality (susceptible species), whereas others may be tolerant, sustain sublethal infections and act as reservoirs of the pathogen (tolerant species; van Rooij et al. 2015). Similarly, different populations of the same species may exhibit different responses to infection; some populations allow long-term pathogen persistence, while others may experience pathogen fade-out or host decline, and even extinction (Briggs et al. 2010, Tobler and Schmidt 2010, Vredenburg et al. 2010, Tobler et al. 2012).

Host Susceptibility and Defense Mechanisms

Highly Susceptible Amphibian Taxa

Bd infection risk may differ at the family, genus and species level (e.g., Bancroft et al. 2011, Baláž et al. 2014, Martel et al. 2014, Bacigalupe et al. 2017; but see Brenes 2013, Greenberg et al. 2017). Even though *Bd* has been found to infect at least 1,000 known hosts from all three amphibian orders (Castro Monzon et al. 2020), severe impacts have been more frequently associated with anurans than urodeles, where the lethal *Bd*-related chytridiomycosis has been reported in a much smaller number of species (e.g., Bosch and Martínez-Solano 2006, Lips et al. 2006, Thien et al. 2013, van Rooij et al. 2015).

A global analysis reported that over 500 amphibians experienced *Bd*-associated declines, with most of the species belonging to the genera *Atelopus*, *Craugastor*, and *Telmatobius* (Scheele et al. 2019a). *Atelopus* is by far one of the most threatened amphibian groups, with around 80% of its species assessed as Endangered or Critically Endangered, including four extinct species and at least 40 that have disappeared from their known localities in the past 20 years (La Marca et al. 2005, Lips 2016, IUCN 2021). Moreover, other highly susceptible species to chytridiomycosis can be found within the families Alytidae, Bombinatoridae, Dendrobatidae, Bufonidae, Ranidae, Pelodryadidae, and Hylidae (e.g., Bosch et al. 2001, Berger et al. 2005, Brem and Lips 2008, Briggs et al. 2010, Harmos et al. 2021). In Europe, mortalities associated with chytridiomycosis have been observed, among others, in all five species of the *Alytes* genus (Bosch et al. 2001, Bosch et al. 2013, Doddington et al. 2013, Rosa et al. 2013, Thumsová et al. 2022). Despite the low tolerance of *A. obstetricans* to *Bd* (e.g., Bosch et al. 2001, Rosa et al. 2013), no negative effects of the fungus were observed in some populations after reaching the enzootic stage (Tobler et al. 2012). However, the rapid spread of *Bd* throughout the whole distribution range of *Alytes dickhilleni* (Thumsová et al. 2021), together with the recently observed mortalities in wild populations of *Bombina variegata* (Harmos et al. 2021) and *Alytes maurus* (Thumsová et al. 2022), support the idea that *Bd* impact in Europe and North Africa may be worse than proposed by Scheele et al. (2019a).

The recent emergence of *Bsal* in Europe has raised concerns about its impact if the pathogen expands throughout the globe as *Bd* did. So far, mortality events in the wild have been limited to a few European urodeles: *Salamandra salamandra*, *Triturus marmoratus* and *Ichthyosaura alpestris* (Martel et al. 2013, Spitzen-van der Sluijs et al. 2016, Dalbeck et al. 2018, Martel et al. 2020, Schmeller et al. 2020). The high susceptibility of these species has also been supported by experimental exposures in the laboratory (Martel et al. 2014, Gilbert et al. 2020, Bosch et al. 2021c). Yet, laboratory trials have also shown *Bsal* to be lethal to newt species in North America, and nearly all urodeles in Europe, North Africa and the Middle East (e.g., Martel et al. 2014, Carter et al. 2020, Gilbert et al. 2020, Bosch et al. 2021c). Therefore, once spreading to new areas, *Bsal* is likely to cause a dramatic impact on the native populations (Gilbert et al. 2020).

Despite the ability to infect a very broad range of ectothermic vertebrates, different strains of ranaviruses vary in their degree of generalism (Duffus et al. 2015, Price et al. 2017). The most studied ranaviruses (i.e., the FV3-clade) have been detected in a great number of wild and captive anuran and urodele species throughout the Americas, Africa, Asia and Europe (Duffus et al. 2015). However, increased susceptibility to infections by this clade appears to be limited to ranids (but see Hoverman et al. 2011). This has been particularly well illustrated in the UK, where an FV3-like *Ranavirus* has caused local declines in common frogs (*R. temporaria*), despite being able to infect other species, such as the common toad (*Bufo bufo*) (Duffus et al. 2014a,b). Regardless of the high number of mortality events worldwide, this case seems to be the only clear evidence of population declines due to FV3 ranaviruses to date (Teacher et al. 2010).

In the wild, ATV-like forms, initially described as the cause of mass mortality of larval tiger salamanders (*Ambystoma tigrinum*), seem to be restricted to North America. Laboratory trials have suggested ATV as a urodele specialist (e.g., Jancovich et al. 2001), however, a following study has shown that anurans may develop infection and lethal disease when experimentally infected with strain from this clade (Schock et al. 2008). While mortality events can be catastrophic, associated population declines have never been detected (Brunner et al. 2004, Price et al. 2017).

No other group of pathogens has been so impactful to amphibian assemblages in Europe as the lethal CMTV-like ranaviruses. Outbreaks have been detected in at least 20 species of 12 genera, i.e., over 20% of the total European amphibian biodiversity (Temple and Cox 2009, Price et al. 2014, Rijks et al. 2016, Price et al. 2017, Rosa et al. 2017, Bosch et al. 2020a, von Essen et al. 2020), suggesting CMTV-like ranaviruses have extremely broad host range (Price et al. 2017). The most severe population declines due to these viral forms were reported in the Picos de Europa National Park (northern Spain). Several species across multiple sites were involved, with *A. obstetricans*, closely followed by *I. alpestris* and *B. spinosus*, as the worst-affected species (Price et al. 2014). This also seems to be the most likely scenario behind the decline of the endemic and threatened *Rana pyrenaica* in the Ordesa National Park (Spain). Extremely high infection loads in both living and dead individuals suggest *R. pyrenaica* as probably the most susceptible species to CMTV-like ranaviruses (Bosch et al. 2020b).

Resistance and Tolerance

Host defense mechanisms are frequently classified as contributing to pathogen avoidance, resistance, and tolerance (Medzhitov et al. 2012). While avoidance works through behavioral modifications to evade and prevent pathogen contact (see below; Kavaliers et al. 2004, Medzhitov et al. 2012), resistance and tolerance concern strategies that deal with pathogen colonization. Resistance reduces the pathogen burden, whereas tolerance is the ability to mitigate the detrimental effects of pathogen colonization, without directly affecting the intensity of infection (Schneider and Ayres 2008, Råberg et al. 2009).

While traits that allow preventing, or rapidly clearing infection are present in resistant species, they are missed in those tolerant (Schneider and Ayres 2008). Even though tolerant hosts may become persistently infected, they are able to reduce the fitness consequences of infection and do not succumb. It follows that while the tolerance mechanism may have a positive effect on the pathogen, the resistance mechanism acts in favor of the host (Roy and Kirchner 2000, Restif and Koella 2004, van Rooij et al. 2015). However, in the literature, tolerant species are often termed and described as resistant species (e.g., Eskew et al. 2018), suggesting a different use of both concepts varying according to the author (van Rooij et al. 2015).

Resistance to lethal chytridiomycosis varies according to the pathogen; while *Bsal* infects predominantly urodeles, *Bd* seems to find higher resistance in this group. Despite the North American plethodontid salamanders having been suggested as susceptible species to *Bd* (Vazquez et al. 2009), the European clade (*Speleomantes* spp.) is a notable example of true resistance to *Bd*. Individuals of *S. strinatii* experimentally exposed to *Bd* were able to clear the infection within 7–14 days, and skin secretions of seven *Speleomantes* species were revealed to be capable of killing *Bd* within 24 hours (Pasmans et al. 2013). Moreover, no *Bd*-infected individuals have been detected during the extensive sampling efforts of all European *Speleomantes* wild populations, even when the pathogen is present within their distribution range (Bovero et al. 2008, Pasmans et al. 2013, Schulz et al. 2021). Yet, the overall proportion of species with a truly developed resistance to *Bd* seems to be scarce (van Rooij et al. 2015).

Some plethodontids have shown resistance to *Bsal* as well, alongside other urodeles experimentally exposed, including ambystomids and some Asian hynobiids (Martel et al. 2014). In Europe, the Palmate newt (*Lissotriton helveticus*) has been primarily considered resistant to *Bsal* (Martel et al. 2014), despite low-intensity infected individuals found in the wild (Dalbeck et al. 2018).

As tolerant species may harbor the pathogen without discernible clinical disease, they serve as reservoirs of the infection. For example, *Bsal* has been found to cause asymptomatic infection in anurans and less susceptible urodele species such as East-Asian salamanders, its presumed original host (Laking et al. 2017, Stegen et al. 2017). The perfect examples of *Bd* carrier species are the American bullfrog (*Lithobates catesbeianus*) and the African clawed frog (*Xenopus laevis*). Both are often exported as part of the global amphibian trade and, therefore, are considered to have played an important role in the dissemination of *Bd* worldwide (Hanselmann et al. 2004, Weldon et al. 2004, Fisher and Garner 2007, Vredenburg et al. 2013). Yet, the two species show different dynamics: while detected *Bd* intensities in *X. leavis* are usually low (Solís et al. 2009), *L. catesbeianus*—together with other species such as *Pseudacris regilla* or *Litoria wilcoxi*—may carry extremely high levels of *Bd*, which can be commonly lethal for other species (Retallick et al. 2004, Garner et al. 2006, Kriger et al. 2007, Reeder et al. 2012). Because they can contribute to a high environmental spread of *Bd* they are often called 'super spreaders' of the infection. Super spreaders are, by definition, species or individuals that disproportionately facilitate the spread of the infection by increasing the contacts within the community (McCaig et al. 2011, DiRenzo et al. 2014). Therefore, they do not include only the species tolerant to extremely high infection intensities, but also the highly susceptible species which are able to produce high infection intensities over several weeks prior to death (for example *Atelopus zeteki*; DiRenzo et al. 2014).

Life-History and Ecology of Susceptible Species

Many aspects of species biology, including their physiology, ecology, life history and behavior may be predictors of species' susceptibility to infection and disease-induced declines (Bielby et al. 2008, Hoverman et al. 2011, Brenes 2013, Scheele et al. 2019a, Valenzuela-Sánchez et al. 2021). The novel synthesis emphasizes the importance of deeper integration of life-history theory into disease ecology and suggests that host response can be determined by the position of host species along the classical slow-fast life-history continuum (Valenzuela-Sánchez et al. 2021). For example, fast-living species tend to favor less costly components such as innate immunity or mechanisms of resistance/tolerance. Therefore, they may be more prone to disease-induced mortality than slow-living species (Gervasi et al. 2017, Valenzuela-Sánchez et al. 2021). Also, compensatory recruitment seems to be more common in fast life-history strategies (Valenzuela-Sánchez et al. 2021). It was demonstrated in the southern Darwin's frog (*Rhinoderma darwinii*) populations. The high *Bd* prevalence population exhibited the highest male reproductive effort and the highest population recruitment. It has led to a growing population trend despite the high mortality of adults (Valenzuela-Sánchez et al. 2022).

Body size has also been linked to susceptibility and disease risk. A recent meta-analysis revealed that large-bodied anurans have experienced more severe declines (Scheele et al. 2019a). This may be a result of increased susceptibility given the larger skin surface area, which can lead to a larger infection and death (Kuris et al. 1980). This hypothesis is supported by the increased infection risk and intensity resulting in lower survival in older frogs (as a proxy for larger body size; Bradley et al. 2019). Yet, the results partially contrast with Greenberg et al. (2017) models that predict larger species to face increased infection risk but lower infection intensities, proposing that large bodies could instead facilitate higher resistance to *Bd* infection.

Population and range size has often been a determining risk factor, making restricted and/or small populations (with reduced levels of genetic diversity) more prone to disease-induced extinction (de Castro and Bolker 2005). Even though wide-range species may get to be infected and suffer local population decline (e.g., Bosch and Martínez-Solano 2006), they are expected to recover more quickly or, at least, persist through immigration, recolonization, and recruitment than small-range species (Bielby et al. 2008).

The *Bd*-associated decline severity was found greatest for species occurring consistently in wet regions, and for those associated with perennial aquatic habitats. Moreover, risks are increasingly higher for species with narrow elevation ranges, with populations occurring at higher elevations having a lower probability of recovery (Scheele et al. 2019a). In both temperate and tropical regions, hundreds of montane aquatic-breeding frog species have experienced *Bd*-associated mortalities and population declines (Bosch et al. 2001, Bosch and Martínez-Solano 2006, Pearl et al. 2007, Lips et al. 2008, Carvalho et al. 2017). Although aquatic and terrestrial direct-developers are expected to show opposite patterns of infection (Lips et al. 2003, Bielby et al. 2008), contrasting results have been reported by some studies (Mesquita et al. 2017, Zumbado-Ulate et al. 2019, Lambertini et al. 2021). As direct-developing species rarely venture into aquatic habitats, they are less frequently exposed to *Bd* and therefore the adaptive responses to this pathogen may be lacking (Mesquita et al. 2017). This may help explain the higher infection intensities in terrestrial species—despite lower prevalence—than in aquatic species (Lambertini et al. 2021). An exception to the rule is the *Craugastor punctariolus* clade (robber frogs) from Costa Rica. While frogs of the *Craugastor* genus are typically direct-developing and with terrestrial reproduction, species of this clade spend a vast majority of their life cycle along fast-flowing streams, an environment highly suitable for *Bd*. This trait leads to a greater susceptibility of the species to *Bd*, resulting in higher infection intensities and eventual declines (Lips et al. 2003, Zumbado-Ulate et al. 2019).

As for *Ranavirus*, while Hoverman et al. (2011) found a phylogenetic signature to pathogen susceptibility, Brenes (2013) failed to confirm a positive relationship between host phylogenies and susceptibility to ranavirus. Nevertheless, due to biological trade-offs, life-history traits such as fast development are frequently associated with higher susceptibility to ranaviral infections (Hoverman et al. 2011). Both studies proposed that host species from stable habitats do not experience variation in infection status. This is the case for permanent wetlands where ranaviruses and amphibians have more opportunities to co-evolve and, as a result, exhibit lower susceptibility to infection (Hoverman et al. 2011, Brenes 2013).

Variation of Host Response through a Life-History Stage

The life-history stage is also an important trait of a host species which can affect infection dynamics and outcome (Haislip et al. 2011, Warne et al. 2011). Because *Bd* infects the keratinized parts of amphibians, the infection in anuran larvae occurs only on mouthparts (Marantelli et al. 2004). Infection in larval amphibians is usually associated with mild symptoms and very rare mortality (Blaustein et al. 2005, Garner et al. 2009a, Hanlon et al. 2015), which makes tadpoles a good reservoir of infection (e.g., Daszak et al. 2003, Blaustein et al. 2005). Some populations have a prolonged larval stage and can overwinter for multiple years (e.g., Walsh et al. 2016). Therefore, they have been implicated in the persistence of *Bd* throughout the year (e.g., Narayan et al. 2014, Bielby et al. 2022), but as well identified as an important source of spill-over to other hosts (e.g., Medina et al. 2015, Clare et al. 2016).

Even though *Bd* infects larvae and adults, mass mortalities in temperate zones usually occur at the end of metamorphosis (e.g., Bosch et al. 2001, Rosa et al. 2013). At metamorphosis, the larval epidermis starts to keratinize and stratify, allowing *Bd* to extend throughout the body (Garner et al. 2009a). Moreover, the immune system suffers dramatic reorganization during the metamorphosis, and the defenses are still not mature in early metamorphosed individuals (Chapter 4). An elevated level of hormones during the metamorphosis triggers the host's immune suppression increasing its susceptibility to infection (Rollins-Smith et al. 2011, Rollins-Smith 2020).

In contrast to *Bd*, all life stages of amphibians have been observed to be severally impacted by ranaviruses (Price et al. 2017). Most of the mortalities caused by North American FV3-like ranaviruses were especially found in late-stage tadpoles and recent metamorphs (Green et al. 2002, Duffus et al. 2015). Many authors have reported hundreds of thousands of dead larvae in just one day (e.g., Green et al. 2002, Wheelwright et al. 2014). On the other hand, *Ranavirus*-driven mortality in the UK has primarily affected adult individuals (Duffus et al. 2013). As for the European CMTV-like ranaviruses, mass mortality events impact all life-history stages across the extremely broad host range (Price et al. 2014, Rijks et al. 2016, Rosa et al. 2017).

Behavior

Aggregation, Host Movements and Behavioral Fever

Host behavior can produce differences in susceptibility to a pathogen. From intentional avoidance to limit contacts (Kavaliers et al. 2004) to abiotic-induced behavior (Rosa et al. 2019), the ways a host interacts with others and the environment affects the risk of infection (Moore 2002). The avoidance hypothesis assumes that there should be fitness advantages to a host that can detect the risk of pathogen exposure before being infected (e.g., Medzhitov et al. 2012). Yet, "ancient" behaviors that have persisted for millions of years may have the opposite effect and be detrimental in novel situations (Blaustein and Bancroft 2007). This is the case for larval aggregation, often displayed during foraging, thermoregulation, and/or antipredator defense (reviewed by Moore 2002). Highly social tadpoles may end up increasing this schooling behavior when infected with *Bd* (Han et al. 2008), which may lead to the increment of infection parameters making gregarious species (or life stages) more at infection risk than singly occurring ones (Venesky et al. 2011). On the other hand, it has been shown that individuals can minimize the likelihood of being infected by reducing time in aquatic habitats (e.g., Ruano-Fajardo et al. 2016, Daversa et al. 2018a, 2021).

Host movements, comprising seasonal migratory behavior, can also induce changes in within-individual infection (e.g., Altizer et al. 2011, Daversa et al. 2018b). Both, migratory individuals leaving high-risk sites before contracting infections (migratory escape), and infection-induced mortality during migration (migratory culling), may drive reductions in infection prevalence (Altizer et al. 2011). Alternatively, a boost in infection prevalence may be observed following the return of hosts to pathogen-rich habitats (van Dijk et al. 2014). A perfect example for studying seasonal changes in infection dynamics is provided by spiny common toads (*B. spinosus*), a susceptible species to *Bd* that migrate annually from terrestrial habitats—used for foraging and winter hibernation—to ponds—used for breed during the summer months—(Daversa et al. 2012). Daversa et al. (2018b) found that individuals of *B. spinosus* did not consistently sustain infection across annual cycles, but they gained and lost the infection from year to year. Radio-tracked individuals that sustained *Bd* during the breeding season lost their infections when migrated to terrestrial habitats, likely due to differences in biotic and abiotic conditions characteristic of each habitat (Daversa et al. 2018b). Additionally, sexual differences in host phenology can affect host infection dynamics. The large number of females of Boscas' newt (*Lissotriton boscai*) remaining in the aquatic habitat after the breeding season induced female-biased mortality by *Ranavirus* (Rosa et al. 2019).

Several studies have shown that terrestrial habitats may provide potential refuges from *Bd* (e.g., Daskin et al. 2011, Puschendorf et al. 2011, Daversa et al. 2012). For example, the lack of canopy cover in high-elevated dry sclerophyll forests enables Australian treefrogs (*Litoria lorica* and *Litoria nannotis*) basking behavior on warm rocks which probably prevents pathogen growth (Puschendorf et al. 2011). The same basking behavior seems to be key for other species such as *Rana iberica*, which never suffered from chytridiomycosis during severe outbreaks even when infected by *Bd* (Bosch et al. 2018). Warm-adapted frogs that experience daily heat influxes when basking are less likely to exceed a critical infection threshold (Greenspan et al. 2017b).

As a response to pathogen exposure, some amphibians can intentionally choose particular microclimates to regulate their body temperature (Kluger 1977, Huey 1991). The phenomenon is called behavioral fever, where individuals are able to avoid or reduce the severity of the infection as a result of increased body temperature by seeking warmer spots (Sherman 2008, Rakus et al. 2017). This fever allows the host to improve its immune function (Evans et al. 2015) while serving as a mechanism to limit pathogen growth rate and survival (Richards-Zawacki 2010). Yet, the study of this behavioral alteration in the context of *Bd* epidemiology has produced mixed conclusions: while Richards-Zawacki (2010) and Karavlan and Venesky (2016) reported on toad metamorphs raising their body temperatures to decrease infection risk, another study failed to detect this behavioral strategy in tadpoles of four frog species (Han et al. 2008). As a response to ranaviral infection, Sauer et al. (2019) found a change in thermal preference where individuals of *Anaxyrus americanus* would also display behavioral fever.

Antipredator Behavior

Predators can significantly alter the transmission dynamics within a system through different mechanisms: (1) reducing host density; (2) reducing the proportion of infected prey; and (3) inducing morphological, physiological or behavior changes in the host (Packer et al. 2003, Ostfeld and Holt 2004). The latter has been observed in infected *Rana pipiens* where tadpoles decreased their activity levels and maintained a greater distance from predators when exposed to visual and chemical cues of a fish predator (Parris et al. 2006). Contrarily, Han et al. (2011) showed that *Bd*-exposed tadpoles of one species (out of the four) increased their activity rates and refuge use in the presence of chemical cues of a salamander predator. Even though the physiological mechanisms that are changing host behavior were not tested within those studies, stereotypical behavioral responses by hosts are expected because *Bd* requires host survival for its transmission (discussed in Parris et al. 2006 and Han et al. 2011).

Host Community and Transmission Dynamic

Dilution and Amplification Effects

Community composition can significantly influence infection and disease risk through ecological phenomena such as pathogen dilution or amplification (Searle et al. 2011, Becker et al. 2014, Venesky et al. 2014, Muths et al. 2020, Lambertini et al. 2021). The two concepts rely on the simple relationship between species richness and disease risk. While a low species diversity may amplify the infection, higher species richness may have a diluting effect, resulting in lower disease risk (Keesing et al. 2006, 2010, Halliday et al. 2017). However, both phenomena are strongly context-dependent and as it has been suggested, whether a dilution or amplification of the pathogens occurs, will most likely depend more on host community composition, competence, and densities than on simple species richness (Randolph and Dobson 2012, Venesky et al. 2014, Luis et al. 2018).

The dilution effect on an amphibian-*Bd* system has been experimentally demonstrated by several studies. For example, Searle et al. (2011) have shown that the addition of Cascades frog (*Rana cascadae*) tadpoles leads to a reduction of *Bd* infection severity and prevalence in Western toad (*Anaxyrus boreas*) tadpoles, even when accounting for the density effect. On the other hand, a few other studies have shown complex results. While *Gastrophryne* spp. diluted the risk of chytridiomycosis for tadpoles of *Bufo terrestris* and *Hyla cinerea*, *B. terrestris* amplified infections for the other species (Venesky et al. 2014). Moreover, whilst North et al. (2015) acknowledge the positive effect of alternative hosts in reducing ranavirosis prevalence (i.e., dilution effect), field data showed that the presence of *B. bufo* was linked to an increased likelihood of ranavirosis occurrence in more susceptible *R. temporaria*. The authors suggest that the lower competency of *B. bufo* as the host may result in reduced transmission efficiency but lead to a more virulent outcome.

Even though higher biodiversity may dilute disease risk at the local scale, it has been predicted to have the opposite effect at a large spatial scale (Halliday and Rohr 2019). Consistent with this, Lambertini et al. (2021) found that infection rates increased with increasing species richness across their broad latitudinal transect. The authors suggested that the high availability of susceptible hosts—and the so-called amplification hosts—throughout the year, could explain this finding. Amplification hosts, including, but not limited to, super spreaders, make a pathogen more abundant and more likely to persist. They also increase infection parameters within the community and the risk of disease emergence in the sympatric more susceptible species (Begon 2008, DiRenzo et al. 2014, Fernández-Beaskoetxea et al. 2016). In agreement with this, Scheele et al. (2016) demonstrated that *Bd* prevalence in susceptible northern corroboree frog (*Pseudophryne pengilleyi*) was higher at sites where a *Bd* tolerant reservoir common eastern froglet (*Crinia signifera*) was present and that the increased abundance of *C. signifera* was a good predictor of chytridiomycosis-associated declines of *P. pengilleyi*.

Transmission and Epidemiological Dynamic

Density-dependent transmission suggests the linear positive relationship between the rate of contacts per host and density. It can be observed when individuals move and contact each other randomly within a community. On the other hand, frequency-dependent transmission assumes that as host density changes, the rate of contacts per individual remains constant. This can increase if contacts only occur within a social group, where abundance does not change with density (McCallum et al. 2001, Begon et al. 2002).

While a significant effect of density for *Bd* has been reported in some studies (Rachowicz and Briggs 2007, Briggs et al. 2010, Wilber et al. 2017), others have not found this pattern (Searle et al. 2011, Fernández-Beaskoetxea et al. 2016, Bosch et al. 2020c). A positive correlation between higher densities of Mountain yellow-legged frog (*Rana muscosa*) and increased infection, mortality and higher number of environmental zoospores has been detected in the Sierra Nevada mountains of California (Rachowicz and Briggs 2007, Briggs et al. 2010). The same pattern was obtained using experimental designs where density-dependent transmission function better described the *Bd* infection dynamics than a frequency-dependent model (Wilber et al. 2017, Bosch et al. 2020c). Yet, in the latter study, the density had a significant effect on *Bd* infection intensity only for groups of *S. salamandra* larvae housed at very low densities.

Epidemiological models support the idea that *Bsal*-associated disease outbreaks can occur even at very low host densities, far lower than host densities commonly observed in the wild (Schmidt et al. 2017, Tompros et al. 2021). Moreover, transmission can saturate and be functionally frequency-dependent, which was suggested by a > 90% infection rate in eastern newts (*Notophthalmus viridescens*) across an experimental range of host densities (Tompros et al. 2021). Higher densities have also been shown to be detrimental to hosts in ranaviral dynamics (e.g., Echaubard et al. 2010). Nonetheless, while frog density was the best predictor of a high ranavirus prevalence in wild populations of ranids (*Lithobates* spp.; Youker-Smith et al. 2018), transmission rapidly reaches a saturation threshold as density increases (Brunner et al. 2017).

It is clear that the network of transmission within a community of multiple species, and its species composition, play a fundamental role in driving a pathogen's spread and persistence at the community level (Fenton and Pederson 2005, Keesing et al. 2006, Fenton et al. 2015). The basic reproductive number (R_0) of a parasite is a key measure in quantifying a species or individual's role in the transmission and maintenance of infection (Anderson and May 1992) and allows us to better understand the circumstances of disease persistence in the environment (Fenton and Pederson 2005, Fenton et al. 2015). While a $R_0 < 1$ indicates that the infection will tend to fade out, a $R_0 > 1$ indicates a high likelihood of pathogen persistence and potential spread (Fine et al. 2011). However, the R_0 may suffer many variations according to several biotic and abiotic variables to which a parasite is exposed. To account for these complexities, Fenton et al. (2015) developed a mathematical model that partitions the contribution each host species makes to the overall R_0 of the pathogen, allowing this way natural systems to be considered. When running the model on a community of five species, Bielby et al. (2021) found that different species may play the same role in *Bd* and ranavirus persistence: while *A. obstetricans* acted as the maintaining host for both pathogens at one site, *Ichthyosaura alpestris* fulfilled this role at another site.

Non-Amphibian Hosts and Vectors

Identification of non-amphibian vectors or hosts, and characterization of the role they play in disease dynamics is important from an epidemiology perspective and may even shape conservation strategies. While vectors can only mediate pathogens' mechanical transmission, abiotic reservoir is an alternative host which moreover allows pathogens to complete their lifecycle and persist outside of their natural host (Haydon et al. 2002, Wilson et al. 2017). Indeed, the presence of an alternative host in the aquatic ecosystem can be a principal cause of amphibian unsuccessful reintroduction efforts (Mitchell et al. 2008, McMahon et al. 2013).

Even though there are many studies suggesting potential carriers and alternative hosts of chytrids or ranavirus, little is known (Brunner et al. 2015, van Rooij et al. 2015). Reptiles (Kilburn et al. 2011, Pontes et al. 2018), birds (Garmyn et al. 2012, Burrowes and De la Riva 2017), nematodes (Shapard et al. 2012), crayfish (McMahon et al. 2013), zebrafish (Liew et al. 2017) or mosquitoes (Gould et al. 2019, Toledo et al. 2021) have been reported to potentially carry *Bd*. However, of these, only crayfish and zebrafish have been demonstrated to maintain the pathogen (McMahon et al. 2013, Oficialdegui et al. 2019, Liew et al. 2017), and only crayfish was able to infect amphibian hosts (McMahon et al. 2013). Despite other studies have reported *Bd* growth on sterile bird feathers (Johnson and Speare 2005), boiled snake skin (Symonds et al. 2008), and sterile toe scale of waterfowl (Garmyn et al. 2012), there is no evidence that the living specimens with functioning immune system can facilitate *Bd* maintenance or spread active zoospores over a long time period.

Amphibian ranaviruses are able to infect a wide range of reptile and fish species (Duffus et al. 2015), but the transmission routes in the wild and implications of non-ectotherm species are very poorly understood (Price et al. 2017). Although North American FV3-like ranaviruses seem to exploit primarily amphibian hosts, spilling over into other ectothermic vertebrate classes is possible (Brenes et al. 2014). Indeed, FV3-like-associated infections in wild North American fish are rare, and if one occurs, it will be probably through very close contact with infected amphibians (Waltzek et al. 2014). Lethal ranavirosis has been observed in chelonians and squamate reptiles infected by FV3-like strains (Belzer and Seiber 2011, Duffus et al. 2015, Kimble et al. 2015), however, initial infection in chelonians likely resulted from amphibian mortality outbreaks (Belzer and Seiber 2011). Kimble et al. (2015) have detected FV3 DNA in mosquitoes sampled during a ranavirus outbreak in semi-captive eastern box turtles. This led the authors to suggest mosquitoes as a potential vector of ranavirus among ectothermic vertebrates. Nevertheless, the viability of the virions was not assessed, leaving the role of mosquitoes in the ranavirus transmission dynamic yet to be determined.

CMTV-like and ATV-like viruses have been detected infecting both marine and freshwater fishes for decades (Jensen et al. 1979, Jancovich et al. 2015, Ariel et al. 2016, Holopainen et al. 2016, Subramaniam et al. 2016, Price et al. 2017). Even though CMTV-like ranaviruses can affect the full breadth of the aquatic vertebrates (reptiles and fish included), ATV-like variants are fish specialists with only one member of the ATV clade infecting amphibians. Two models of ancestral host identification in the amphibian-associated ranaviruses have been proposed: while the first involved an ancestral fish host and multiple subsequent jumps into amphibians, the second model involved an ancestral amphibian host which jumped into fish (Jancovich et al. 2010). When considering the amphibian-specific ATV-like ranaviruses, Price et al. (2017) suggested that ancestral fish jumping into amphibians may have led to subsequent specialization and adaptation of the ATV strain to ambystomatid salamanders.

Co-Infections

As the great majority of hosts may carry several pathogens, co-infections within a host represent a common context in disease dynamics (Pedersen and Fenton 2007). The interactions between pathogens infecting the same host may produce different outcomes. On the one hand, via cross-immunity or trophic interactions, one pathogen may have a negative effect on the development of the other (antagonism), resulting in a negative relationship between infection levels of pathogens. On the other hand, by diverting or suppressing further immune response of the host, one pathogen may promote the development of the other (synergism), resulting in a positive relationship between infection levels of the pathogens in question (Pedersen and Fenton 2007). It has been noticed that epidemics do not act in isolation either. The term 'synzootic' has been proposed to describe a worse health outcome of infected animals as a result of interactions of two or more enzootic or epizootic disease processes (Sweeny et al. 2021). This recently described concept offers new valuable insight into the complex infection outcome and may explain disease emergence in many studied cases (e.g., Whitfield et al. 2013, McDonald et al. 2020).

Co-infection by *Bd* and ranavirus (e.g., Whitfield et al. 2013, Warne et al. 2016, Olori et al. 2018, Ayres et al. 2020, Bosch et al. 2020a) and by *Bd* and *Bsal* (Lötters et al. 2018) have been documented. Nevertheless, most of the studies simply focus on reporting pathogens' co-occurrence and do not directly address their potential interaction and effects. For example, Whitfield et al. (2013) found a positive association between *Bd* and ranavirus infection, suggesting that infection by one pathogen could promote higher susceptibility of the host to the other. Yet, another study in Serra da Estrela (Portugal) found no strong evidence of cumulative or amplified effects caused by the asynchronous emergence of the two lethal pathogens (Rosa et al. 2017). In fact, ranavirus drove the declines of host assemblages and changed host community composition and structure, even at sites where *Bd* was not having deleterious impacts (Rosa et al. 2013, 2017). The results are supported by Bosch et al. (2020a) where the infection with one pathogen did not increase the probability of co-infection in two susceptible species (*A. obstetricans* and *I. alpestris*). Moreover, the negative association between pathogens was observed when the presence of one was linked to a lower infection intensity in the other. The ecology of hosts and the distinct temperature requirements of both pathogens (often leading to asynchronous emergences) could help explain these findings.

Antagonistic interactions have also been documented involving amphibian pathogens. Consistent outcomes were found across anuran species when these were infected with trematode (*Echinoparyphium* spp.) following ranavirus exposure. The macroparasite infection resulted in lower infection intensity of ranavirus and increased survival of coinfected individuals (Wuerthner et al. 2017). A similar species-depend interaction was also reported between *Bd* and *Bsal*; even though the pre-exposure to low virulent isolates of *Bd* had no effect on *Bsal*-induced mortality or infection course in *S. salamandra*, reduced *Bsal*-induced mortality was observed in *T. marmoratus* and *Pleurodeles waltl* (Greener et al. 2020). On the other hand, simultaneous co-infection by *Bd* and *Bsal* has resulted, in individuals of *N. viridescens*, in the clearance of *Bd* and a high mortality rate driven by *Bsal* persistence (Longo et al. 2019).

Interactions between pathogens within a host, including increases in their effect, may greatly depend on the specifics of the host, the pathogens, and the ecological and environmental conditions in which they occur (e.g., Bosch et al. 2020a). Therefore, experimentally controlled, or natural systems studies of co-infections, have considerable limitations that may hinder the generality of their findings and require cautious interpretation (Fenton et al. 2014, McArdle et al. 2018).

Transkingdom and Interspecies Interactions

Commensal Bacteria and Fungi

Innate mechanisms prevent the colonization of amphibian skin by pathogens (Rollins-Smith et al. 2005) and the major histocompatibility complex (MHC) heterozygosity is a significant predictor of survival across populations (Savage and Zamudio 2011). Additionally, antimicrobial peptides (AMPs) produced by amphibian hosts can contribute to the suppression of disease with mixed results (Rollins-Smith 2009, Briggs et al. 2010, McMahon et al. 2014). However, the skin of amphibians also hosts a complex microbial assemblage, including taxa from both bacterial and fungi communities (Rebollar et al. 2020). While some of them have evolved to survive on amphibian skin, others may be obtained from the environment and have a transient character (Muletz et al. 2012). These communities of microbes (microbiome) play relevant functions for host health and survival (Berg 2009, Ottman et al. 2012, Ross et al. 2019), including host susceptibility to a pathogen and disease (e.g., Ford and King 2016, Warne et al. 2019).

Microbiomes may support host resistance through the production of compounds with antimicrobial properties, or by limiting pathogen adhesion to host cells (e.g., Hooper et al. 2012, Kamada et al. 2012, Buffie and Pamer 2013). The structure and maintenance of symbiont´s diversity on amphibian skin depend on a group of biotic and abiotic factors such as species development (e.g., Griffiths et al. 2018, Prest et al. 2018), presence or abundance of the pathogen (e.g., Jani et al. 2017, Bates et al. 2018, Harrison et al. 2019), and the characteristic of the environment at a landscape (e.g., Bates et al. 2018, Kueneman et al. 2019) or microhabitat scale (e.g., Rebollar et al. 2016, Garcia-Recinos et al. 2019). Any disruption or imbalance caused in the structure and composition of these bacterial assemblages may then lead to increasing host susceptibility to a disease (e.g., Brown et al. 2012, Harrison et al. 2019).

Amphibian skin microbiomes have also been shown to change over time and host life stages (e.g., Kueneman et al. 2014, Bates et al. 2018). Within the same populations, Bates et al. (2018) found few shared taxa between larvae and metamorphic midwife toads. Moreover, the authors noted a signature of the population infection history on the host's skin microbiota. In populations exhibiting epizootic disease dynamics, the bacterial alpha diversity was significantly lower compared to those populations presenting enzootic dynamics (Bates et al. 2018). Harrison et al. (2019) have shown that ranavirus may produce changes in microbiome structure. Even though these alterations were irrespective of total microbial diversity, individuals with higher microbiome diversity before ranavirus exposure, appeared to exhibit higher survival. On the other hand, in the study of Jervis et al. (2021), the increased probability of infection in stream-breeding amphibians was associated with increased abundance and diversity of non-*Batrachochytrium* chytrid fungi in the skin and environmental microbiome. The increased alpha diversity and relative abundance of fungi were higher in the skin microbiome of pond-breeding adult amphibians than in stream-breeding adult amphibians. Nevertheless, this association has not been observed for bacteria. Finally, pond tadpoles exhibited higher proportions of predicted protective microbial taxa than stream tadpoles, suggesting increased resistance to infection (Jervis et al. 2021).

As an often-common result, these studies suggest that higher diversity of microbiome communities offers enhanced protection against pathogens, where skin-associated bacteria or fungi can be more effective in competition than the invading pathogen (Ma et al. 2016). Another possible explanation relies on the presence of skin bacteria metabolites, which can have anti-fungal properties and thus an important component of extended immune defense. When bacteria are co-cultured, the production of anti-pathogen metabolites increases, making them more effective in limiting the colonization by pathogens or in reducing their growth (Becker et al. 2009, Lam et al. 2010, Muletz-Wolz et al. 2017). Nevertheless, and despite the ubiquitous presence of bacteria with anti-fungal capacity in all culturable phyla, only a small proportion exhibit broad-spectrum inhibition towards *Bd* (Antwis et al. 2015, Bletz et al. 2017, Catenazzi et al. 2018, Rebollar et al. 2020). Therefore, a bacteria consortium is likely to provide a greater antimicrobial capacity (Antwis et al. 2015) and thus higher biotic resistance (Woodhams et al. 2007a, Lam et al. 2010, Burkart et al. 2017, Jervis et al. 2021).

Microscopic and Small Aquatic Predators

Microscopic and small predators are an integral part of any freshwater environment, including planktonic crustaceans, rotifers and other free-swimming protozoans. These aquatic predators provide an important role in nutrient recycling, consuming organic detritus, algae, and other microbes (Gliwicz 2008). This way they can be seen as an environmental barrier that pathogens must face before contact with uninfected individuals. Several laboratory-based studies have focused on the role some of these micro creatures may play in *Bd* or ranavirus epidemiology. The first evidence of *Bd* zoospores' consumption came from *Daphnia* (Buck et al. 2011, Hamilton et al. 2012). This predator is not only capable of dramatically decreasing *Bd* abundance in the water column but leads to a reduction in infection intensity in tadpoles (Hamilton et al. 2012, Searle et al. 2013). Yet, these findings seem to be ecological context-dependent, varying with host species, *Daphnia* density and algal concentration (Searle et al. 2013).

In addition to planktonic crustaceans, several microorganisms, including ciliates (*Paramecium caudatum* and *P. aurelia*) and rotifers (*Notommatidae* spp. and *Lecane stichaea*) have been shown to be effective consumers of *Bd*'s free-living infectious zoospores in Pyrenean mountain lakes (Schmeller et al. 2014). Furthermore, infection prevalence and intensity in larvae of *Alytes obstetricans* and *Discoglossus scovazzi* have been negatively correlated with the presence of these microorganisms in water (Schmeller et al. 2014). The higher abundance of *Bd*-predators has been shown to inhibit aquatic zoospore survival also in bromeliad water across Central America, resulting in five times lower *Bd* prevalence than in stream microhabitats with a large portion of the protist taxa (Blooi et al. 2017). Nevertheless, even though the protists *P. caudatum* and *P. aurelia* in Schmeller et al. (2014) study have been linked with a decrease of *Bd* zoospores, it seems that a large portion of the protist taxa has not this ability likely due to their different feeding strategy or autotrophic style (Blooi et al. 2017).

In contrast to *Bd*, fewer studies have been conducted to determine the importance of aquatic predators in ranavirus epidemiology. *Daphnia* has been shown to reduce the persistence of ranavirus in the environment by inactivation of its viral particles, suggesting a possible impact on the ranavirus dynamic as well (Johnson and Brunner 2014). Another study investigated the role of scavenging invertebrates in reducing the transmission of ranavirus (Le Sage et al. 2019). While carcasses can be very an important route for ranavirus transmission (Pearman et al. 2004, Brunner et al. 2007), scavenging invertebrates can dramatically reduce ranaviral transmission. The hypothesis that scavengers could increase transmission when moving infectious tissues throughout the environment has not been confirmed. In contrast, scavenging results in negligible water-borne transmission (Le Sage et al. 2019).

Mitigating the Impacts of Emerging Infectious Diseases

Understanding the factors limiting infection risk and disease progression is essential for the development of an effective strategy to maintain amphibian diversity locally and globally. Unfortunately, the list of studies addressing ecological drivers of emerging pathogens outweighs the very limited number of available disease control attempts (reviewed by Woodhams et al. 2011, Garner et al. 2016, Scheele et al. 2019b, Thomas et al. 2019). Current mitigation measures involve either environment or host manipulation (Table 2).

Table 2. Suggested mitigation tools for fighting pathogens *in situ* and *ex situ*.

Mitigation action	Description	References
Trade restrictions and import controls	Establishing import bans/restrictions of amphibian vectors of pathogens, mandatory quarantine and strengthened biosecurity measures	EFSA 2018, Stark et al. 2018, Thomas et al. 2019
Increasing resistance to infections	Selective breeding or genetic engineering	Garner et al. 2016, Thomas et al. 2019
	Immunization and vaccination	Ramsey et al. 2010, Woodhams et al. 2011, McMahon et al. 2014
	Bioaugmentation by inoculating beneficial probiotics to the animal host or into the habitat	Woodhams et al. 2007b, Muletz et al. 2012, Bletz et al. 2013, 2018, Knapp et al. 2022
Biocontrol	Using of aquatic predators for the pathogen removal from the environment	Johnson and Brunner 2014, Schmeller et al. 2014, Blooi et al. 2017
Culling	Reduction host abundance or density	Woodhams et al. 2011, Canessa et al. 2018, Bosch et al. 2020c,
Manipulation of salinity levels	Increasing salinity level of the aquatic environment	Heard et al. 2014, Stockwell et al. 2015
Fungicides application	Fungicides treatment of the host and/or application in the habitat	Garner et al. 2009b, Bosch et al. 2015, Hudson et al. 2016, Geiger et al. 2017, Knapp et al. 2022
Temperature manipulation	Individuals exposed to elevated temperatures, elevating the temperature in the environment	Geiger et al. 2011, Scheele et al. 2015, Hettyey et al. 2019, Price et al. 2019
Water draining	Drying breeding ponds	Bosch et al. 2015, Fernández-Loras et al. 2019b
No action	Unmitigated course of pathogens in a population during the epidemic phase	Canessa et al. 2018, Bozzuto et al. 2020

Nevertheless, despite some of these methods being successfully tested under controlled laboratory conditions, attempts to apply them in wild populations remain very limited. This is particularly true for *Bd*, but even more marked in the case of *Bsal* and ranavirus, whose mitigation actions are neglectable or inexistent. Indeed, certain methods are either impractical outside the laboratory, require a disproportionate effort, are associated with immense financial costs, or can be harmful to the environment (Hettyey et al. 2019). Therefore, all these disadvantages make their implementation very complex or risky, especially when their reliability is not guaranteed.

But even so, the few methods that have been tested in the field have not yielded positive results. For example, while reducing host density is frequently suggested, Bosch et al. (2020c) demonstrated that only heavy culling efforts—simply by removing hosts from the system—will probably succeed. Unfortunately, this approach is not possible in many natural systems either due to the existence of other reservoirs or simply because of the large abundance of hosts or their difficulty to be captured. Similar unsatisfactory results were again obtained when testing severe breeding habitat intervention (complete drying and fencing for the whole breeding season): Fernández-Loras et al. (2019b) found that the intervention only achieved temporary success. Immediate *Bd* load reduction succeeded in another field experiment which included the antifungal drug itraconazole in populations of the endangered *Rana muscosa* during, or immediately after, chytridiomycosis epidemics (Knapp et al. 2022). Even though the treatment increased adult frog survival, subadult *Bd* load and survival returned to pre-treatment levels in less than one year. Neither augmentation of the skin microbiome of subadults with a commensal bacterium with antifungal properties was successful. Concentrations of this bacterium on frogs declined within one month and did not result in a protective effect against *Bd*. This suggests that *Bd* cannot be eliminated without controlling for its (multiple) reservoirs in the surrounding breeding sites (Bosch et al. 2015). That was, in fact, the only successful elimination of *Bd* from a natural system, probably due to the combination of water draining with complete environmental chemical disinfection (including surrounding areas to breeding pools). Although the direct application of chemicals to the environment remains very controversial and is not suitable for every natural system, this could be the unique possible way to combat fungal diseases in nature.

It is clear that the complexity of new and emerging pathogens, and the current epidemiological gaps in the knowledge, make the implementation of conservation strategies difficult and challenging (Bienentreu and Lesbarrères 2020). Unfortunately, the lack of published studies reporting on successful and failed attempts to mitigate diseases in nature may be indicating that field trials are not a priority of many researchers (Garner et al. 2016). Nonetheless, without effective methods to combat these diseases *in situ*, this 'emerging pathogen crisis' will continue impairing global amphibian biodiversity.

Acknowledgement

BT was supported by a 'Doctorados Industriales de la Comunidad de Madrid' grant (Spain, ref. IND2020/AMB-17438). GMR and JB were supported by FCT (Fundação para a Ciência e a Tecnologia) through R&D project funding (PTDC/

BIA-CBI/2434/2021). We thank Benedikt R. Schmidt, Joice Ruggeri and Phillip Jervis for providing insightful comments on earlier versions of this text.

References

Altizer, S., R. Bartel and B.A. Han. 2011. Animal migration and infectious disease risk. Science 331: 296–302.

Altizer S., R.S. Ostfeld, P.T. Johnson, S. Kutz and C.D. Harvell. 2013. Climate change and infectious diseases: from evidence to a predictive framework. Science 341: 514–519.

An, D. and B. Waldman. 2016. Enhanced call effort in Japanese tree frogs infected by amphibian chytrid fungus. Biol. Lett. 12: 20160018.

Anderson, R.M. and R.M. May. 1992. Infectious Diseases of Humans: Dynamics and Control. Oxford University Press, Oxford, UK.

Anlauf, K.J. and C.M. Moffitt. 2008. Models of stream habitat characteristics associated with tubificid populations in an intermountain watershed. Hydrobiologia 603: 147–158.

Antwis, R.E., R.F. Preziosi, X.A. Harrison and T.W.J. Garner. 2015. Amphibian symbiotic bacteria do not show universal ability to inhibit growth of the global pandemic lineage of *Batrachochytrium dendrobatidis*. Appl. Environ. Microbiol. 18: 3706–3711.

Ariel, E., J. Kielgast, H.E. Svart, K. Larsen, H. Tapiovaara, B.B. Jensen et al. 2009a. Ranavirus in wild edible frogs *Pelophylax* kl. *esculentus* in Denmark. Dis. Aquat. Organ. 85: 7–14.

Ariel, E., N. Nicolajsen, M.-B. Christophersen, R. Holopainen, H. Tapiovaara and B.B. Jensen. 2009b. Propagation and isolation of ranaviruses in cell culture. Aquaculture 294: 159–164.

Ariel, E., N.K. Steckler, K. Subramaniam, N.J. Olesen and T.B. Waltzek. 2016. Genomic sequencing of ranaviruses isolated from turbot (*Scophthalmus maximus*) and atlantic cod (*Gadus morhua*). Genome Announc. 4: e01393-16.

Ayres, C., I. Acevedo, C. Monsalve-Carcaño, B. Thumsová and J. Bosch. 2020. Triple dermocystid-chytrid fungus-ranavirus co-infection in a *Lissotriton helveticus*. Eur. J. Wildl. Res. 66: 2018–2020.

Bacigalupe, L.D., C. Soto-Azat, C. García-Vera, I. Barría-Oyarzo and E.L. Rezende. 2017. Effects of amphibian phylogeny, climate and human impact on the occurrence of the amphibian-killing chytrid fungus. Glob. Change Biol. 23: 3543–3553.

Baláž, V., J. Vörös, P. Civiš, J. Vojar, A. Hettyey, E. Sós et al. 2014. Assessing risk and guidance on monitoring of *Batrachochytrium dendrobatidis* in Europe through identification of taxonomic selectivity of infection. Conserv. Biol. 28: 213–223.

Bancroft, B.A., N.J. Baker and A.R. Blaustein. 2008. A meta-analysis of the effects of ultraviolet B radiation and its synergistic interactions with pH, contaminants, and disease on amphibian survival. Conserv. Biol. 22: 987–996.

Bancroft, B.A., B.A. Han, C.L. Searle, L.M. Biga, D.H. Olson, L.B. Kats et al. 2011. Species-level correlates of susceptibility to the pathogenic amphibian fungus *Batrachochytrium dendrobatidis* in the United States. Biodivers. Conserv. 20: 1911–1920.

Bates, K.A., F.C. Clare, S. O'Hanlon, J. Bosch, L. Brookes, K. Hopkins et al. 2018. Amphibian chytridiomycosis outbreak dynamics are linked with host skin bacterial community structure. Nat. Commun. 9: 693.

Battaglin, W.A., K.L. Smalling, C. Anderson, D. Calhoun, T. Chestnut and E. Muths. 2016. Potential interactions among disease, pesticides, water quality and adjacent land cover in amphibian habitats in the United States. Sci. Total Environ. 566: 320–332.

Becker, C.G., D. Rodriguez, A.V. Longo, A.L. Talaba and K.R. Zamudio. 2012. Disease risk in temperate amphibian populations is higher at closed-canopy sites. PLoS ONE 7: e48205.

Becker, C.G., D. Rodriguez, L.F. Toledo, A.V. Longo, C. Lambertini, D.T. Corrêa et al. 2014. Partitioning the net effect of host diversity on an emerging amphibian pathogen. Proc. R. Soc. B 281: 20141796.

Becker, M.H., R.M. Brucker, C.R. Schwantes, R.N. Harris and K.P.C. Minbiole. 2009. The bacterially produced metabolite violacein is associated with survival of amphibians infected with a lethal fungus. Appl. Environ. Microbiol. 75: 6635–6638.

Begon, M., M. Bennett, R.G. Bowers, N.P. French, S.M. Hazel and J. Turner. 2002. A clarification of transmission terms in host–microparasite models: numbers, densities and areas. Epidemiol. Infect. 129: 147–153.

84 *Evolutionary Ecology of Amphibians*

Begon, M. 2008. Effects of host diversity on disease dynamics. pp. 12–29. *In*: Ostfeld, R.S., F.F. Keesing and V. Eviner [eds.]. Infectious Disease Ecology: Effects of Ecosystems on Disease and of Disease on Ecosystems. Princeton University Press, Princeton, USA.

Bellard, C., C. Bertelsmeier, P. Leadley, W. Thuiller and F. Courchamp. 2012. Impacts of climate change on the future of biodiversity. Ecol. Lett. 15: 365–377.

Belzer, W. and S. Seiber. 2011. A natural history of *Ranavirus* in an eastern box turtle population. Turt. Tortoise Newsl. 15: 18–25.

Bennett, E., S. Carpenter and N. Caraco. 2001. Human impact on erodable phosphorus and eutrophication: A global perspective. BioScience 51: 227–234.

Berg, G. 2009. Plant-microbe interactions promoting plant growth and health: Perspectives for controlled use of microorganisms in agriculture. Appl. Microbial. Biotechnol. 84: 11–18.

Berger, L., R. Speare, P. Daszak, D.E. Green, A.A. Cunningham, C.L. Goggin et al. 1998. Chytridiomycosis causes amphibian mortality associated with population declines in the rain forests of Australia and Central America. Proc. Natl. Acad. Sci. U.S.A. 95: 9031–9036.

Berger, L., R. Speare, H.B. Hines, G. Marantelli, A.D. Hyatt, K.R. McDonald et al. 2004. Effect of season and temperature on mortality in amphibians due to chytridiomycosis. Aust. Vet. J. 82: 434–439.

Berger, L., A.D. Hyatt, R. Speare and J.E. Longcore. 2005. Life cycle stages of the amphibian chytrid *Batrachochytrium dendrobatidis*. Dis. Aquat. Organ. 68: 51–63.

Beyer, S.E., C.A. Phillips and R.L. Schooley. 2015. Canopy cover and drought influence the landscape epidemiology of an amphibian chytrid fungus. Ecosphere 6: 78.

Bielby, J., N. Cooper, A.A. Cunningham, T.W.J. Garner and A. Purvis. 2008. Predicting susceptibility to future declines in the world's frogs. Conserv. Lett. 1: 82–90.

Bielby, J., S.J. Price, C. Monsalve-Carcaño and J. Bosch. 2021. Host contribution to parasite persistence is consistent between parasites and over time, but varies spatially. Ecol. Appl. 31: e02256.

Bielby. J., C. Sausor, C. Monsalve-Carcaño and J. Bosch. 2022. Temperature and duration of exposure drive infection intensity with the amphibian pathogen *Batrachochytrium dendrobatidis*. PeerJ 10: e12889.

Bienentreu, J.F. and D. Lesbarrères. 2020. Amphibian disease ecology: Are we just scratching the surface? Herpetologica 76: 153–166.

Blaustein, A.R., J.M. Kiesecker, D.P. Chivers and R.G. Anthony. 1997. Ambient UV-B radiation causes deformities in amphibian embryos. Proc. Natl. Acad. Sci. U.S.A. 94: 13735–13737.

Blaustein, A.R., J.M. Romansic, E.A. Scheessele, B.A. Han, A.P. Pessier and J.E. Longcore. 2005. Interspecific variation in susceptibility of frog tadpoles to the pathogenic fungus *Batrachochytrium dendrobatidis*. Conserv. Biol. 19: 1460–1468.

Blaustein, A.R. and B.A. Bancroft. 2007. Amphibian population declines: evolutionary considerations. BioScience 57: 437–444.

Bletz, M.C., A.H. Loudon, M.H. Becker, S.C. Bell, D.C. Woodhams, K.P.C. Minbiole et al. 2013. Mitigating amphibian chytridiomycosis with bioaugmentation: Characteristics of effective probiotics and strategies for their selection and use. Ecol. Lett. 16: 807–820.

Bletz, M.C., J. Myers, D.C. Woodhams, F.C.E. Rabemananjara, A. Rakotonirina, C. Weldon et al. 2017. Estimating herd immunity to amphibian chytridiomycosis in Madagascar based on the defensive function of amphibian skin bacteria. Front. Microbiol. 8: 1751.

Bletz, M.C., M. Kelly, J. Sabino-Pinto, E. Bales, S. van Praet, W. Bert, F. Boyen et al. 2018. Disruption of skin microbiota contributes to salamander disease. Proc. R. Soc. B 285: 20180758.

Blooi, M., A.E. Laking, A. Martel, F. Haesebrouck, M. Jocque, T. Brown et al. 2017. Host niche may determine disease-driven extinction risk. PLoS ONE 12: e0181051.

Bollinger, T.K., J. Mao, D. Schock, R.M. Brigham and V.G. Chinchar. 1999. Pathology, isolation and preliminary molecular characterization of a novel iridovirus from tiger salamanders in Saskatchewan. J. Wildl. Dis. 35: 413–429.

Bosch, J., I. Martínez-Solano and M. García-París. 2001. Evidence of a chytrid fungus infection involved in the decline of the common midwife toad (*Alytes obstetricans*) in protected areas of Central Spain. Biol. Conserv. 97: 331–337.

Bosch, J. and I. Martínez-Solano. 2006. Chytrid fungus infection related to unusual mortalities of *Salamandra salamandra* and *Bufo bufo* in the Peñalara Natural Park, Spain. Oryx. 40: 84–89.

Bosch, J., L.M. Carrascal, L. Durán, S. Walker and M.C. Fisher. 2007. Climate change and outbreaks of amphibian chytridiomycosis in a montane area of Central Spain; is there a link? Proc. R. Soc. B 274: 253–260.

Bosch, J., D. García-Alonso, S. Fernández-Beaskoetxea, M.C. Fisher and T.W.J. Garner. 2013. Evidence for the introduction of lethal chytridiomycosis affecting wild betic midwife toads (*Alytes dickhilleni*). EcoHealth 10: 82–89.

Bosch, J., E. Sanchez-Tomé, A. Fernández-Loras, J.A. Oliver, M.C. Fisher and T.W.J. Garner. 2015. Successful elimination of a lethal wildlife infectious disease in nature. Biol. Lett. 11: 20150874.

Bosch, J., S. Fernández-Beaskoetxea, T.W.J. Garner and L.M. Carrascal. 2018. Long-term monitoring of an amphibian community after a climate change- and infectious disease-driven species extirpation. Glob. Change Biol. 24: 2622–2632.

Bosch, J., C. Monsalve-Carcaño, S.J. Price and J. Bielby. 2020a. Single infection with *Batrachochytrium dendrobatidis* or *Ranavirus* does not increase probability of co-infection in a montane community of amphibians. Sci. Rep. 10: 21115.

Bosch, J., B. Thumsová, R. Velarde and A. Martínez-Silvestre. 2020b. Diferente susceptibilidad de *Rana pyrenaica* a dos enfermedades emergentes de anfibios. VI Jornada de Investigación. Parque Nacional de Ordesa y Monte Perdido.

Bosch, J., L.M. Carrascal, A. Manica and T.W.J. Garner. 2020c. Significant reductions of host abundance weakly impact infection intensity of *Batrachochytrium dendrobatidis*. PLoS ONE 15: e0242913.

Bosch, J., S. Elvira, C. Sausor, J. Bielby, I. González-Fernández, R. Alonso et al. 2021a. Increased tropospheric ozone levels enhance pathogen infection levels of amphibians. Sci. Total Environ. 759: 143461.

Bosch, J., B. Thumsová, N. López-Rojo, J. Pérez, A. Alonso, M.C. Fisher et al. 2021b. Microplastics increase susceptibility of amphibian larvae to the chytrid fungus *Batrachochytrium dendrobatidis*. Sci. Rep. 11: 22438.

Bosch, J., A. Martel, J. Sopniewski, B. Thumsová, C. Ayres, B.C. Scheele et al. 2021c. *Batrachochytrium salamandrivorans* threat to the iberian urodele hotspot. J. Fungi 7: 1–16.

Bovero, S., G. Sotgiu, C. Angelini, S. Doglio, E. Gazzaniga, A.A. Cunningham et al. 2008. Detection of chytridiomycosis caused by *Batrachochytrium dendrobatidis* in the endangered sardinian newt (*Euproctus platycephalus*) in Southern Sardinia, Italy. J Wildl. Dis. 44: 712–715.

Bozzuto, C., B.R. Schmidt and S. Canessa. 2020. Active responses to outbreaks of infectious wildlife diseases: objectives, strategies and constraints determine feasibility and success. Proc. R. Soc. B 287: 20202475.

Bradley, P.W., P.W. Snyder and A.R. Blaustein. 2019. Host age alters amphibian susceptibility to *Batrachochytrium dendrobatidis*, an emerging infectious fungal pathogen. PLoS ONE 14: e0222181.

Brand, M.D., R.D. Hill, R. Brenes, J.C. Chaney, R.P. Wilkes, L. Grayfer et al. 2016. Water temperature affects susceptibility to *Ranavirus*. EcoHealth 13: 350–359.

Brem, F.M.R. and K.R. Lips. 2008. *Batrachochytrium dendrobatidis* infection patterns among Panamanian amphibian species, habitats and elevations during epizootic and enzootic stages. Dis. Aquat. Organ. 81: 189–202.

Brenes, R. 2013. Mechanisms contributing to the emergence of ranavirus in ectothermic vertebrate communities. Ph.D. Thesis, University of Tennessee, Knoxville, TN, USA.

Brenes, R., M.J. Gray, T.B. Waltzek, R.P. Wilkes and D.L. Miller. 2014. Transmission of *Ranavirus* between ectothermic vertebrate hosts. PLoS ONE 9: e92476.

Bridges, C.M. and R.D. Semlitsch. 2000. Variation in pesticide tolerance of tadpoles among and within species of ranidae and patterns of amphibian decline. Conserv. Biol. 14: 1490–1499.

Briggs, C.J., R.A. Knapp and V.T. Vredenburg. 2010. Enzootic and epizootic dynamics of the chytrid fungal pathogen of amphibians. Proc. Natl. Acad. Sci. USA 107: 9695–9700.

Brown, K., D. DeCoffe, E. Molcan and D.L. Gibson. 2012. Diet-induced dysbiosis of the intestinal microbiota and the effects on immunity and disease. Nutrients 4: 1095–1119.

Brühl, C., T. Schmidt, S. Pieper and A. Alscher. 2013. Terrestrial pesticide exposure of amphibians: An underestimated cause of global decline? Sci. Rep. 3: 1135.

Brunner, J.L., D.M. Schock, E.W. Davidson and J.P. Collins. 2004. Intraspecific reservoirs: complex life history and the persistence of a lethal ranavirus. Ecology 85: 560–566.

Brunner, J.L., D.M. Schock and J.P. Collins. 2007. Transmission dynamics of the amphibian ranavirus *Ambystoma tigrinum* virus. Dis. Aquat. Organ. 77: 87–95.

Brunner, J.L., A. Storfer, M.J. Gray and J.T. Hoverman. 2015. Ranavirus ecology and evolution: from epidemiology to extinction. pp. 71–104. *In*: Gray, M.J. and V.G. Chinchar [eds.]. Ranaviruses: Lethal Pathogens of Ectothermic Vertebrates. Springer International Publishing, Berlin, Germany.

Brunner, J.L., L. Beaty, A. Guitard and D. Russell. 2017. Heterogeneities in the infection process drive ranavirus transmission. Ecology 98: 576–582.

Brunner, J.L. and C.M. Yarber. 2018. Evaluating the importance of environmental persistence for *Ranavirus* transmission and epidemiology. Adv. Virus Res. 101: 129–148.

Buck, J.C., L. Truong and A.R. Blaustein. 2011. Predation by zooplankton on *Batrachochytrium dendrobatidis*: Biological control of the deadly amphibian chytrid fungus? Biodivers. Conserv. 20: 3549–3553.

Buffie, C.G. and E.G. Pamer. 2013. Microbiota-mediated colonization resistance against intestinal pathogens. Nat. Rev. Immunol. 13: 790–801.

Burkart, D., S.V. Flechas, V.T. Vredenburg and A. Catenazzi. 2017. Cutaneous bacteria, but not peptides, are associated with chytridiomycosis resistance in Peruvian marsupial frogs. Anim. Conserv. 20: 483–491.

Burrowes, P.A. and I. De la Riva. 2017. Detection of the amphibian chytrid fungus *Batrachochytrium dendrobatidis* in museum specimens of andean aquatic birds: Implications for pathogen dispersal. J. Wildl. Dis. 53: 349–355.

Canessa, S., C. Bozzuto, E.H.C. Grant, S.S. Cruickshank, M.C. Fisher, J.C. Koella et al. 2018. Decision making for mitigating wildlife diseases: from theory to practice for an emerging fungal pathogen of amphibians. J. Appl. Ecol. 55: 1987–1996.

Cardinale, B.J., J.E. Duffy, A. Gonzalez, D.U. Hooper, C. Perrings, P. Venail et al. 2012. Biodiversity loss and its impact on humanity. Nature 486: 59–67.

Carter, E.D., D.L. Miller, A.C. Peterson, W.B. Sutton, J.P.W. Cusaac, J.A. Spatz et al. 2020. Conservation risk of *Batrachochytrium salamandrivorans* to endemic lungless salamanders. Conserv. Lett. 13: e12675.

Carvalho T., C.G. Becker and L.F. Toledo. 2017. Historical amphibian declines and extinctions in Brazil linked to chytridiomycosis. Proc. R. Soc. B 284: 20162254.

Castro Monzon, F., M.O. Rödel and J.M. Jeschke. 2020. Tracking *Batrachochytrium dendrobatidis* infection across the globe. EcoHealth 17: 270–279.

Catenazzi, A., E. Lehr, L.O. Rodriguez and V.T. Vredenburg. 2011. *Batrachochytrium dendrobatidis* and the collapse of anuran species richness and abundance in the upper Manu National Park, southeastern Peru. Conserv. Biol. 25: 382–391.

Catenazzi, A., S.V. Flechas, D. Burkart, N.D. Hooven, J. Townsend and V.T. Vredenburg. 2018. Widespread elevational occurrence of antifungal bacteria in Andean amphibians decimated by disease: A complex role for skin symbionts in defense against chytridiomycosis. Front. Microbiol. 9: 465.

Clare, F.C., J.B. Halder, O. Daniel, J. Bielby, M.A. Semenov, T. Jombart et al. 2016. Climate forcing of an emerging pathogenic fungus across a montane multihost community. Philos. Trans. R. Soc. B 371: 20150454.

CLRTAP. 2017. Mapping Critical Levels for Vegetation. Manual on Methodologies and Criteria for Modelling and Mapping Critical Loads and Levels and Air Pollution Effects, Risks and Trends. Umwelbundesamt, Berlin, Germany.

Clulow, S., J. Gould, H. James, M. Stockwell, J. Clulow and M. Mahony. 2017. Elevated salinity blocks pathogen transmission and improves host survival from the global amphibian chytrid pandemic: Implications for translocations. J. Appl. Ecol. 55: 830–840.

Cohen, J.M., D. Venesky and E.L. Sauer. 2017. The thermal mismatch hypothesis explains host susceptibility to an emerging infectious disease. Ecol. Lett. 20: 184–193.

Cohen, J.M., D.J. Civitello, M.D. Venesky, T.A. McMahon and J.R. Rohr. 2018. An interaction between climate change and infectious disease drove widespread amphibian declines. Glob. Change Biol. 25: 927–937.

Cohen, J.M., T.A. McMahon, C. Ramsay, E.A. Roznik, E.L. Sauer, S. Bessler et al. 2019. Impacts of thermal mismatches on chytrid fungus *Batrachochytrium dendrobatidis* prevalence are moderated by life stage, body size, elevation and latitude. Ecol. Lett. 22: 817–825.

Collins, J.P. and A. Storfer. 2003. Global amphibian declines: sorting the hypotheses. Divers. Distrib. 9: 89–98.

Conlon, J.M., J. Kolodziejek and N. Nowotny. 2004. Antimicrobial peptides from ranid frogs: taxonomic and phylogenetic markers and a potential source of new therapeutic agents. Biochim. Biophys. Acta 1696: 1–14.

Corn, P.S. 2005. Climate change and amphibians. Anim. Biodivers. Conserv. 28: 59–67.

Craft, M.E., P.L. Hawthorne, C. Packer and A.P. Dobson. 2008. Dynamics of a multihost pathogen in a carnivore community. J. Anim. Ecol. 77: 1257–1264.

Crispo, E. 2007. The Baldwin effect and genetic assimilation: Revisiting two mechanisms of evolutionary change mediated by phenotypic plasticity. Evolution 61: 2469–2479.

Cunningham, A.A., T.E.S. Langton, P.M. Bennett, J.F. Lewin, S.E.N. Drury, R.E. Gough et al. 1996. Pathological and microbiological findings from incidents of unusual mortality of the common frog (*Rana temporaria*). Philos. Trans. R. Soc. B 351: 1539–1557.

Cunningham, A.A., P. Daszak and J.L.N. Wood. 2017. One health, emerging infectious diseases and wildlife: two decades of progress? Philos. Trans. R. Soc. B 372: 20160167.

Cusaac, J.P.W., E.D. Carter, D.C. Woodhams, J. Robert, J.A. Spatz, J.L. Howard et al. 2021. Emerging pathogens and a current-use pesticide: Potential impacts on eastern hellbenders. J. Aquat. Anim. Health 33: 24–32.

Dalbeck, L., H. Düssel-Siebert, A. Kerres, K. Kirst, A. Koch, S. Lötters et al. 2018. Die Salamanderpest und ihr Erreger *Batrachochytrium salamandrivorans* (*Bsal*): aktueller Stand in Deutschland. Z. Feldherpetol. 25: 1–22.

Daskin, J.H., R.A. Alford and R. Puschendorf. 2011. Short-term exposure to warm microhabitats could explain amphibian persistence with *Batrachochytrium dendrobatidis*. PLoS ONE 6: e26215.

Daszak, P., A.A. Cunningham and A.D. Hyatt. 2000. Emerging infectious diseases of wildlife—threats to biodiversity and human health. Science 287: 443–449.

Daszak, P., A.A. Cunningham and A.D. Hyatt. 2003. Infectious disease and amphibian population declines. Divers. Disturb. 9: 141–150.

Daversa, D.R., E. Muths and J. Bosch. 2012. Terrestrial movement patterns of the common toad (*Bufo bufo*) in Central Spain reveal habitat of conservation importance. J. Herpetol. 46: 658–664.

Daversa, D.R., A. Manica, J. Bosch, J.W. Jolles and T.W.J. Garner. 2018a. Routine habitat switching alters the likelihood and persistence of infection with a pathogenic parasite. Funct. Ecol. 32: 1262–1270.

Daversa, D.R., C. Monsalve-Carcaño, L.M. Carrascal and J. Bosch. 2018b. Seasonal migrations, body temperature fluctuations, and infection dynamics in adult amphibians. PeerJ 6: e4698.

Daversa, D.R., A. Manica, H. Bintanel Cenis, P. Lopez, T.W.J. Garner and J. Bosch. 2021. Alpine newts (*Ichthyosaura alpestris*) avoid habitats previously used by parasite-exposed conspecifics. Front. Ecol. Evol. 9: 636099.

Davidson, C., M.F. Benard, H.B. Shaffer, J. Parker, C. O'Leary, J.M. Conlon et al. 2007. Effects of chytrid and carbaryl exposure on survival, growth, and skin peptide defenses in foothill yellow-legged frogs. Environ. Sci. Technol. 41: 1771–1776.

Davis, A.K., M.J. Yabsley, M.K. Keel and J.C. Maerz. 2007. Discovery of a novel alveolate pathogen affecting southern leopard frogs in Georgia: description of the disease and host effects. EcoHealth 4:310–317.

de Castro, F. and B. Bolker. 2005. Mechanism of disease-induced extinction. Ecol. Lett. 8: 117–126.

Deutsch, C.A., J.J. Tewksbury, R.B. Huey, K.S. Sheldon, C.K. Ghalambor, D.C. Haak et al. 2008. Impacts of climate warming on terrestrial ectotherms across latitude. Proc. Natl. Acad. Sci. USA 105: 6668–6672.

DiRenzo, G.V., P.F. Langhammer, K.R. Zamudio and K.R. Lips. 2014. Fungal infection intensity and zoospore output of *Atelopus zeteki*, a potential acute chytrid supershedder. PLoS ONE 9: e93356.

Doddington, B.J., J. Bosch, J.A. Oliver, N.C. Grassly, G. Garcia, B.R. Schmidt et al. 2013. Context dependent amphibian host population response to an invading pathogen. Ecology 94: 1795–1804.

Dohm, M.R., W.J. Mautz, P.G. Looby, K.S. Gellert and J.A. Andrade. 2001. Effects of ozone on evaporative water loss and thermoregulatory behavior of marine toads (*Bufo marinus*). Environ. Res. 86: 274–286.

Dohm, M.R., W.J. Mautz, J.A. Andrade, K.S. Gellert, L.J. Salas-Ferguson, N. Nicolaisen et al. 2005. Effects of ozone exposure on nonspecific phagocytic capacity of pul- monary macrophages from an amphibian, *Bufo marinus*. Environ. Toxicol. Chem. 24: 205–210.

Duffus, A.L.J., R.A. Nichols and T.W.J. Garner. 2013. Investigations into the life history stages of the common frog (*Rana temporaria*) affected by an amphibian ranavirus in the United Kingdom. Herpetol. Rev. 44: 260–263.

Duffus, A.L.J., R.A. Nichols and T.W.J. Garner. 2014a. Experimental evidence in support of single host maintenance of a multihost pathogen. Ecosphere 5: 142.

Duffus, A.L.J., R.A. Nichols and T.W.J. Garner. 2014b. Detection of a frog virus 3-like ranavirus in native and introduced amphibians in the United Kingdom in 2007 and 2008. Herpetol. Rev. 45: 608–610.

Duffus, A.L.J., T.B. Waltzek, A.C. Stöhr, M.C. Allender, M. Gotesman, R.J. Whittington et al. 2015. Distribution and host range of ranaviruses. pp. 9–57. *In*: Gray, M.J. and V.G. Chinchar [eds.]. Ranaviruses: Lethal Pathogens of Ectothermic Vertebrates. Springer International Publishing, Berlin, Germany.

EFSA Panel on Animal Health and Welfare (AHAW), S. More, M.A. Miranda, D. Bicout, A. Bøtner, A. Butterworth et al. 2018. Risk of survival establishment, spread of *Batrachochytrium salamandrivorans* (*Bsal*) in the EU. EFSA J. 16: 5259.

Echaubard, P., K. Little, B. Pauli and D. Lesbarrères. 2010. Context-dependent effects of ranaviral infection on northern leopard frog life history traits. PLoS ONE 5: e13723.

Eskew, E.A., B.C. Shock, E.E.B. LaDouceur, K. Keel, M.R. Miller, J.E. Foley et al. 2018. Gene expression differs insusceptible and resistant amphibians exposed to *Batrachochytrium dendrobatidis*. R. Soc. Open Sci. 5: 170910.

Evans, S.S., E.A. Repasky and D.T. Fisher. 2015. Fever and the thermal regulation of immunity: the immune system feels the heat. Nat. Rev. Immunol. 15: 335–349.

Fenton, A. and A.B. Pedersen. 2005. Community epidemiology framework for classifying disease threats. Emerg. Infect. Dis. 11: 1815.

Fenton, A., S.C. Knowles, O.L. Petchey and A.B. Pedersen. 2014. The reliability of observational approaches for detecting interspecific parasite interactions: comparison with experimental results. Int. J. Parasitol. 44: 437–445.

Fenton, A., D.G. Streicker, O.L. Petchey and A.B. Pedersen. 2015. Are all hosts created equal? Partitioning host species contributions to parasite persistence in multihost communities. Am. Nat. 186: 610–622.

Fernández-Beaskoetxea, S., L.M. Carrascal, A. Fernández-Loras, M.C. Fisher and J. Bosch. 2015. Short term minimum water temperatures determine levels of infection by the amphibian chytrid fungus in *Alytes obstetricans* tadpoles. PLoS ONE 10: e0120237.

Fernández-Beaskoetxea, S., J. Bosch and J. Bielby. 2016. Infection and transmission heterogeneity of a multi-host pathogen (*Batrachochytrium dendrobatidis*) within an amphibian community. Dis. Aquat. Organ. 118: 11–20.

Fernández-Loras, A., L. Boyero, F. Correa-Araneda, M. Tejedo, A. Hettyey and J. Bosch. 2019a. Infection with *Batrachochytrium dendrobatidis* lowers heat tolerance of tadpole hosts and cannot be cleared by brief exposure to CTmax. PLoS ONE 14: e0216090.

Fernández-Loras, A., L. Boyero and J. Bosch. 2019b. *In-situ* severe breeding habitat intervention only achieves temporary success in reducing *Batrachochytrium dendrobatidis* infection. Amphibia-Reptilia 41: 261–267.

Fijan, N., Z. Matašin, Z. Petrinec, I. Valpo ć and L.O. Zwillenberg. 1991. Isolation of an iridovirus-like agent from the green frog (*Rana esculenta L.*). Vet. Arh. Zagreb 61: 151–158.

Fine, P., K. Eames and D.L. Heymann. 2011. Herd immunity: A rough guide. Clin. Infect. Dis. 52: 911–916.

Fisher, M.C. and T.W.J. Garner. 2007. The relationship between the emergence of *Batrachochytrium dendrobatidis*, the international trade in amphibians and introduced amphibian species. Fungal Biol. Rev. 21: 2–9.

Fisher, M.C., T.W.J. Garner and S.F. Walker. 2009. Global emergence of *Batrachochytrium dendrobatidis* and amphibian chytridiomycosis in space, time, and host. Annu. Rev. Microbiol. 63: 291–310.

Fisher, M.C. and T.W.J. Garner. 2020. Chytrid fungi and global amphibian declines. Nat. Rev. Microbiol. 18: 332–343.

Fitzpatrick, L.D., F. Pasmans, A. Martel and A.A. Cunningham. 2018. Epidemiological tracing of *Batrachochytrium salamandrivorans* identifies widespread infection and associated mortalities in private amphibian collections. Sci. Rep. 8: 13845.

Ford, S.A. and K.C. King. 2016. Harnessing the power of defensive microbes: evolutionary implications in nature and disease control. PLoS Pathog. 12: e1005465.

Gahl, M.K. and A.J.K. Calhoun. 2010. The role of multiple stressors in ranavirus-caused amphibian mortalities in Acadia National Park wetlands. Can. J. Zool. 88: 108–121.

Gahl, M., B. Pauli and J.E. Houlahan. 2011. Effects of chytrid fungus and glyphosate-based herbicide on survival and growth of wood frogs (*Lithobates sylvaticus*). Ecol. Appl. 21: 2521–2529.

Garcia, T.S., J.M. Romansic and A.R. Blaustein. 2006. Survival of three species of anuran metamorphs exposed to UV-B radiation and the pathogenic fungus *Batrachochytrium dendrobatidis*. Dis. Aquat. Organ. 72:163–169.

Garcia-Recinos, L., P.A. Burrowes and M. Dominguez-Bello. 2019. The skin microbiota of *Eleutherodactylus* frogs: Effects of host ecology, phylogeny, and local environment. Front. Microbiol. 10: 2571.

Garmyn, A., P. van Rooij, F. Pasmans, T. Hellebuyck, W. van Den Broeck, F. Hasebrouck et al. 2012. Waterfowl: potential environmental reservoirs of the chytrid fungus *Batrachochytrium dendrobatidis*. PLoS ONE 7: e35038.

Garner, T.W.J., M.W. Perkins, P. Govindarajulu, D. Seglie, S. Walker, A.A. Cunningham et al. 2006. The emerging amphibian pathogen *Batrachochytrium dendrobatidis* globally infects introduced populations of the North American bullfrog, *Rana catesbeiana*. Biol. Lett. 2: 455–459.

Garner, T.W.J., S. Walker, J. Bosch, S. Leech, J.M. Rowcliffe, A.A. Cunningham et al. 2009a. Life history trade-offs influence mortality associated with the amphibian pathogen *Batrachochytrium dendrobatidis*. Oikos 118: 783–791.

Garner, T.W.J., G. Garcia, B. Carroll and M.C. Fisher. 2009b. Using itraconazole to clear *Batrachochytrium dendrobatidis* infection, and subsequent depigmentation of *Alytes muletensis* tadpoles. Dis. Aquat. Organ. 83: 257–260.

Garner, T.W.J., B.R. Schmidt, A. Martel, F. Pasmans, E. Muths, A.A. Cunningham et al. 2016. Mitigating amphibian chytridiomycosis in nature. Philos. Trans. R. Soc. B 371: 20160207.

Gascon, C., J.P. Collins, R.D. Moore, D.R. Church, J.E. McKay and J.R. Mendelson III. 2007. Amphibian Conservation Action Plan. IUCN/SSC Amphibian Specialist Group, Gland, Switzerland and Cambridge, U.K.

Geiger, C.C., E. Küpfer, S. Schär, S. Wolf and B.R. Schmidt. 2011. Elevated temperature clears chytrid fungus infections from tadpoles of the midwife toad, *Alytes obstetricans*. Amphibia-Reptilia 32: 276–280.

Geiger,C.C., C. Bregnard, E. Maluenda, M.J. Voordouw and B.R. Schmidt. 2017. Antifungal treatment of wild amphibian populations caused a transient reduction in the prevalence of the fungal pathogen, *Batrachochytrium dendrobatidis*. Sci. Rep. 7: 5956.

Gervasi, S.S. and J. Foufopoulos. 2008. Costs of plasticity: Responses to desiccation decrease post-metamorphic immune function in a pond-breeding amphibian. Funct. Ecol. 22:100–108.

Gervasi, S.S., P.R. Stephens, J. Hua, C.L. Searle, G.Y. Xie, J. Urbina et al. 2017. Linking ecology and epidemiology to understand predictors of multi-host responses to an emerging pathogen, the amphibian chytrid fungus. PLoS One 12: e0167882.

Gilbert, M.J., A.M. Spitzen-van der Sluijs, S. Canessa, J. Bosch, A.A. Cunningham, E. Grasselli et al. 2020. Mitigating *Batrachochytrium salamandrivorans* in Europe. *Batrachochytrium salamandrivorans* Action Plan for European urodeles. Nijmegen, the Netherlands.

Gliwicz, Z.M. 2008. Zooplankton. Pp. 461–516. *In*: O'Sullivan, P. and C.S. Reynolds [eds.]. The Lakes Handbook: Limnology and Limnetic Ecology. John Wiley & Sons, Hoboken, NJ, USA.

González-Hernández, M., M. Denoël, A.J. Duffus, T.W.J. Garner, A.A. Cunningham and K. Acevedo-Whitehouse. 2010. Dermocystid infection and associated skin lesions in free-living palmate newts (*Lissotriton helveticus*) from Southern France. Parasitol. Int. 59: 344–350.

Goodman R.M., E.D. Carter and D.L. Miller. 2021. Influence of herbicide exposure and ranavirus infection on growth and survival of juvenile red-eared slider turtles (*Trachemys scripta elegans*). Viruses 13: 1440.

Gould, J., J.W. Valdez, M.P. Stockwell, S. Clulow and M.J. Mahony. 2019. Mosquitoes as a potential vector for the transmission of the amphibian chytrid fungus. Zool. Ecol. 29: 36–42.

Granoff, A., P.E. Came and K.A. Rafferty. 1965. The isolation and properties of viruses from *Rana pipiens*: their possible relationship to the renal adenocarcinoma of the leopard frog. Ann. N. Y. Acad. Sci. 126: 237–255.

Gray, M.J., D.L. Miller and J.T. Hoverman. 2009. Ecology and pathology of amphibian ranaviruses. Dis. Aquat. Organ. 87: 243–266.

Green, D.E., K.A. Converse and A.K. Schrade. 2002. Epizootiology of sixty-four amphibian morbidity and mortality events in the USA, 1996–2001. Ann. N. Y. Acad. Sci. 969: 323–339.

Greenberg, D.A., W.J. Palen and A. Mooers. 2017. Amphibian species traits, evolutionary history and environment predict *Batrachochytrium dendrobatidis* infection patterns, but not extinction risk. Evol. Appl. 10: 1130–1145.

Greener, M.S., E. Verbrugghe, M. Kelly, M. Blooi, W. Beukema, S. Canessa et al. 2020. Presence of low virulence chytrid fungi could protect European amphibians from more deadly strains. Nat. Commun. 11: 1–11.

Greenspan, S.E., D.S. Bower, E.A. Roznik, D.A. Pike, G. Marantelli, R.A. Alford et al. 2017a. Infection increases vulnerability to climate change via effects on host thermal tolerance. Sci. Rep. 7: 9349.

Greenspan, S.E., D.S. Bower, R.J. Webb, E.A. Roznik, L.A. Stevenson, L. Berger et al. 2017b. Realistic heat pulses protect frogs from disease under simulated rainforest frog thermal regimes. Funct. Ecol. 31: 2274–2286.

Griffiths, S.M., X.A. Harrison, C. Weldon, M.D. Wood, A. Pretorius, K. Hopkins et al. 2018. Genetic variability and ontogeny predict microbiome structure in a disease- challenged montane amphibian. ISME J. 12: 2506–2517.

Haislip, N.A., M.J. Gray, J.T. Hoverman and D.L. Miller. 2011. Development and disease: How susceptibility to an emerging pathogen changes through anuran development. PLoS ONE 6: e22307.

Häkkinen, J., S. Pasanen and J.V.K. Kukkonen. 2001. The effects of solar UV-B radiation on embryonic mortality and development in three boreal anurans (*Rana temporaria*, *Rana arvalis* and *Bufo bufo*). Chemosphere 44: 441–446.

Hall, E.M., C.S. Goldberg, J.L. Brunner and E.J. Crespi. 2018. Seasonal dynamics and potential drivers of ranavirus epidemics in Wood Frog populations. Oecologia 188: 1253–1262.

Hall, E.M., J.L. Brunner, B. Hutzenbiler and E.J. Crespi. 2020. Salinity stress increases the severity of ranavirus epidemics in amphibian populations. Proc. R. Soc. B 287: 20200062.

Halliday, F.W., R.W. Heckman, P.A. Wilfahrt and C.E. Mitchell. 2017. A multivariate test of disease risk reveals conditions leading to disease amplification. Proc. R. Soc. B 284: 20171340.

Halliday, F.W. and J.R. Rohr. 2019. Measuring the shape of the biodiversity-disease relationship across systems reveals new findings and key gaps. Nat. Commun. 10: 5032.

Hamilton, P.T., J.M.L. Richardson and B.R. Anholt. 2012. *Daphnia* in tadpole mesocosms: Trophic links and interactions with *Batrachochytrium dendrobatidis*. Freshw. Biol. 57: 676–683.

Han, B.A., P.W. Bradley and A.R. Blaustein. 2008. Ancient behaviors of larval amphibians in response to an emerging fungal pathogen, *Batrachochytrium dendrobatidis*. Behav. Ecol. Sociobiol. 63: 241–250.

Han, B.A., C.L. Searle and A.R. Blaustein. 2011. Effects of an infectious fungus, *Batrachochytrium dendrobatidis*, on amphibian predator-prey interactions. PLoS ONE 6: e16675.

Hanlon, S.M., J.L. Kerby and M.J. Parris. 2012. Unlikely remedy: Fungicide clears infection from pathogenic fungus in larval southern leopard frogs (*Lithobates sphenocephalus*). PLoS One 7: e43573.

Hanlon, S.M., K.J. Lynch, J. Kerby and M.J. Parris. 2015. *Batrachochytrium dendrobatidis* exposure effects on foraging efficiencies and body size in anuran tadpoles. Dis. Aquat. Organ. 112: 237–242.

Hanselmann, R., A. Rodríguez, M. Lampo, L. Fajardo-Ramos, A. Alonso Aguirre, A. Marm Kilpatrick et al. 2004. Presence of an emerging pathogen of amphibians in introduced bullfrogs *Rana catesbiana* in Venezuela. Biol. Conserv. 120: 115–119.

Harmos, K., J. Bosch, B. Thumsová, A. Martínez-Silvestre, R. Velarde and J. Vörös. 2021. Evidence of amphibian mortality caused by chytridiomycosis in Central Europe. Herpetol. Notes 14: 1213–1218.

Harrison, X.A., S.J. Price, K. Hopkins, W.T.M. Leung, C. Sergeant and T.W.J. Garner. 2019. Diversity-stability dynamics of the amphibian skin microbiome and susceptibility to a lethal viral pathogen. Front. Microbiol. 10: 2883.

Harvell, C.D., C.E. Mitchell, J.R. Ward, S. Altizer, A.P. Dobson, R.S. Ostfeld, et al. 2002. Climate warming and disease risks for terrestrial and marine biota. Science 296: 2158–2162.

Haydon, D.T., S. Cleaveland, L.H. Taylor and M.K. Laurenson. 2002. Identifying reservoirs of infection: A conceptual and practical challenge. Emerg. Infect. Dis. 8: 1468–1473.

Hayes, T.B., P. Falso, S. Gallipeau and M. Stice. 2010. The cause of global amphibian declines: a developmental endocrinologist's perspective. J. Exp. Biol. 213: 921–933.

Heard, G.W., M.P. Scroggie, N. Clemann and D.S.L. Ramsey. 2014. Wetland characteristics influence disease risk for a threatened amphibian. Ecol. Appl. 24: 650–662.

Hettyey, A., J. Ujszegi, D. Herczeg, D. Holly, J. Vörös, B.R. Schmidt et al. 2019. Mitigating disease impacts in amphibian populations: Capitalizing on the thermal optimum mismatch between a pathogen and its host. Front. Ecol. Evol. 7: 254.

Hite, J.L., J. Bosch, S. Fernández-Beaskoetxea, D. Medina and S.R. Hall. 2016. Joint effects of habitat, zooplankton, host stage structure and diversity on amphibian chytrid. Proc. R. Soc. B 283: 20160832.

Holopainen, R., K. Subramaniam, N.K. Steckler, S.C. Claytor, E. Ariel and T.B. Waltzek. 2016. Genome sequence of a ranavirus isolated from pike-perch *Sander lucioperca*. Genome Announc. 4: e01295–16.

Hooper, L.V., D.R. Littman and A.J. Macpherson. 2012. Interactions between the microbiota and the immune system. Science 336: 1268–1273.

Hoverman, J.T., M.J. Gray, N.A. Haislip and D.L. Miller. 2011. Phylogeny, life history, and ecology contribute to differences in amphibian susceptibility to ranaviruses. EcoHealth 8: 301–319.

Hoyt, J.R., A.M. Kilpatrick and K.E. Langwig. 2021. Ecology and impacts of white-nose syndrome on bats. Nat. Rev. Genet. 19: 196–210.

Hua, J., R. Cothran, A. Stoler and R.A. Relyea. 2013. Cross-tolerance in amphibians: Wood frog mortality when exposed to three insecticides with a common mode of action. Environ. Toxicol. Chem. 32: 932–936.

Hua, J., D.K. Jones, B.M. Mattes, R.D. Cothran, R.A. Relyea and J.T. Hoverman. 2015. The contribution of phenotypic plasticity to the evolu- tion of insecticide tolerance in amphibian populations. Evol. Appl. 8: 586–596.

Hua, J., V.P. Wuerthner, D.K. Jones, B. Mattes, R.D. Cothran, R.A. Relyea et al. 2017. Evolved pesticide tolerance influences susceptibility to parasites in amphibians. Evol. Appl. 10: 802–812.

Hudson, M.A., R.P. Young, Y. Lopez, L. Martin, C. Fenton, R. McCrea et al. 2016. *In-situ* itraconazole treatment improves survival rate during an amphibian chytridiomycosis epidemic. Biol. Conserv. 195: 37–45.

Hudson, M.A., R.A. Griffiths, L. Martin, C. Fenton, S.-L. Adams, A. Blackman et al. 2019. Reservoir frogs: seasonality of *Batrachochytrium dendrobatidis* infection in robber frogs in Dominica and Montserrat. PeerJ 7: e7021.

Huey, R.B. 1991. Physiological consequences of habitat selection. Am. Nat. 137(Suppl.): S91–S115.

Isidro-Ayza, M., W.J. Barichivich, D.L. Calhoun, D.A. Grear, J.M. Lorch and M. Winzeler. 2017. Pathogenic lineage of Perkinsea associated with mass mortality of frogs across the United States. Sci. Rep. 7: 1–10.

IUCN 2021. The IUCN Red List of Threatened Species. Version 2021-1. https://www.iucnredlist.org. Accessed on 20 of April 2021.

Jancovich, J.K., E.W. Davidson, A. Seiler, B.L. Jacobs and J.P. Collins. 2001. Transmission of the *Ambystoma tigrinum* virus to alternative hosts. Dis. Aquat. Organ. 46: 159–163.

Jancovich, J.K., M. Bremont, J.W. Touchman and B.L. Jacobs. 2010. Evidence for multiple recent host species shifts among the ranaviruses (Family Iridoviridae). J. Virol. 84: 2636–2647.

Jancovich, J.K., N.K. Steckler and T.B. Waltzek. 2015. Ranavirus taxonomy and phylogeny. pp. 59–70. *In*: Gray, M.J. and V.G. Chinchar [eds.]. Ranaviruses: Lethal Pathogens of Ectothermic Vertebrates. Springer International Publishing, Berlin, Germany.

Jani, A.J., R.A. Knapp and C.J. Briggs. 2017. Epidemic and endemic pathogen dynamics correspond to distinct host population microbiomes at a landscape scale. Proc. R. Soc. B 284: 20170944.

Jensen, N.J., B. Bloch and J.L. Larsen. 1979. The ulcus-syndrome in cod (*Gadus morhua*). III. A preliminary virological report. Nord. Vet. Med. 31: 436–442.

Jervis, P., P. Pintanel, K. Hopkins, C. Wierzbicki, J.M.G. Shelton, E. Skelly et al. 2021. Post-epizootic microbiome associations across communities of neotropical amphibians. Mol. Ecol. 30: 1322–1335.

Johnson, A.F. and J.L. Brunner. 2014. Persistence of an amphibian ranavirus in aquatic communities. Dis. Aquat. Organ. 111: 129–138.

Johnson, M.L. and R. Speare. 2003. Survival of *Batrachochytrium dendrobatidis* in water: quarantine and disease control implications. Emerg. Infect. Dis. 9: 922–925.

Johnson, M.L. and R. Speare. 2005. Possible modes of dissemination of the amphibian chytrid *Batrachochytrium dendrobatidis* in the environment. Dis. Aquat. Organ. 65: 181–186.

Jones, K.E., N.G. Patel, M.A. Levy, A. Storeygard, D. Balk, J.L. Gittleman and P. Daszak. 2008. Global trends in emerging infectious diseases. Nature 451: 990–993.

Kamada, N, Y.G. Kim, H.P. Sham, B.A. Vallance, J.L. Puente, E.C. Martens et al. 2012. Regulated virulence controls the ability of a pathogen to compete with the gut microbiota. Science 336: 1325–1329.

Karavlan, S.A. and M.D. Venesky. 2016. Thermoregulatory behavior of *Anaxyrus americanus* in response to infection with *Batrachochytrium dendrobatidis*. Copeia 104: 746–751.

Karraker, N.E., J.P. Gibbs and J.R. Vonesh. 2008. Impacts of road deicing salt on the demography of vernal pool-breeding amphibians. Ecol. Appl. 18: 724–734.

Kärvemo, S., S. Meurling, D. Berger, J. Höglund and A. Laurila. 2018. Effects of host species and environmental factors on the prevalence of *Batrachochytrium dendrobatidis* in northern Europe. PLoS ONE 13: e0199852.

Kavaliers, M., E. Choleris, A. Ågmo and D.W. Pfaff. 2004. Olfactory-mediated parasite recognition and avoidance: Linking genes to behavior. Horm. Behav. 46: 272–283.

Keesing, F., R.D. Holt and R.S. Ostfeld. 2006. Effects of species diversity on disease risk. Ecol. Lett. 9: 485–498.

Keesing, F., L.K. Belden, P. Daszak, A. Dobson, C.D. Harvell, R.D. Holt et al. 2010. Impacts of biodiversity on the emergence and transmission of infectious diseases. Nature 468: 647–652.

Kerr, J.B. and C.T. McElroy. 1993. Evidence for large upward trends of ultraviolet-B radiation linked to ozone depletion. Science 262: 1032–1034.

Kilburn, V., R. Ibáñez and D. Green. 2011. Reptiles as potential vectors and hosts of the amphibian pathogen *Batrachochytrium dendrobatidis* in Panama. Dis. Aquat. Organ. 97: 127–134.

Kimble, S.J.A., A.K. Karna, A.J. Johnson, J.T. Hoverman and R.N. Williams. 2015. Mosquitoes as a potential vector of *Ranavirus* transmission in terrestrial turtles. EcoHealth 12: 334–338.

Kluger, M.J. 1977. Fever in the frog *Hyla cinerea*. J. Therm. Biol. 2: 79–81.

Knapp R.A., M.B. Joseph, T.C. Smith, E.E. Hegeman, V.T. Vredenburg, J.E. Erdman Jr. et al. 2022. Effectivness of antifungal treatments during chytridiomycosis epizootics in populations of an endangered frog. PeerJ 10: e12712.

Knight, J. and S. Harrison. 2013. The impacts of climate change on terrestrial Earth surface systems. Nature Clim. Change 3: 24–29.

Kohli, A.K., A.L. Lindauer, L.A. Brannelly, M.E.B. Ohmer, C. Richards-Zawacki, L. Rollins-Smith et al. 2019. Disease and the drying pond: Examining possible links among drought, immune function, and disease development in amphibians. Physiol. Biochem. Zool. 92: 339–348.

Kolby, J.E., S.D. Ramirez, L. Berger, K.L. Richards-Hrdlicka, M. Jocque and L.F. Skerratt. 2015. Terrestrial Dispersal and potential environmental transmission of the amphibian chytrid fungus (*Batrachochytrium dendrobatidis*). PLoS ONE 10: e0125386.

Kriger, K.M. 2009. Lack of evidence for the drought-linked chytridiomycosis hypothesis. J. Wildlife Dis. 45: 537–541.

Kriger, K.M. and J.-M. Hero. 2007. Large-scale seasonal variation in the prevalence and severity of chytridiomycosis. J. Zool. 271: 352–359.

Kriger, K.M., F. Pereoglou and J.-M. Hero. 2007. Latitudinal variation in the prevalence and severity of chytrid (*Batrachochytrium dendrobatidis*) infection in Eastern Australia. Conserv. Biol. 21: 1280–1290.

Kriger, K.M. and J.-M. Hero. 2008. Altitudinal distribution of chytrid (*Batrachochytrium dendrobatidis*) infection in subtropical Australian frogs. Austral. Ecol. 33: 1022–1032.

Kueneman, J.G., L.W. Parfrey, D.C. Woodhams, H.M. Archer, R. Knight and V.J. McKenzie. 2014. The amphibian skin-associated microbiome across species, space and life history stages. Mol. Ecol. 23: 1238–1250.

Kueneman, J.G., M.C. Bletz, V.J. McKenzie, C.G. Becker, M.B. Joseph., J.G. Abarca et al. 2019. Community richness of amphibian skin bacteria correlates with bioclimate at the global scale. Nature Ecol. Evol. 3: 381–389.

Kuris, A.M., A.R. Blaustein and J.J. Alio. 1980. Hosts as islands. Am. Nat. 116: 570–586.

La Fauce, K., E. Ariel, S. Munns, C. Rush and L. Owens. 2012. Influence of temperature and exposure time on the infectivity of Bohle iridovirus, a ranavirus. Aquaculture 354-355: 64–67.

La Marca, E., K.R. Lips, S. Lötters, R. Puschendorf, R. Ibáñez, J.V. Rueda-Almonacid et al. 2005. Catastrophic population declines and extinctions in neotropical harlequin frogs (Bufonidae: *Atelopus*). Biotropica 37: 190–201.

Lafferty, K.D. and E.A. Mordecai. 2016. The rise and fall of infectious disease in a warmer world. F1000Res. 5: F1000 Faculty Rev-2040. doi: 10.12688/f1000research.8766.1.

Laking, A.E., H.N. Ngo, F. Pasmans, A. Martel and T.T. Nguyen. 2017. *Batrachochytrium salamandrivorans* is the predominant chytrid fungus in Vietnamese salamanders. Sci. Rep. 7: 44443.

Lam, B.A., J.B. Walke, V.T. Vredenburg and R.N. Harris. 2010. Proportion of individuals with anti–*Batrachochytrium dendrobatidis* skin bacteria is associated with population persistence in the frog *Rana muscosa*. Biol. Conserv. 143: 529–531.

Lambert, S. and M. Wagner. 2018. Microplastics are contaminants of emerging concern in fresh water environments: an overview. pp. 1–23. *In*: Wagner, M. and S. Lambert [eds.]. Freshwater Microplastics. Springer Nature, Cham, Switzerland.

Lambertini, C., C.G. Becker, A.M. Belasen, A. Valencia-Aguilar, C.H.L. Nunes-de-Almeida, C.M. Betancourt-Román et al. 2021. Biotic and abiotic determinants of *Batrachochytrium dendrobatidis* infections in amphibians of the Brazilian Atlantic Forest. Fungal Ecol. 49: 100995.

Landsberg, J.H., K.M. Enge, Y. Kiryu, A.P. Pessier, A. Preston, S. Reintjes-Tolen et al. 2013. Co-infection by alveolate parasites and frog virus 3-like ranavirus during an amphibian larval mortality event in Florida, USA. Dis. Aquat. Org. 105: 89–99.

Le Sage, M.J., B.D. Towey and J.L. Brunner. 2019. Do scavengers prevent or promote disease transmission? The effect of invertebrate scavenging on *Ranavirus* transmission. Funct. Ecol. 33: 1342–1350.

Liew, N., M.J.M. Moya, C.J. Wierzbicki, M. Hollinshead, M.J. Dillon, C.R. Thornton et al. 2017. Chytrid fungus infection in zebrafish demonstrates that the pathogen can parasitize non-amphibian vertebrate hosts. Nat. Commun. 8: 15048.

Lindauer, A.L. and J. Voyles. 2019. Out of the frying pan, into the fire? Yosemite toad (*Anaxyrus canorus*) susceptibility to *Batrachochytrium dendrobatidis* after development under drying conditions. Herpetol. Conserv. Bio. 14: 185–198.

Lips, K.R. 1998. Decline of a tropical montane amphibian fauna. Conserv. Biol. 12: 106–117.

Lips, K.R., J.D. Reeve and L.R. Witters. 2003. Ecological traits predicting amphibian population declines in Central America. Conserv. Biol. 17: 1078–1088.

Lips, K.R., F. Brem, R. Brenes, J.D. Reeve, R.A. Alford, J. Voyles et al. 2006. Emerging infectious disease and the loss of biodiversity in a Neotropical amphibian community. Proc. Natl. Acad. Sci. USA 103: 3165–3170.

Lips, K.R., J. Diffendorfer, J.R. Mendelson III and M.W. Sears. 2008. Riding the wave: Reconciling the roles of disease and climate change in amphibian declines. PLoS Biol. 6: e72.

Lips, K.R. 2016. Overview of chytrid emergence and impacts on amphibians. Philos. Trans. R. Soc. B 371: 20150465.

Longcore, J.E., A.P. Pessier and D.K. Nichols. 1999. *Batrachochytrium dendrobatidis* gen. et sp. nov., a chytrid pathogenic to amphibians. Mycologia 91: 219–227.

Longo, A.V., P.A. Burrowes and R.L. Joglar. 2010. Seasonality of *Batrachochytrium dendrobatidis* infection in direct-developing frogs suggests a mechanism for persistence. Dis. Aquat. Organ. 92: 253–260.

Longo, A.V., R.J. Ossiboff, K.R. Zamudio and P.A. Burrowes. 2013. Lability in host defenses: terrestrial frogs die from chytridiomycosis under enzootic conditions. J. Wildl. Dis. 49: 197–199.

Longo, A.V., R.C. Fleischer and K.R. Lips. 2019. Double trouble: co-infections of chytrid fungi will severely impact widely distributed newts. Biol. Invasions 21: 2233–2245.

Lötters, S., N. Wagner, A. Kerres, M. Vences, S. Steinfartz, J. Sabino-Pinto et al. 2018. First report of host co-infection of parasitic amphibian chytrid fungi. Salamandra 54: 287–290.

Lötters, S., M. Veith, N. Wagner, A. Martel and F. Pasmans. 2020. The amphibian pathogen *Batrachochytrium salamandrivorans* in the hotspot of its European invasive range: past—present—future. Salamandra 56: 173–188.

Luis, A.D., A.J. Kuenzi and J.N. Mills 2018. Species diversity concurrently dilutes and amplifies transmission in a zoonotic host–pathogen system through competing mechanisms. Proc. Natl. Acad. Sci. USA 115: 7979–7984.

Luquet, E., T.W.J. Garner, J.-P. Léna, C. Bruel, P. Joly, T. Lengagne et al. 2012. Genetic erosion in wild populations makes resistance to a pathogen more costly. Evolution 66: 1942–1952.

Luttershmidt, W.I. and V.H. Hutchison. 1997. The critical thermal maximum: history and critique. Can. J. Zool. 75: 1561–1574.

Ma, C., S.-P. Li, Z. Pu, J. Tan, M. Liu, J. Zhou et al. 2016. Different effects of invader-native phylogenetic relatedness on invasion success and impact: A meta-analysis of Darwin's naturalization hypothesis. Proc. R. Soc. B 283: 20160663.

Marantelli, G., L. Berger, R. Speare and L. Keegan. 2004. Distribution of the amphibian chytrid *Batrachochytrium dendrobatidis* and keratin during tadpole development. Pac. Conserv. Biol. 10: 173–179.

Márquez, R., J.L. Olmo and J. Bosch. 1995. Recurrent mass mortality of larval midwife toads Alytes obstetricans in a lake in the Pyrenean mountains. Herp. J. 5: 287–289.

Martel, A., C. Adriaensen, S. Bogaerts, R. Ducatelle, H. Favoreel, S. Crameri et al. 2012a. Novel Chlamydiaceae disease in captive salamanders. Emerg. Infect. Dis. 18: 1020–1022.

Martel, A., C. Adriaensen, M. Sharifian-Fard, M. Vandewoestyne, D. Deforce D., H. Favoreel et al. 2012b. The novel '*Candidatus* Amphibiichlamydia ranarum' is highly prevalent in invasive exotic bullfrogs (*Lithobates catesbeianus*). Environ. Microbiol. Rep. 5: 105–108.

Martel, A., A. Spitzen-van der Sluijs, M. Blooi, W. Bert, R. Ducatelle, M.C. Fisher et al. 2013. *Batrachochytrium salamandrivorans* sp. nov. causes lethal chytridiomycosis in amphibians. Proc. Natl. Acad. Sci. USA 110: 15325–15329.

Martel, A., M. Blooi, C. Adriaensen, P. van Rooij, W. Beukema, M.C. Fisher et al. 2014. Recent introduction of a chytrid fungus endangers Western Palearctic salamanders. Science 346: 630–631.

Martel, A., M. Vila-Escale, D. Fernández-Giberteau, A. Martinez-Silvestre, S. Canessa, S. van Praet et al. 2020. Integral chain management of wildlife diseases. Conserv. Lett. 13: 1–6.

Matson, P.A., W.J. Parton, A.G. Power and M.J. Swift. 1997. Agricultural intensification and ecosystem properties. Science 277: 504–509.

Mautz, W.J. and M.R. Dohm. 2004. Respiratory and behavioral effects of ozone on a lizard and a frog. Comp. Biochem. Phys. A 139: 371–377.

Mazzoni, R., A.J. de Mesquita, L.F.F., Fleury, W.M.E.D. de Brito, I.A. Nunes, J. Robert et al. 2009. Mass mortality associated with a frog virus 3-like Ranavirus infection in farmed tadpoles Rana catesbeiana from Brazil. Dis. Aquat. Org. 86: 181–191.

McArdle, A.J., A. Turkova and A.J. Cunnington. 2018. When do co-infections matter? Curr. Opin. Infect. Dis. 31: 209.

McCaig, C., M. Begon, R. Norman and C. Shankland. 2011. A symbolic investigation of superspreaders. Bull. Math. Biol. 73: 777–794.

McCallum, H., N. Barlow and J. Hone. 2001. How should pathogen transmission be modelled? Trends Ecol. Evol. 16: 295–300.

McCoy, K.A. and A.L. Peralta. 2018. Pesticides could alter amphibian skin microbiomes and the effects of *Batrachochytrium dendrobatidis*. Front. Microbiol. 9: 748.

McDonald, C.A., A.V. Longo, K.R. Lips and K.R. Zamudio. 2020. Incapacitating effects of fungal coinfection in a novel pathogen system. Mol. Ecol. 29: 3173–3186.

McKenzie, R.L., P.J. Aucamp, A.F. Bais, L.O. Björn, M. Ilyas and S. Madronich S. 2011. Ozone depletion and climate change: impacts on UV radiation. Photochem. Photobiol. Sci. 10: 182–98.

McMahon, T.A., L.A. Brannelly, M.W.H. Chatfield, P.T.J. Johnson, M.B. Joseph, V.J. McKenzie et al. 2013. Chytrid fungus *Batrachochytrium dendrobatidis* has nonamphibian hosts and releases chemicals that cause pathology in the absence of infection. Proc. Natl. Acad. Sci. USA 110: 210–215.

McMahon, T.A., B.F. Sears, M.D. Venesky, S.M. Bessler, J.M. Brown, K. Deutsch et al. 2014. Amphibians acquire resistance to live and dead fungus overcoming fungal immunosuppression. Nature 511: 224–227.

Medina, D., T.W.J. Garner, L.M. Carrascal and J. Bosch. 2015. Delayed metamorphosis of amphibian larvae facilitates *Batrachochytrium dendrobatidis* transmission and persistence. Dis. Aquat. Organ. 117: 85–92.

Medzhitov, R., D.S. Schneider and M.P. Soares. 2012. Disease tolerance as a defense strategy. Science 335: 936–941.

Mesquita, A.F.C., C. Lambertini, M. Lyra, L.R. Malagoli, T.Y. James, L.F. Toledo et al. 2017. Low resistance to chytridiomycosis in direct-developing amphibians. Sci. Rep. 7: 16605.

Miller, D., M. Gray and A. Storfer. 2011. Ecopathology of ranaviruses infecting amphibians. Viruses 3: 2351–2373.

Mitchell, K.M., T.S. Churcher, T.W.J. Garner and M.C. Fisher. 2008. Persistence of the emerging pathogen *Batrachochytrium dendrobatidis* outside the amphibian host greatly increases the probability of host extinction. Proc. R. Soc. B 275: 329–334.

Moore, J. 2002. Parasites and the Behaviour of Animals. Oxford University Press, Oxford, UK.

Moura-Campos, D., S.E. Greenspan, G.V. DiRenzo, W.J. Neely, L.F. Toledo and C.G. Becker. 2021. Fungal disease cluster in tropical terrestrial frogs predicted by low rainfall. Biol. Conserv. 261: 109246.

Muletz, C.R., J.M. Myers, R.J. Domangue, J.B. Herrick and R.N. Harris. 2012. Soil bioaugmentation with amphibian cutaneous bacteria protects amphibian hosts from infection by *Batrachochytrium dendrobatidis*. Biol. Conserv. 152: 119–126.

Muletz-Wolz, C.R., J.G. Almario, S.E. Barnett, G.V. DiRenzo, A. Martel, F. Pasmans et al. 2017. Inhibition of fungal pathogens across genotypes and temperatures by amphibian skin bacteria. Front. Microbiol. 8: 1551.

Munro, J., A.E. Bayley, N.J. McPherson and S.W. Feist. 2016. Survival of frog virus 3 in freshwater and sediment from an English lake. J. Wildl. Dis. 52: 138–142.

Murray, K.A., D. Rosauer, H. McCallum and L.F. Skerratt. 2011. Integrating species traits with extrinsic threats: closing the gap between predicting and preventing species declines. Proc. R. Soc. B 278: 1515–1523.

Murray, K.A., L.F. Skerratt, S. Garland, D. Kriticos and H. McCallum. 2013. Whether the weather drives patterns of endemic amphibian chytridiomycosis: A pathogen proliferation approach. PLoS ONE 8: e61061.

Muths, E., B.R. Hossack, E.H. Campbell Grant, D.S. Pilliod and B.A. Mosher. 2020. Effects of snowpack, temperature, and disease on demography in a wild population of amphibians. Herpetologica 76: 132–143.

Nagl, A.M. and R. Hofer. 1997. Effects of ultraviolet radiation on early larval stages of the Alpine newt, *Triturus alpestris*, under natural and laboratory conditions. Oecologia 110: 514–519.

Narayan, E.J., C. Graham, H. McCallum and J.-M. Hero. 2014. Over-wintering tadpoles of *Mixophyes fasciolatus* act as reservoir host for *Batrachochytrium dendrobatidis*. PLoS ONE 9: e92499.

Nava-González, B.A., I. Suazo-Ortuño, G. Parra-Olea, L. López-Toledo and J. Alvarado-Díaz. 2019. *Batrachochytrium dendrobatidis* infection in amphibians from a high elevation habitat in the trans-Mexican volcanic belt. Aquat. Ecol. 54: 75–87.

Nazir, J., M. Spengler and R.E. Marschang. 2012. Environmental persistence of amphibian and reptilian ranaviruses. Dis. Aquat. Organ. 98: 177–184.

Nguyen, T.T., T.V. Nguyen, T. Ziegler, F. Pasmans and A. Martel. 2017. Trade in wild anurans vectors the urodelan pathogen *Batrachochytrium salamandrivorans* into Europe. Amphibia-Reptilia 38: 554–556.

North, A.C., D.J. Hodgson, S.J. Price and A.G.F. Griffiths. 2015. Anthropogenic and ecological drivers of amphibian disease (ranavirosis). PLoS ONE 10: e0127037.

Nowakowski, A.J., S.M. Whitfield, E.A. Eskew, M.E. Thompson, J.P. Rose, B.L. Caraballo et al. 2016. Infection risk decreases with increasing mismatch in host and pathogen environmental tolerances. Ecol. Lett. 19: 1051–1061.

O'Hanlon, S.J., A. Rieux, R.A. Farrer, G.M. Rosa, B. Waldman, A. Bataille et al. 2018. Recent Asian origin of chytrid fungi causing global amphibian declines. Science 360: 621–627.

Oficialdegui, F.J., M.I. Sánchez, C. Monsalve-Carcaño, L. Boyero and J. Bosch. 2019. The invasive red swamp crayfish (*Procambarus clarkii*) increases infection of the amphibian chytrid fungus (*Batrachochytrium dendrobatidis*). Biol. Invasions 21: 3221–3231.

Olori, J.C., R. Netzband, N. McKean, J. Lowery, K. Parsons and S.T. Windstam. 2018. Multi-year dynamics of ranavirus, chytridiomycosis, and co-infections in a temperate host assemblage of amphibians. Dis. Aquat. Organ. 130: 187–197.

Olson, D.H. and K.L. Ronnenberg. 2014. Global *Bd* mapping project: 2014 update. FrogLog 22: 17–21.

Ortiz-Santaliestra, M.E., M.C. Fisher, S. Fernández-Beaskoetxea, M.J. Fernández-Benéitez and J. Bosch. 2011. Ambient ultraviolet B radiation and prevalence of infection by *Batrachochytrium dendrobatidis* in two amphibian species. Conserv. Biol. 25: 975–982.

Ostfeld, R.S. and R.D. Holt 2004. Are predators good for your health? Evaluating evidence for top-down regulation of zoonotic disease reservoirs. Front. Ecol. Environ. 2: 13–20.

Ottman, N., H. Smidt, W.M. de Vos and C. Belzer. 2012. The function of our microbiota: who is out there and what do they do? Front. Cell. Infect. Microbiol. 2: 104.

Packer, C., R.D. Holt, P.J. Hudson, K.D. Lafferty and A.P. Dobson. 2003. Keeping the herds healthy and alert: implications of predator control for infectious disease. Ecol. Lett. 6: 797–802.

Parris, M.J., E. Reese and A. Storfer. 2006. Antipredator behavior of chytridiomycosis-infected northern leopard frog (*Rana pipiens*) tadpoles. Can. J. Zool. 84: 58–65.

Pasmans, F., P. van Rooij, M. Blooi, G. Tessa, S. Bogaerts, G. Sotgiu et al. 2013. Resistance to chytridiomycosis in European plethodontid salamanders of the genus *Speleomantes*. PLoS ONE 8: e63639.

Patton, A.H., M.F. Lawrance, M.J. Margres, C.P. Kozakiewicz, R. Hamede, M.A. Ruiz-Aravena et al. 2020. A transmissible cancer shifts from emergence to endemism in Tasmanian devils. Science 370: 6522.

Pearl, C.A., E.L. Bull, D.E. Green, J. Bowerman, M.J. Adams, A. Hyatt et al. 2007. Occurrence of the amphibian pathogen *Batrachochytrium dendrobatidis* in the Pacific Northwest. J. Herpetol. 41: 145–149.

Pearman, P.B., T.W.J. Garner, M. Straub and U.F. Greber. 2004. Response of the Italian agile frog (*Rana latastei*) to a *Ranavirus*, frog virus 3: a model for viral emergence in naive populations. J. Wildl. Dis. 40: 660–669.

Pedersen, A.B. and A. Fenton. 2007. Emphasizing the ecology in parasite community ecology. Trends Ecol. Evol. 22: 133–139.

Penczykowski, R.M., S.R. Hall, D.J. Civitello and M.A. Duffy. 2014. Habitat structure and ecological drivers of disease. Limnol. Oceanogr. 59: 340–348.

Pessier, A.P., D.K. Nichols, J.E. Longcore and M.S. Fuller. 1999. Cutaneous chytridiomycosis in poison dart frogs (*Dendrobates* spp.) and White's tree frogs (*Litoria caerulea*). J. Vet. Diagn. Invest. 11: 194–199.

Phillott, A.D., L.F. Grogan, S.D. Cashins, K.R. McDonald, L. Berger and L.F. Skerratt. 2013. Chytridiomycosis and seasonal mortality of tropical stream-associated frogs 15 years after introduction of *Batrachochytrium dendrobatidis*. Conserv. Biol. 27: 1058e1068.

Piotrowski, J.S., S.L. Annis and J.E. Longcore. 2004. Physiology of *Batrachochytrium dendrobatidis*, a chytrid pathogen of amphibians. Mycologia 96: 9–15.

Pochini, K.M. and J.T. Hoverman. 2017. Reciprocal effects of pesticides and pathogens on amphibian hosts: The importance of exposure order and timing. Environ. Pollut. 221: 359–366.

Pongsiri, M.J., J. Roman, V.O. Ezenwa, T.L. Goldberg, H.S. Koren, S.C. Newbold et al. 2009. Biodiversity loss affects global disease ecology. BioScience 59: 945–954.

Pontes, M., G. Augusto-Alves, C. Lambertini and L. Toledo. 2018. A lizard acting as carrier of the amphibian-killing chytrid *Batrachochytrium dendrobatidis* in southern Brazil. Acta Herpetol. 13: 201–205.

Pounds, J.A., M.R. Bustamante, L.A. Coloma, J.A. Consuegra, M.P.L. Fogden, P.N. Foster et al. 2006. Widespread amphibian extinctions from epidemic disease driven by global warming. Nature 439: 161–167.

Prest, T.L., A.K. Kimball, J.G. Kueneman and V.J. McKenzie. 2018. Host- associated bacterial community succession during amphibian development. Mol. Ecol. 27: 1–15.

Price, S.J., T.W.J. Garner, R.A. Nichols, F. Balloux, C. Ayres, A. Mora-Cabello de Alba et al. 2014. Collapse of amphibian communities due to an introduced ranavirus. Curr. Biol. 24: 2586–2591.

Price, S.J., E. Ariel, A. Maclaine, G.M. Rosa, M.J. Gray, J.L. Brunner et al. 2017. From fish to frogs and beyond: Impact and host range of emergent ranaviruses. Virology 511: 272–279.

Price, S.J., W.T.M. Leung, C.J. Owen, C. Sergeant, A.A. Cunningham, F. Balloux et al. 2019. Effects of historic and projected climate change on the range and impacts of an emerging wildlife disease. Glob. Change Biol. 25: 2548–2660.

Puschendorf, R., C.J. Hoskin, S.D. Cashins, K. Mcdonald, L.F. Skerratt, J. Vanderwal et al. 2011. Environmental refuge from disease-driven amphibian extinction. Conserv. Biol. 25: 956–964.

Råberg, L., A.L. Graham and A.F. Read. 2009. Decomposing health: tolerance and resistance to parasites in animals. Philos. Trans. R. Soc. B 364: 37–49.

Rachowicz, L.J., R.A. Knapp, J.A.T. Morgan, M.J. Stice, V.T. Vredenburg, J.M. Parker et al. 2006. Emerging infectious disease as a proximate cause of amphibian mass mortality. Ecology 87: 1671–1683.

Rachowicz, L.J. and C.J. Briggs. 2007. Quantifying the disease transmission function: effects of density on *Batrachochytrium dendrobatidis* transmission in the mountain yellow-legged frog *Rana muscosa*. J. Anim. Ecol. 76: 711–721.

Raffel, T.R., J.R. Rohr, J.M. Kiesecker and P.J. Hudson. 2006. Negative effects of changing temperature on amphibian immunity under field conditions. Funct. Ecol. 20: 819–828.

Raffel, T.R., J.M. Romansic, N.T. Halstead, T.A. McMahon, M.D. Venesky and J.R. Rohr. 2012. Disease and thermal acclimation in a more variable and unpredictable climate. Nat. Clim. Chang. 3: 146–151.

Raffel, T.R., N.T. Halstead, T.A. McMahon, A.K. Davis and J.R. Rohr. 2015. Temperature variability and moisture synergistically interact to exacerbate an epizootic disease. Proc. R. Soc. B 282: 20142039.

Rakus, K., M. Ronsmans and A. Vanderplasschen. 2017. Behavioral fever in ectothermic vertebrates. Dev. Comp. Immunol. 66: 84–91.

Ramsey, J.P., L.K. Reinert, L.K. Harper, D.C. Woodhams and L.A. Rollins-Smith. 2010. Immune defenses against *Batrachochytrium dendrobatidis*, a fungus linked to global amphibian declines, in the South African Clawed Frog, *Xenopus laevis*. Infect. Immun. 78: 3981–3992.

Randolph, S.E. and A.D. Dobson. 2012. Pangloss revisited: A critique of the dilution effect and the biodiversity-buffers-disease paradigm. Parasitology 139: 847–863.

Rebollar, E.A., M.C. Hughey, D. Medina, R.N. Harris, R. Ibáñez and L.K. Belden. 2016. Skin bacterial diversity of Panamanian frogs is associated with host susceptibility and presence of *Batrachochytrium dendrobatidis*. ISME J. 10: 1682–1695.

Rebollar, E.A., E. Martínez-Ugalde and A.H. Orta. 2020. The amphibian skin microbiome and its protective role against chytridiomycosis. Herpetologica 76: 167–177.

Reeder, N.M.M., A.P. Pessier and V.T. Vredenburg. 2012. A reservoir species for the emerging amphibian pathogen *Batrachochytrium dendrobatidis* thrives in a landscape decimated by disease. PLoS ONE 7: e33567.

Restif, O. and J. Koella. 2004. Concurrent evolution of resistance and tolerance to pathogens. Am. Nat. 164: E90–102.

Retallick, R.W.R., H. McCallum and R. Speare. 2004. Endemic infection of the amphibian chytrid fungus in a frog community post-decline. PLoS Biol. 2: 1965–1971.

Richards-Zawacki, C.L. 2010. Thermoregulatory behaviour affects prevalence of chytrid fungal infection in a wild population of Panamanian golden frogs. Proc. R. Soc. B 277: 519–528.

Rijks, J.M., B. Saucedo, A. Spitzen-van der Sluijs, G.S. Wilkie, A.J.A.M. Asten, J. van Broek et al. 2016. Investigation of amphibian mortality events in wildlife reveals an on-going *Ranavirus* epidemic in the north of the Netherlands. PLoS ONE 11: e0157473.

Rohr, J.R., J. Brown, W.A. Battaglin, T.A. McMahon and R.A. Relyea. 2017. A pesticide paradox: fungicides indirectly increase fungal infections. Ecol. Appl. 27: 2290–2302.

Rohr, J.R. and J.M. Cohen. 2020. Understanding how temperature shifts could impact infectious disease. PLoS Biol. 18: e3000938.

Rollins-Smith, L.A., L.K. Reinert, C.J. O'Leary, L.E. Houston and D.C. Woodhams. 2005. Antimicrobial peptide defenses in amphibian skin. Integr. Comp. Biol. 45: 137–142.

Rollins-Smith, L.A. 2009. The role of amphibian antimicrobial peptides in protection of amphibians from pathogens linked to global amphibian declines. Biochim. Biophys. Acta 1788: 1593–1599.

Rollins-Smith, L.A., J.P. Ramsey, J.D. Pask, L.K. Reinert and D.C. Woodhams. 2011. Amphibian immune defenses against chytridiomycosis: Impacts of changing environments. Integr. Comp. Biol. 51: 552–562.

Rollins-Smith, L.A. 2020. Global amphibian declines, disease, and the ongoing battle between *Batrachochytrium* fungi and the immune system. Herpetologica 76: 178–188.

Rosa, G.M., I. Anza, P.L. Moreira, J. Conde, F. Martins, M.C. Fisher et al. 2013. Evidence of chytrid-mediated population declines in common midwife toad in Serra da Estrela, Portugal. Anim. Conserv. 16: 306–315.

Rosa, G.M., J. Sabino-Pinto, T.G. Laurentino, A. Martel, F. Pasmans, R. Rebelo et al. 2017. Impact of asynchronous emergence of two lethal pathogens on amphibian assemblages. Sci. Rep. 7: 43260.

Rosa, G.M., J. Bosch, A. Martel, F. Pasmans, R. Rebelo, R.A. Griffiths et al. 2019. Sex-biased disease dynamics increase extinction risk by impairing population recovery. Anim. Conserv. 22: 579–588.

Ross, A.A., A. Rodrigues Hoffmann and J.D. Neufeld. 2019. The skin microbiome of vertebrates. Microbiome 7: 79.

Roy, B.A. and J.W. Kirchner. 2000. Evolutionary dynamics of pathogen resistance and tolerance. Evolution 54: 51–63.

Roznik, E.A., S.J. Sapsford, D.A. Pike, L. Schwarzkopf and R.A. Alford. 2015. Natural disturbance reduces disease risk in endangered rainforest frog populations. Sci. Rep. 5: 13472.

Ruano-Fajardo, G., LK.F. Toledo and T. Mott. 2016. Jumping into a trap: high prevalence of chytrid fungus in the preferred microhabitats of a bromeliad-specialist frog. Dis. Aquat. Org. 121: 223–232.

Ruggeri, J., L.F. Toledo and S.P. de Carvalho-e-Silva. 2018b. Stream tadpoles present high prevalence but low infection loads of *Batrachochytrium dendrobatidis* (Chytridiomycota). Hydrobiologia 806: 303–311.

Ruggeri, J., S.P. De Carvalho-E-silva, T.Y. James and L.F. Toledo. 2018a. Amphibian chytrid infection is influenced by rainfall seasonality and water availability. Dis. Aquat. Org. 127: 107–115.

Ryan M.E., W.J. Palen, M.J. Adams and R.M. Rochefort. 2014. Amphibians in the climate vise: loss and restoration of resilience of montane wetland ecosystems in the western US. Front. Ecol. Environ. 12: 232–240.

Santini, A., A. Liebhold, D. Migliorini and S. Woodward. 2018. Tracing the role of human civilization in the globalization of plant pathogens. ISME J. 12: 647–652.

Sapsford, S.J., R.A. Alford and L. Schwarzkopf. 2013. Elevation, temperature, and aquatic connectivity all influence the infection dynamics of the amphibian chytrid fungus in adult frogs. PLoS ONE 8: e82425.

Saucedo, B., T.W.J. Garner, N. Kruithof, S.J.R. Allain, M.J. Goodman, R.J. Cranfield et al. 2019. Common midwife toad ranaviruses replicate first in the oral cavity of smooth newts (*Lissotriton vulgaris*) and show distinct strain-associated pathogenicity. Sci. Rep. 9: 4453.

Sauer, E.L., R.C. Fuller, C.L. Richards-Zawacki, J. Sonn, J.H. Sperry and J.R. Rohr. 2018. Variation in individual temperature preferences, not behavioural fever, affects susceptibility to chytridiomycosis in amphibians. Proc. R. Soc. B 285: 20181111.

Sauer, E.L., N. Trejo, J.T. Hoverman and J.R. Rohr. 2019. Behavioural fever reduces ranaviral infection in toads. Funct. Ecol. 33: 2172–2179.

Sauer, E.L., J.M. Cohen, M.J. Lajeunesse, T.A. McMahon, D.J. Civitello, S.A. Knutie et al. 2020. A meta-analysis reveals temperature, dose, life stage, and taxonomy influence host susceptibility to a fungal parasite. Ecology 101: e02979.

Savage, A.E. and K.R. Zamudio. 2011. MHC genotypes associate with resistance to a frog-killing fungus. Proc. Natl. Acad. Sci. USA 108: 16705–16710.

Scheele, B.C., D.A. Driscoll, J. Fischer, A.W. Fletcher, J. Hanspach, J. Vörös et al. 2015. Landscape context influences chytrid fungus distribution in an endangered European amphibian. Anim. Conserv. 18: 480–488.

Scheele, B.C., D.A. Hunter, L.A. Brannelly, L.F. Skerratt and D.A. Driscoll. 2016. Reservoir-host amplification of disease impact in an endangered amphibian. Conserv. Biol. 31: 592–600.

Scheele, B.C., F. Pasmans, L.F. Skerratt, L. Berger, A. Martel, W. Beukema et al. 2019a. Amphibian fungal panzootic causes catastrophic and ongoing loss of biodiversity. Science 363: 1459–1463.

Scheele, B.C., C.N. Foster, D.A. Hunter, D.B. Lindenmayer, B.R. Schmidt and G.W. Heard. 2019b. Living with the enemy: facilitating amphibian coexistence with disease. Biol. Conserv. 236: 52–59.

Schloegel, L.M., A.M. Picco, A.M. Kilpatrick, A.J. Davies, A.D. Hyatt and P. Daszak. 2009. Magnitude of the US trade in amphibians and presence of *Batrachochytrium dendrobatidis* and ranavirus infection in imported North American bullfrogs (*Rana catesbeiana*). Biol. Conserv. 142: 1420–1426.

Schloegel, L.M., L.F. Toledo, J.E. Longcore, S.E. Greenspan, C.A. Vieira, M. Lee et al. 2012. Novel, panzootic and hybrid genotypes of amphibian chytridiomycosis associated with the bullfrog trade. Mol. Ecol. 21: 5162–5177.

Schmeller, D.S., M. Blooi, A. Martel, T.W.J. Garner, M.C. Fisher, F. Azemar et al. 2014. Microscopic aquatic predators strongly affect infection dynamics of a globally emerged pathogen. Curr. Biol. 24: 176–180.

Schmeller, D.S., R. Utzel, F. Pasmans and A. Martel. 2020. *Batrachochytrium salamandrivorans* kills alpine newts (*Ichthyosaura alpestris*) in southernmost Germany. Salamandra 56: 230–232.

Schmidt, B.R. 2004. Pesticides, mortality and population growth rate. Trends Ecol. Evol. 19: 459–460.

Schmidt, B.R., C. Bozzuto, S. Lötters and S. Steinfartz. 2017. Dynamics of host populations affected by the emerging fungal pathogen *Batrachochytrium salamandrivorans*. R. Soc. Open Sci. 4: 160801.

Schneider, D.S. and J.S. Ayres. 2008. Two ways to survive infection: what resistance and tolerance can teach us about treating infectious diseases. Nat. Rev. Immunol. 8: 889–895.

Schock, D.M., T.K. Bollinger, V.G. Chinchar, J.K. Jancovich and J.P. Collins. 2008. Experimental evidence that amphibian ranaviruses are multi-host pathogens. Copeia 1: 133–143.

Schulz, V., A. Schulz, M. Klamke, K. Preissler, J. Sabino-Pinto, M. Müsken et al. 2020. *Batrachochytrium salamandrivorans* in the Ruhr District, Germany: history, distribution, decline dynamics and disease symptoms of the salamander plague. Salamandra 56: 189–214.

Schulz, V., P. Gerhardt, D. Stützer, U. Seidel and M. Vences. 2021. Lungless salamanders of the genus speleomantes in the solling, germany: Genetic identification, *Bd/Bsal*-screening, and introduction hypothesis. Herpetol. Notes 14: 421–429.

Searle, C.L., L.K. Belden, B.A. Bancroft, B.A. Han, L.M. Biga and A.R. Blaustein. 2010. Experimental examination of the effects of ultraviolet-B radiation in combination with other stressors on frog larvae. Oecologia 162: 237–245.

Searle, C.L., L.M. Biga, J.W. Spatafora and A.R. Blaustein. 2011. A dilution effect in the emerging amphibian pathogen *Batrachochytrium dendrobatidis*. Proc. Natl. Acad. Sci. USA 108: 16322–16326.

Searle, C.L., J.R. Mendelson, L.E. Green and M.A. Duffy. 2013. *Daphnia* predation on the amphibian chytrid fungus and its impacts on disease risk in tadpoles. Ecol. Evol. 3: 4129–4138.

Shapard, E.J., A.S. Moss and M.J. San Francisco. 2012. *Batrachochytrium dendrobatidis* can infect and cause mortality in the nematode *Caenorhabditis elegans*. Mycopathologia 173: 121–126.

Sherman, E. 2008. Thermal biology of newts (*Notophthalmus viridescens*) chronically infected with a naturally occurring pathogen. J. Therm. Biol. 33: 27–31.

Skerratt, L.F., L. Berger, R. Speare, S. Cashins, K.R. McDonald, A.D. Phillott et al. 2007. Spread of chytridiomycosis has caused the rapid global decline and extinction of frogs. EcoHealth 4: 125–134.

Skerratt, L.F., K.R. McDonald, H.B. Hines, L. Berger, D. Mendez, A.D. Phillott et al. 2010. Application of the survey protocol for chytridiomycosis to Queensland, Australia. Dis. Aquat. Organ. 92: 117–129.

Solís, R., G. Lobos, S.F. Walker, M.C. Fisher and J. Bosch. 2009. Presence of *Batrachochytrium dendrobatidis* in feral populations of *Xenopus laevis* in Chile. Biol. Invasions 12: 1641–1646.

Sonn, J.M., S. Berman and C.L. Richards-Zawacki. 2017. The influence of temperature on chytridiomycosis *in vivo*. EcoHealth 14: 762–770.

Spitzen-van der Sluijs, A., F. Spikmans, W. Bosman, M. de Zeeuw, T. van der Meij, E. Goverse et al. 2013. Rapid enigmatic decline drives the re salamander (*Salamandra salamandra*) to the edge of extinction in the Netherlands. Amphibia-Reptilia 34: 233–239.

Spitzen-van der Sluijs, A., A. Martel, J. Asselberghs, E.K. Bales, W. Beukema, M.C. Bletz et al. 2016. Expanding distribution of lethal amphibian fungus *Batrachochytrium salamandrivorans* in Europe. Emerg. Infect. Dis. 22: 1286–1288.

Stark T., A. Martel, F. Pasmans, V. Thomas, M. Gilbert and A. Spitzen-van der Sluijs. 2018. A European early warning system for a deadly salamander pathogen. Amphibian Ark Newsletter 42: 12–14.

Stegen, G., F. Pasmans, B.R. Schmidt, L.O. Rouffaer, S. van Praet, M. Schaub et al. 2017. Drivers of salamander extirpation mediated by *Batrachochytrium salamandrivorans*. Nature 544: 353–356.

Stockwell, M.P., J. Clulow and M.J. Mahony. 2012. Sodium chloride inhibits the growth and infective capacity of the amphibian chytrid fungus and increases host survival rates. PLoS ONE 7: e36942.

Stockwell, M.P., L.J. Storrie, C.J. Pollard, J. Clulow and M.J. Mahony. 2015. Effects of pond salinization on survival rate of amphibian hosts infected with the chytrid fungus. Conserv. Biol. 29: 391–399.

Stöhr, A.C., A. López-Bueno, S. Blahak, M.F. Caeiro, G.M. Rosa, A.P. Alves de Matos et al. 2015. Phylogeny and differentiation of reptilian and amphibian Ranaviruses detected in Europe. PLoS ONE 10: e0118633.

Stoler, A.B., K.A. Berven and T.R. Raffel. 2016. Leaf litter inhibits growth of an amphibian fungal pathogen. EcoHealth 13: 392–404.

Strauss, A. and K.G. Smith. 2013. Why does amphibian chytrid (*Batrachochytrium dendrobatidis*) not occur everywhere? An exploratory study in Missouri ponds. PLoS ONE 8: e76035.

Subramaniam, K., A. Toffan, E. Cappellozza, N.K. Steckler, N.J. Olesen, E. Ariel et al. 2016. Genomic sequence of a *Ranavirus* isolated from short-finned eel (*Anguilla australis*). Genome Announc. 4: e00843–16.

Sutton, M.A., C.M. Howard, J.W. Erisman, G. Billen, A. Bleeker, P. Grennfelt et al. 2011. The European Nitrogen Assessment. Cambridge University Press, Cambridge, UK.

Sweeny, A.R., G.F. Albery, D.J. Becker, E.A. Eskew and C.J. Carlson. 2021. Synzootics. J. Anim. Ecol. 90: 2744–2754.

Symonds, E.P., D.J. Trott, P.S. Bird and P. Mills. 2008. Growth characteristics and enzyme activity in *Batrachochytrium dendrobatidis* isolates. Mycopathologia 166: 143–147.

Teacher, A.G.F., A.A. Cunningham and T.W.J. Garner. 2010. Assessing the long-term population impact of *Ranavirus* infection in wild common frog populations. Anim. Conserv. 13: 514–522.

Temple, H.J. and N.A. Cox. 2009. European Red List of Amphibians. Office for Official Publications of the European Communities, Luxembourg.

Terrell, V.C.K., N.J. Engbrecht, A.P. Pessier and M.J. Lannoo. 2014. Drought reduces chytrid fungus (*Batrachochytrium dendrobatidis*) infection intensity and mortality but not prevalence in adult crawfish frogs (*Lithobates areolatus*). J. Wildl. Dis. 50: 56–62.

Tevini, M. 1993. UV-B Radiation and Ozone Depletion: Effects on Humans, Animals, Plants, Microorganisms, and Materials. Lewis, Boca Raton, Florida.

Thien, T.N., A. Martel, M. Brutyn, S. Bogaerts, M. Sparreboom, F. Haesebrouck et al. 2013. A survey for *Batrachochytrium dendrobatidis* in endangered and highly susceptible Vietnamese salamanders (*Tylototriton* spp.). J. Zoo. Wildl. Med. 44: 627–633.

Thomas, C.D., A. Cameron, R.E. Green, M. Bakkenes, L.J. Beaumont, Y.C. Collingham et al. 2004. Extinction risk from climate change. Nature 427: 145–148.

Thomas, V., Y. Wang, P. van Rooij, E. Verbrugghe, V. Baláž and J. Bosch. 2019. Mitigating *Batrachochytrium salamandrivorans* in Europe. Amphibia-Reptilia 40: 265–290.

Thumsová, B., D. Donaire-Barroso, E.H. El Mouden and J. Bosch. 2022. Fatal chytridiomycosis in the Moroccan midwife toad *Alytes maurus* and potential distribution of *Batrachochytrium dendrobatidis* across Morocco. Afr. J. Herpetol. 71: 72–82.

Thumsová, B., E. González-Miras, S.C. Faulkner and J. Bosch. 2021. Rapid spread of a virulent amphibian pathogen in nature. Biol. Invasions 23: 3151–3160.

Tobler, U. and B.R. Schmidt. 2010. Within- and among-population variation in chytridiomycosis-induced mortality in the toad *Alytes obstetricans*. PLoS ONE 5: e10927.

Tobler, U., A. Borgula and B.R. Schmidt. 2012. Populations of a susceptible amphibian species can grow despite the presence of a pathogenic chytrid fungus. PLoS ONE 7: e34667.

Toledo, L.F., J. Ruggeri, L. Leite Ferraz de Campos, M. Martins, S. Neckel-Oliveira and C.P.B. Breviglieri. 2021. Midges not only sucks, but may carry lethal pathogens to wild amphibians. Biotropica 53: 722–725.

Tompros, A., A.D. Dean, A. Fenton, M.Q. Wilber, E.D. Carter and M.J. Gray. 2021. Frequency-dependent transmission of *Batrachochytrium salamandrivorans* in eastern newts. Transbound Emerg. Dis. 69: 731–741.

Valencia-Aguilar, A., L.F. Toledo, M.V.C. Vital and T. Mott. 2016. Seasonality, environmental factors, and host behavior linked to disease risk in stream-dwelling tadpoles. Herpetologica 72: 98–106.

Valenzuela-Sánchez, A., M.Q. Wilber, S. Canessa, L.D. Bacigalupe, E. Muths, B.R. Schmidt et al. 2021. Why disease ecology needs life-history theory: a host perspective. Ecol. Lett. 24: 876–890.

Valenzuela-Sánchez, A., C. Azat, A.A. Cunningham, S. Delgado, L.D. Bacigalupe, J. Beltrand et al. 2022. Interpopulation differences in male reproductive effort drive the population dynamics of a host exposed to an emerging fungal pathogen. J. Anim. Ecol. 91: 308–319.

van Dijk, J.G.B., B.J. Hoye, J.H. Verhagen, B.A. Nolet, R.A.M. Fouchier and M. Klaassen. 2014. Juveniles and migrants as drivers for seasonal epizootics of avian influenza virus. J. Anim. Ecol. 83: 266–275.

van Rooij, P., A. Martel, F. Haesebrouck and F. Pasmans. 2015. Amphibian chytridiomycosis: A review with focus on fungus-host interactions. Vet. Res. 46: 1–22.

Vazquez, V.M., B.B. Rothermel and A.P. Pessier. 2009. Experimental infection of North American plethodontid salamanders with the fungus *Batrachochytrium dendrobatidis*. Dis. Aquat. Organ. 84: 1–7.

Venesky, M.D., J.L. Kerby, A. Storfer and M.J. Parris. 2011. Can differences in host behavior drive patterns of disease prevalence in tadpoles? PLoS ONE 6: e24991.

Venesky, M.D., X. Liu, E.L. Sauer and J.R. Rohr. 2014. Linking manipulative experiments to field data to test the dilution effect. J. Anim. Ecol. 83: 557–565.

von Essen, M., W.T.M. Leung, J. Bosch, S. Pooley, C. Ayres and S.J. Price. 2020. High pathogen prevalence in an amphibian and reptile assemblage at a site with risk factors for dispersal in Galicia, Spain. PLoS ONE 15: e0236803.

Voyles, J., L. Berger, S. Young, R. Speare, R. Webb, J. Warner et al. 2007. Electrolyte depletion and osmotic imbalance in amphibians with chytridiomycosis. Dis. Aquat. Organ. 77: 113–118.

Voyles, J., S. Young, L. Berger, C. Campbell, W.F. Voyles, A. Dinudom et al. 2009. Pathogenesis of chytridiomycosis, a cause of catastrophic amphibian declines. Science 326: 582–585.

Voyles, J., L.R. Johnson, J. Rohr, R. Kelly, C. Barron, D. Miller et al. 2017. Diversity in growth patterns among strains of the lethal fungal pathogen *Batrachochytrium dendrobatidis* across extended thermal optima. Oecologia 184: 363–373.

Vredenburg, V.T., R.A. Knapp, T.S. Tunstall and C.J. Briggs. 2010. Dynamics of an emerging disease drive large-scale amphibian population extinctions. Proc. Natl. Acad. Sci. USA 107: 9689–9694.

Vredenburg, V.T., S.A. Felt, E.C. Morgan, S.V.G. McNally, S. Wilson and S.L. Green. 2013. Prevalence of *Batrachochytrium dendrobatidis* in *Xenopus* Collected in Africa (1871–2000) and in California (2001–2010). PLoS ONE 8: 63791.

Walker, S.F., J. Bosch, V. Gomez, T.W.J. Garner, A.A. Cunningham, D.S. Schmeller et al. 2010. Factors driving pathogenicity vs. prevalence of amphibian panzootic chytridiomycosis in Iberia. Ecol. Lett. 13: 372–382.

Walsh, P.T., J.R. Downie and P. Monaghan. 2016. Factors affecting the overwintering of tadpoles in a temperate amphibian. J. Zool. 298: 183–190.

Waltzek, T.B., D.L. Miller, M.J. Gray, B. Drecktrah, J.T. Briggler, B. MacConnell et al. 2014. New disease records for hatchery-reared sturgeon. I. Expansion of frog virus 3 host range into *Scaphirhynchus albus*. Dis. Aquat. Organ. 111: 219–227.

Warne, R.W., E.J. Crespi and J.L. Brunner. 2011. Escape from the pond: Stress and developmental responses to ranavirus infection in wood frog tadpoles. Funct. Ecol. 25: 139–146.

Warne, R.W., B. LaBumbard, S. LaGrange, V.T. Vredenburg and A. Catenazzi. 2016. Co-infection by chytrid fungus and ranaviruses in wild and harvested frogs in the tropical Andes. PLoS ONE 11: e0145864.

Warne, R.W., L. Kirschman and L. Zeglin. 2019. Manipulation of gut microbiota during critical developmental windows affect host physiological performance and disease susceptibility across ontogeny. J. Anim. Ecol. 88: 845–856.

Welbergen, J.A., S.M. Klose, N. Markus and P. Eby. 2008. Climate change and the effects of temperature extremes on Australian flying-foxes. Proc. R. Soc. B 275: 419–425.

Weldon, C., L.H. Du Preez, A.D. Hyatt, R. Muller and R. Speare. 2004. Origin of the amphibian chytrid fungus. Emerg. Infect. Dis. 10: 2100–2105.

Werner, E.E. and K.S. Glennemeier. 1999. Influence of forest canopy cover on the breeding pond distributions of several amphibian species. Copeia 1999: 1–12.

Wheelwright, N.T., M.J. Gray, R.D. Hill and D.L. Miller. 2014. Sudden mass die-off of a large population of wood frog (*Lithobates sylvaticus*) tadpoles in Maine, USA, likely due to *Ranavirus*. Herpetol. Rev. 45: 240–242.

Whitfield, S.M., E. Geerdes, I. Chacon, E. Ballestero Rodriguez, R.R. Jimenez, M.A. Donnelly et al. 2013. Infection and co-infection by the amphibian chytrid fungus and ranavirus in wild Costa Rican frogs. Dis. Aquat. Organ. 104: 173–178.

Whitmore, T.C. 1998. An Introduction to Tropical Rain Forests. 2nd edn. Oxford University Press, New York, USA.

WHO Ebola Response Team. 2014. Ebola virus disease in West Africa—the first 9 months of the epidemic and forward projections. N. Engl. J. Med. 371: 1481–1495.

WHO. 2020. Coronavirus disease 2019 (COVID-19) situation report. https://www.who.int/emergencies/diseases/novel-coronavirus-2019. Accessed on 20 of April 2021.

Wilber, M.Q., R.A. Knapp, M. Toothman and C.J. Briggs. 2017. Resistance, tolerance and environmental transmission dynamics determine host extinction risk in a load-dependent amphibian disease. Ecol. Lett. 30: 1169–1181.

Wilson, A.J., E.R. Morgan, M. Booth, R. Norman, S.E. Perkins, H.C. Hauffe et al. 2017. What is a vector? Philos. Trans. R. Soc. B 372: 20160085.

Woodhams, D.C., J. Voyles, K.R. Lips, C. Carey and L.A. Rollins-Smith. 2006. Predicted disease susceptibility in a Panamanian amphibian assemblage based on skin peptide defenses. J. Wildl. Dis. 42: 207–218.

Woodhams, D.C., K. Ardipradja, R.A. Alford, G. Marantelli, L.K. Reinert and L.A. Rollins-Smith. 2007a. Resistance to chytridiomycosis varies among amphibian species and is correlated with skin peptide defenses. Anim. Conserv. 10: 409–417.

Woodhams, D.C., V.T. Vredenburg, M.-A. Simon, D. Billheimer, B. Shakhtour, Y. Shyr, et al. 2007b. Symbiotic bacteria contribute to innate immune defenses of the threatened mountain yellow-legged frog, *Rana muscosa*. Biol. Conserv. 138: 390–398.

Woodhams, D.C., J. Bosch, C.J. Briggs, S. Cashins, L.R. Davis, A. Lauer et al. 2011. Mitigating amphibian disease: strategies to maintain wild populations and control chytridiomycosis. Front. Zool. 8: 8.

Wright, R.K. and E.L. Cooper. 1981. Temperature effects on ectotherm immune responses. Dev. Comp. Immunol. 5: 117–122.

Wuerthner, V.P., J. Hua and J.T. Hoverman. 2017. The benefits of coinfection: Trematodes alter disease outcomes associated with virus infection. J. Anim. Ecol. 86: 921–931.

Youker-Smith, T.E., P.H. Boersch-Supan, C.M. Whipps and S.J. Ryan. 2018. Environmental drivers of *Ranavirus* in free-living amphibians in constructed ponds. EcoHealth 15: 608–618.

Zumbado-Ulate, H., K.N. Nelson, A. García-Rodríguez, G. Chaves, E. Arias, F. Bolaños et al. 2019. Endemic infection of *Batrachochytrium dendrobatidis* in Costa Rica: Implications for amphibian conservation at regional and species level. Diversity 11: 129.

CHAPTER 6

Thermoregulation and Hydric Balance in Amphibians

Zaida Ortega,[1,2,*] *Carolina Cunha Ganci*[2,3] *and Marga L. Rivas*[4]

INTRODUCTION

Research on wildlife ecophysiology is accelerating as the urgency of the global biodiversity crisis is recognized (Bovo et al. 2018). Amphibians are an extraordinarily diverse group of tetrapods—with more than 8,000 species living from the sea level to mountaintops up to 5,000 m asl—that managed to adapt to different terrestrial and freshwater environments across the globe (Duellman and Trueb 1994). Despite some differences between the three orders of amphibians, they share important morphological and physiological traits that make them key models to understand central issues of physiological ecology. Salamanders, anurans, and caecilians share a particular link to water (de Andrade et al. 2016). Only through the lens of the interrelations between temperature and humidity can we try to understand the complexity of amphibian ecological physiology. Moreover, as climate change advances, this understanding is crucial to assess amphibian vulnerability and plan their conservation. In this chapter we aim to briefly explain the state of the art on amphibian thermoregulation and hydric balance, their interconnections, and the relation with main threats imposed on these animals in the Anthropocene.

Ecophysiology aims to understand how animals deal with problems and exploit the opportunities offered by environmental conditions (Willmer et al. 2009). The most prominent environmental conditions driving the ecology and physiology of amphibians are temperature and water or moisture regimes. Temperature, because

[1] Departamento de Zoología, Facultad de Ciencias, Universidad de Granada 18071, Granada, Spain.
[2] Programa de Pós-Graduação em Ecologia e Conservação, Instituto de Biociências, Universidade Federal de Mato Grosso do Sul, 79070–900, Campo Grande, MS, Brazil.
[3] Biology Department, University of Massachusetts Dartmouth 02740, Massachusetts, USA.
  Email: carolinacganci@gmail.com
[4] Departamento de Biología, Facultad de Ciencias del Mar y Ambientales, Universidad de Cádiz, 11510, Cádiz, Spain.
  Email: mlrivas@uca.es
* Corresponding author: zaidaortega@usal.es

amphibians are ectotherms—i.e., animals that do not produce enough metabolic heat to maintain constant body temperatures. And since all cellular processes are temperature-dependent, maintaining a constant internal temperature enhances physiological performance and individuals' fitness (Angilletta 2009, Taylor et al. 2020). Water, mainly because of two facts: larval states are aquatic, and the skin of adult amphibians is highly permeable, turning them profoundly dependable on water and humidity to live and reproduce (Tracy 1976). Moreover, temperature and humidity are tightly related, both environmentally and with regards to amphibian's physiology (Rozen-Rechels et al. 2019).

Despite being interconnected phenomena, thermal and hydric ecology have been traditionally studied separately. Thus, we will begin this chapter by reviewing amphibians' thermal and hydric ecology, with an emphasis on how these animals regulate their body temperature and water balance, to be able to go into depth on both processes separately. Then we will assess the interactions between temperature and humidity, introducing the thermo-hydroregulation framework, and their effects on the biology and ecology of amphibians. Finally, we will briefly comment on the main impacts of human activities in relation with amphibian ecophysiology (more details on amphibian ecophysiology in chapter 3).

Thermal Ecology

As ectotherms, amphibians are highly sensitive to the effects of environmental temperature. Their habitats often experience substantial daily and annual thermal fluctuations (Huey 1982, Huey and Berrigan 2001). Species' environmental physiology can vary considerably, with important consequences for performance and fitness under different environments (Huey and Stevenson 1979, Sinclair et al. 2016). Behavioral thermoregulation is the main strategy of amphibians to buffer climatic variations (Huey et al. 2012, Sunday et al. 2014, Diele-Viegas and Rocha 2018). Thus, we can understand the thermal ecology of one species or population by assessing two main aspects: thermal sensitivity of performance and thermoregulation (Angilletta 2009).

Thermal Sensitivity

Although this chapter focuses on amphibian's thermoregulation, we will briefly introduce thermal sensitivity (or thermosensitivity) to better understand how thermal regulation functions. Thermal sensitivity describes how performance of any biological function, including behavior and relative fitness, depends on environmental temperature. Thermal sensitivity places species and populations in a non-static gradient between thermal generalists—performing under a wide range of environmental temperatures—and thermal specialists—performing under a narrow thermal range (Angilletta 2009). The Australian frog *Geocrinia alba*, for example, shows narrow thermal (and hydric) limits that, living in a drying and warming habitat, make it highly vulnerable to climate change (Hoffmann et al. 2021). At the other extreme of the thermal sensitivity gradient would be, for example, *Bufo bufo*, whose generalist thermal physiology could provide the species a remarkable resilience to climate change (Rosset and Oertli 2011). There is a general trend of specialists reaching higher performance rates than generalists (the "hotter is better" hypothesis), probably because having a biochemical machinery that performs under different temperature ranges is costly (Angilletta et al. 2010, Richter-Boix et al. 2015).

A function defining thermal sensitivity is called a thermal reaction norm. Some thermal reaction norms, the thermal performance curves (TPC) are commonly obtained to establish and compare the thermal sensitivity of different populations or species. For adult amphibians, it is frequent to make TPCs of locomotion or calling frequency (Bevier 2016), while growth or survival rates are mainly estimated for larval stages (Navas et al. 2008). Locomotion TPCs commonly assess the thermal dependence of swimming burst speed or jumping performance, for aquatic and terrestrial species, respectively (e.g., Careau et al. 2014, Lai et al. 2018, Greenberg and Palen 2021). A typical TPC is left-skewed and bell-shaped, with a minimum temperature value defined as critical thermal minimum (CT_{min}) and a maximum value defined as a critical thermal maximum (CT_{max}). There is an optimal temperature (T_{opt}) for which performance is maximized and a performance breadth that usually covers the temperatures that limit the top 80% of the performance curve (Fig. 1) (Huey and Stevenson 1979, Angilletta 2006). In terms of thermal physiology, temperatures that exceed an ectotherm's critical thermal maximum (CT_{max}) or thermal optimum (T_{opt}) would impair fitness, limit activity, or induce mortality, hence informing on the absolute thermal tolerance range of the species (Sunday et al. 2014). For this reason, and because they are frequently estimated, TPCs are helpful to compare thermal biology of amphibian species and their vulnerability to climate change (Rosset and Oertli 2011, Taylor et al. 2020).

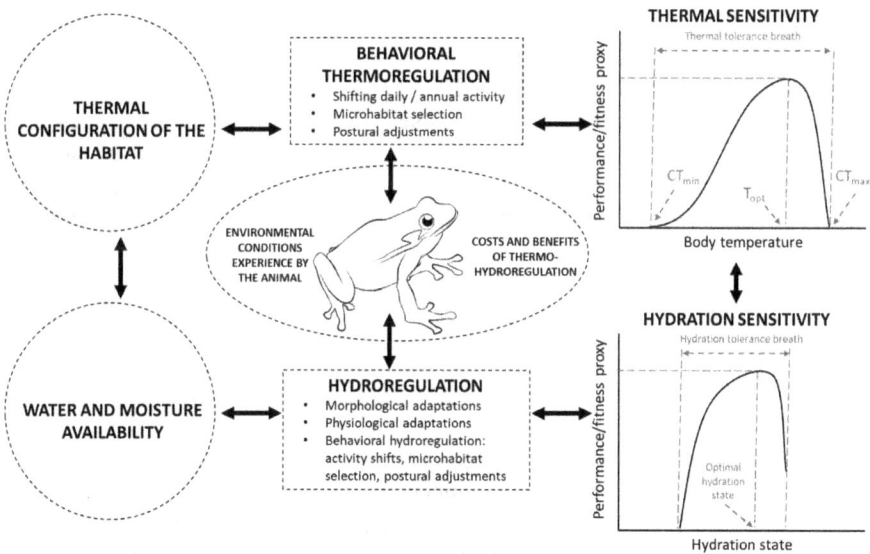

Figure 1. Integrative scheme on how the environment and the organism interact with respect to thermal ecology and water balance of amphibians. The hypothetical situations would be quite different depending on the habitat (aquatic, semi-aquatic or terrestrial) and period of activity (diurnal or nocturnal) species, being the most complex situation that of a terrestrial diurnal amphibian. Abbreviations: CT_{min} = critical thermal minimum, CT_{max} = critical thermal maximum, T_{opt} = optimal temperature.

Variation on thermal sensitivity of amphibian populations can be due to evolution or phenotypic plasticity. These processes interact and, in many cases, it is difficult to assess whether thermal sensitivity differs due to one or the other (Bodensteiner et al. 2021). Critical temperatures usually evolve differently, with CT_{min} evolving at a faster rate than CT_{max}, probably because thermoregulation through microhabitat selection allows animals to actively avoid hot places. This way, thermoregulation may slow evolutionary adaptation of some physiological parameters, a process known as the "Bogert effect" (Huey et al. 2003), as suggested to happen in plethodontid salamanders (Farallo et al. 2018). Although generally evolving slowly, the parameters at the hot part of the TPC, as the T_{opt} and CT_{max} tend to coevolve, allowing thermal adaptation to new conditions. For example, tadpoles of *Lithobates sylvaticus* rapidly evolved, increasing their CT_{max}, when invader beavers modified thermal conditions of their habitats (Skelly and Freidenburg 2000). Moreover, amphibians frequently show events of microevolution and local adaptation to microclimatic conditions, probably due to their low dispersal capacity (Bodensteiner et al. 2021). For example, genetically connected populations of *Rana arvalis* diverged on larval life history and locomotor performance depending on the thermal conditions of ponds (Richter-Boix et al. 2015).

Phenotypic plasticity can also be reflected into ontological differences on thermal sensitivity or thermal acclimation. Eggs or tadpoles may show different TPCs than adult amphibians, as found in four anuran species—*Hypsiboas crepitans*, *Engytomops pustulosus*, *Rhinella humboldti* and *Espadarama prosoblepon*—in Colombia (Turriago et al. 2015). How other intrinsic (e.g., sex, nutritional state) or extrinsic factors (e.g., altitude, water availability, predation pressure) interact with phenotypical plasticity of thermal sensitivity of amphibians will allow us to better understand the ecophysiology of this group and their responses to human impacts. For instance, research on the treefrog *Hyla versicolor* can illustrate the complexity of plastic responses on amphibian thermal sensitivity, where exposure to predatory chemical cues and pesticides induced interactive effects on the TCP of tadpoles (Katzenberger et al. 2014). In any case, it is important to consider the different aspects of the thermal sensitivity of an amphibian species or population to better understand how they regulate their body temperature.

Thermoregulation

The second key aspect of thermal ecology is temperature regulation or thermoregulation, i.e., the ability to maintain body temperatures with a certain mean and variability, responding to conditions of the external environment. Thus, we can imagine a continuum of strategies from thermoconformer animals—whose body temperatures reflect environmental temperatures—to perfect thermoregulators—whose body temperatures are constant, despite environmental fluctuations. Thermoregulation would evolve to enhance performance on thermally variable environments (Angilletta 2009). Many amphibians are aquatic and/or nocturnal, and, thus, live in quite thermally homogeneous environments with limited opportunities for behavioral thermoregulation. That would also be the case of caecilians that, having less permeable skin than other amphibians and living in moist soils or water, are thought to be quite thermoconformers (Measey and Van Dongen 2006). Diurnal species living in tropical rainforests also experience quite thermally homogeneous environments, even at high elevation (e.g., Navas 1997). However, diurnal amphibians more linked to terrestrial habitats, as open areas, rely on behavioral adjustments to thermoregulate, allowing them more time for feeding, reproduction, or growth (Brattstrom 1979).

For ectotherms, heat exchange relies on solar radiation (direct and reflected on objects), conduction, convection, and evaporation. Specifically, amphibians would use solar radiation, conduction from the substrate or the water, and convection from the air to gain heat; and would rely on evaporation, conduction and radiation to the substrate, water and/or air, and convection to air to cool themselves. The combination of these sources and mechanisms of heat exchange will depend on the habitat (terrestrial, semi-aquatic, aquatic) and period of activity (diurnal vs nocturnal), as well as the presence of morphological adaptations (Brattstrom 1963) (Fig. 2).

Thermoregulation is usually conceived as a balance of costs and benefits (Huey and Slatkin 1976, Vickers et al. 2011). Benefits are obvious, since achieving a body temperature within a range that maximizes performance will enhance physiological performance of the individual. However, thermoregulation also entail costs, which will depend on habitat traits, thermoregulation strategies and interactions with other organisms (Sears and Angilletta 2015). We can divide costs into two types: energetic and non-energetic costs. Energetic costs consider that energy used for thermoregulation is not allocated to other activities, such as forage or mate. Non-energetic costs include the risk of predation associated with thermoregulation strategies, as well as the loss of opportunities for doing other activities while thermoregulating (Angilletta 2009). For aquatic amphibians, for example, selecting

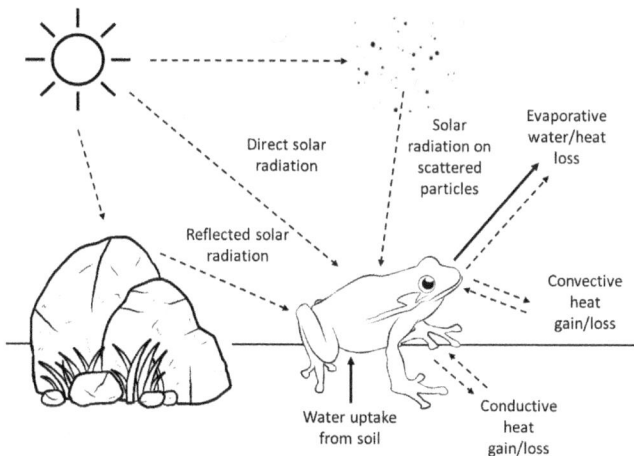

Figure 2. Main sources of heat and water flux for a hypothetical diurnal amphibian. Dashed lines depict heat exchange and continuous lines water exchange. For an aquatic animal (tadpole or adult), heat exchange would be mainly due to convection with the surrounding water, with a possible contribution of solar radiation at the surface of the water. For a nocturnal terrestrial amphibian, heat exchange would rely on convection with air, conduction with the substrate and evaporative cooling, while water exchange would be qualitative similar than this depicted in the figure (evaporative water loss and water uptake from moist soil).

warmer waters may imply less oxygen availability, thus hindering respiration and osmoregulation (Brattstrom 1979). As with most aspects of the behavioral ecology of amphibians, thermoregulatory strategies are tightly linked to their dependence on water and atmospheric humidity. For aquatic amphibians—or aquatic larval stages of terrestrial species—thermoregulation is limited and mainly behavioral, by means of habitat selection. For terrestrial amphibians, thermoregulation may combine physiology—by means of evaporative cooling—and behavior—by modulating activity, microhabitat selection and postural adjustments (Figs. 1 and 2) (Brattstrom 1963). This way, the inevitable water loss resulting from dermal respiration of terrestrial amphibians is also an important cooling mechanism, and thermoregulation will be only prioritized under suitable hydration or water availability situations (Brattstrom 1979).

Heliothermy (i.e., quickly heat up through solar radiation combining microhabitat selection and postural adjustments) is the main thermoregulatory strategy of many diurnal ectotherms, as lizards. However, the trade-off between cooling through evapotranspiration and maintaining suitable hydration would turn heliothermy too costly for most amphibian species. Nonetheless, some anurans can use heliothermy by being in direct contact with water or by decreasing skin permeability. Some temperate species of *Rana, Bufo* and *Hyla* are known to bask under the sun with the ventral part of their bodies in contact with a moist substrate that they actively selected (Brattstrom 1963). Moreover, *Lithobates catesbeianus* combines basking with postural adjustments to keep activity body temperatures within a relatively narrow range (Lillywhite 1970). Different morphological adaptations can allow basking anurans to maintain water balance, such as skin grooves or mucous secretions (Brattstrom 1979). Some amphibians, as *Anaxyrus boreas*, even bask on dry soil, perhaps due to a high water-storage capacity (Lillywhite et al. 1973).

Microhabitat selection and modulation of daily and annual activity patterns are the other two main mechanisms of behavioral thermoregulation (Fig. 1). Plethodontid salamanders are also known to use microhabitat selection to thermoregulate while maintaining water balance (Farallo et al. 2018). While thermal conditions of tropical areas are quite temporally stable, considerable thermal shifts on daily or seasonal activity patterns would be important thermoregulatory strategies for temperate amphibians (Lillywhite 2016). Many frogs and salamanders inhabiting high latitudes or altitudes hibernate when the habitat is below-zero, and some of them have anti-freezing substances in their blood, as is the case of northern populations of *L. sylvaticus* (Costanzo et al. 2013). At not-so-extreme conditions, the toad *A. boreas* is known to thermoregulate by being more diurnal or nocturnal depending on the thermal conditions of the environment (Carey 1978). High mountain Andean populations of *Rhinella spinulosa* use a combination of thigmotermy (i.e., heating up by conduction from the substrate), basking, regulating daily activity hours and microhabitat selection (Sinsch 1989). Thermoregulation through microhabitat selection is not limited to adult amphibians. On the contrary, thermoregulation can begin as early as with the selection of thermally suitable oviposition sites for some species, as documented on *L. sylvaticus* (Freidenburg 2017). For example, tadpoles of *Pseudacris regilla* and *A. boreas* select warmer water both in laboratory and field conditions (Bancroft et al. 2008). Furthermore, tadpoles of high altitude *R. spinulosa* populations use grouping behavior, or aggregation, to achieve higher temperatures (Espinoza and Quinteros 2008). As we can see, morphological adaptations related to water balance and thermo-hydroregulation behaviors are usually combined in amphibians.

After almost a century of research, we are beginning to understand the complex mechanisms driving the thermal ecology of amphibians. Next steps will elucidate how other important extrinsic and intrinsic factors, such as oxygen availability or personality, interact and shape amphibian ecophysiology. There is also a male bias in studies of thermal traits, that needs to be addressed in the next decades to fully understand the evolutionary dynamics and consequences of climate change to these threatened vertebrates (Bodensteiner et al. 2021). If the reader wants to study thermal ecology of amphibians, or to know more about the available methodologies, we refer to a thorough review in Taylor et al. (2020).

Hydric Balance

As many other vertebrates, Amphibians are composed of 70–80% water and need to maintain a stable internal environment for chemical reactions inside cells. The word 'amphibian' translates from Greek as 'both lives', due to their interphase between aquatic and terrestrial habitats. Amphibians are particular among tetrapods due to their permeable skin, considered an adaptation to terrestrial life (Shoemaker et al. 1992, Toledo and Jared 1993). This, added to the fact that oviposition and larval phases are generally aquatic, makes amphibians a special case regarding their dependence on water. This happens even to the point that the availability of water is an essential driver of the distribution of amphibians, probably even more than temperature (Greenberg and Palen 2021). Aquatic and fossorial species present reduced skin permeability, but terrestrial amphibians possess a skin that is highly permeable to water, equally in both directions (Brattstrom 1963). Main functions of the amphibian skin are osmoregulation and respiration—to the point that some groups, such as plethodontid salamanders, lack lungs. This skin allows water uptake directly from the water, or from moist substrates, but not from air (Brattstrom 1963). Finally, this permeable skin also contributes to thermoregulation via evaporation (Fig. 2) (Toledo and Jared 1993).

The hydration state of the organism also influences performance (Mitchell and Bergmann 2016). Thus, we can think on a 'hydration sensitivity' of all biological functions (and, ultimately, fitness) that would be analogous to thermal sensitivity. The relation between hydration state and performance would be more asymmetrical than temperature since performance would drop off sharply under the dehydration extreme of the axis (Fig. 1). This hydration sensitivity would be more evident for semi-aquatic or terrestrial amphibians, since most aquatic phases or species, under normal conditions, would not experience dehydration situations in their habitats. Finally, as different species show a variety of morphological and physiological adaptations to prevent water loss and enhance water uptake, the shape of the hydration sensitivity functions would vary accordingly. In any case, it makes more sense to think of hydration sensitivity as the third dimension of a thermal-hydroregulation framework than to consider it separately (see next section and Rozen-Rechels et al. 2019).

Analogously as we saw for thermal ecology, the other axis of amphibians' hydric ecology would be hydroregulation, or regulation of the hydric balance. In addition to behavioral adjustments, some of them similar to those used for thermoregulation— e.g., postural adjustments, being active when humidity is suitable for the species and select humid microhabitats or shelters—many terrestrial amphibians also rely on morphological adaptations to regulate their hydric balance (Fig. 1; also see Chapter 7). Amphibians do not drink in normal conditions; instead, water enters through the skin. Granular skin is more vascularized, allowing rapid water uptake (Canziani and Cannata 1980). Thus, while aquatic species generally have homogeneously smooth skin, terrestrial anurans and salamanders show some parts of granular skin, and the degree of skin smoothness even change within species, being lower on those inhabiting drier environments. Cutaneous grooves are the other most common adaptation enhancing water intake of anurans and salamanders, carrying the water from the moist surface throughout the amphibian body by capillarity. A particular adaptation for water absorbance is the 'pelvic patch', an area concentrating granular skin and/ or cutaneous grooves for rehydration. Other modifications of the skin are adapted to decrease the animal's water loss, such as epidermis with higher lipid content, iridophores, mucus, cocoons that cover the whole animal preventing desiccation on dry seasons, co-ossification and osteoderms of some hylid frogs living inside bromeliad leaves, or a calcified dermal layer (Toledo and Jared 1993). In addition, some behavioral adaptations may be exclusive to prevent water loss during the driest periods, as burying for some species of *Bufo*, who bury themselves deeper the drier is the soil (Ruibal and Hillman 1981). Finally, another morphological adaptation to enhance water balance is to use the bladder for water storage in terrestrial anurans (e.g., Ruibal 1962) and salamanders (e.g., Brown et al. 1977).

Main behavioral hydroregulation strategies imply adjusting daily or seasonal activity periods, looking for moist shelters, and using water-conserving and water-absorbing postures (Lillywhite 2016). These behavioral adjustments are usually combined with morphological and physiological ones (Fig. 1). For example, the desert-dwelling toad *Anaxyrus punctatus* absorbs water by pressing its pelvic patch with the moist substrate (Brekke et al. 1991). Hylid frogs covering themselves with lipids to reduce evaporative water loss show a typical wiping behavior to spread the secretions over the body surface. After wiping, they generally remain immobile for a while, on water-conserving postures, with some species even reducing their metabolism to torpor states (Blaylock et al. 1976, Gomez et al. 2006, Barbeau and Lillywhite 2005). The desert-dwelling frog *Phyllomedusa sauvagei* wipes cutaneous lipid secretions until environmental temperature reaches 30°C, and then, skin water evaporation raises, preventing the animal from overheating (Brattstrom 1979). This example of the use of water evaporation through the skin as a thermoregulatory strategy is widespread among amphibians and links thermoregulation and hydric balance, as we see in more detail in the next section.

Interactions Between Thermoregulation and Hydric Balance

At the planetary level, water availability and temperature are correlated on space and time. At the individual level, body temperature drives biochemical reactions and water balance determines cell metabolism. Therefore, body temperature and hydration state are correlated physiological axes that will interactively shape performance and fitness (Fig. 1) (Greenberg and Palen 2021). Despite much research on cases where thermo- and hydroregulation interact, an integrative conceptual framework for thermo-hydroregulation is recent (Rozen-Rechels et al. 2019). This framework establishes an integrative panorama where the thermal and water landscapes interact to shape the biophysical properties and behavior of animals, ultimately driving their performance, population dynamics and range distributions (Rozen-Rechels et al. 2019). The adaptation and application of this framework of thermo-hydroregulation for the particular biology of amphibians will be of great interest in future years. Thermo-hydroregulation is defined as an interactive set of behavioral and physiological processes maintaining the organismal thermal and water balance, and thus optimizing performance, survival, and reproduction (Rozen-Rechels et al. 2019). This is obvious for amphibians, whose permeable skin and larval dependence on water bodies make them particularly sensitive to water (Anderson and Andrade 2017). We already explained the main connections between thermoregulation and water balance along the other sections, but here we will briefly summarize the most relevant shared physiological and behavioral mechanisms found in amphibians, to help the reader visualize the integrated framework.

Physiological mechanisms of thermo-hydroregulation are common in amphibians. As evaporation through the permeable skin reduces body temperature, cutaneous adaptations to prevent water loss (see previous section) interact with thermal ecology. Cutaneous resistance to water loss allows arboreal basking frogs to warm up while avoiding hydric stress, and larger frog species (lower surface/volume ratio) may enhance their ability to be away from water sources for longer time than smaller ones (Tracy et al. 2010). Basking *Hyla arenicolor* is known to adapt its evaporative water loss to enhance thermoregulation (Snyder and Hammerson 1993). At the same time, water deprivation can lead amphibians to select for lower body temperatures to avoid further water stress (Anderson and Andrade 2017, Hoffmann et al. 2021). Critical thermal limits are known to change depending on the hydration state of the organism, reducing performance and thermal tolerance as the individual dehydrates (Rozen-Rechels et al. 2019, Hoffmann et al. 2021). In fact, Rozen-Rechels and collaborators (2019) propose to modify thermal performance curves to accommodate hydration state as a third dimension, since there is enough evidence supporting that temperature plus water explain amphibian's performance and fitness better than only temperature (Anderson and Andrade 2017). Although future research will shed more light, and specific differences may exist, the effects of water and temperature on performance are probably interactive rather than additive (Rozen-Rechels et al. 2019).

Main behavioral mechanisms for thermo- and hydroregulation are shared, as the seasonal and daily shifts on activity patterns to match suitable conditions of environmental temperature and water/moisture availability, postural adjustments (that can be different depending on the animal's priority for regulating temperature or water balance) and microhabitat selection (Fig. 1) (Rozen-Rechels et al. 2019). Plethodontid salamanders change the rate of dehydration—by absorbing water from the soil—depending on body size, air humidity and environmental temperature, and showed different behavioral adaptations to thermo-hydroregulate. For example, *Plethodon ouachitae* and *P. caddoensis* burrowed themselves during hot dry summers (Spotila 1972). In the forest frog *Eleutherodactylus coqui*, postural adjustments throughout the day would lead to optimize thermoregulation while minimizing water loss at the same time (Pough et al. 1983). With all of this in mind, the model of costs and benefits of thermoregulation needs to be updated to consider the effect of water availability in the habitat. The costs and benefits of hydroregulation may be delayed with respect to those of thermoregulation, since water uptake and loss tend to be slower than heat exchange. In addition, costs of hydroregulation may be more asymmetric, since they would be more associated with a lack of water than an excess of it (Rozen-Rechels et al. 2019). For instance, *Rana temporaria* selects moist and cool habitats in the field (Köhler et al. 2011), and other amphibians seem to also select moist microhabitats over optimal thermal conditions (Greenberg and Palen 2021). In addition, *Lithobates clamitans* that, as many other aquatic frogs, uses jumping as a main mechanism for predator avoidance—and while temperature and moisture interact to shape jumping performance—prioritizes hydroregulation over thermoregulation (Mitchell and Bergmann 2016). Assessing the relative contribution of each environmental variable and their interactions on microhabitat selection of each species will further elucidate how amphibians make these decisions (Ortega et al. 2019).

Relations Between Human Impacts and Amphibian Ecophysiology

Anthropogenic threats, such as climate change (Velasco et al. 2021), habitat destruction (Cordier et al. 2021), and diseases (Brannelly et al. 2021) are causing a global extinction trend in amphibians. Approximately 40% of amphibian species may be close to extinction (Ceballos et al. 2020).

Temperature of the Earth's surface has been changing since the periods of the industrial revolution. Global surface temperature is increasing at the rate of $0.2° \pm 0.1°C$ per decade since then and is projected to reach 1.5°C above the pre-industrial period between 2030 and 2052 (Allen et al. 2018). Actions such as pollution, burning of fossil fuels and deforestation have an unequivocal direct influence on climate change (IPCC 2021). Like a cascade, climate change affects several other systems, such as rainfall flow, wildland fire regimes and changes in daily temperature fluctuations (Cheng et al. 2007, Flannigan et al. 2009, Vázquez et al. 2015, Sun et al. 2018). This is not only due to a direct effect on environmental temperature, but also because climate change alters other factors that also impact amphibians' thermo-hydroregulation. For example, frog populations of the Savannah River Site (South Carolina, USA) are declining by the reduction in the amount of rain per year (Daszak et al. 2005).

Vulnerability to climate change depends on physiological traits and environmental conditions (Scheffers et al. 2014). While climate change threatens amphibian species, it remains unclear how they could cope with increasing temperatures, seasonal shifts, and increasing frequency of extreme climatic events (Huey et al. 2012). Climate change episodes may vary among locations, and some populations might be more affected by climatic events than others (Boyer et al. 2021). For example, temperatures are predicted to increase at higher rates in tropical locations than in the temperate zones. A study carried out in the Brazilian Atlantic Forest showed the increasing occurrence of frosts contributing to the decline of the frog community (Heyer et al. 1988). In addition, Pounds and Crump (1994) noted that a population drop in Costa Rican *Ollotis periglenes* in 1987, which was attributed to the irregular climate, was more likely induced by global warming. Species adapted to cold or extreme environments, for example polar or high-altitude environments, are even more worrying cases (Parmesan 2006, Navas and Otani 2007).

Due to their strong physiological dependence on thermo-hydroregulation, changes in environmental temperature and precipitation deeply affects amphibian's ecophysiology (Li et al. 2013). The effect of climate warming on the toad *Anaxyrus boreas* is known to depend on the interaction of temperature and atmospheric moisture (Bartelt et al. 2010). In addition, the limited dispersal capability of amphibians make them particularly vulnerable to habitat modifications (Hoffmann et al. 2021). Species depending on seasonal streams are already experiencing niche contractions due to dryness associated with climate change (Hoffmann et al. 2021). In addition, studying the effect of water availability and temperature on embryonic development is key to assess the impact of climate change (Taylor et al. 2020). Dehydration risk may be the primary driver of amphibian activity and environmental restriction, entailing high fitness costs, and thus being another main driver of amphibians' climate vulnerability (Greenberg and Palen 2021). There is potential for local adaptation of amphibians to adapt to increasing water stress, but we still ignore if this will be enough to face the pace of climate change (Hoffmann et al. 2021). Acclimation and/or plasticity of performance curves will enable some adaptation of amphibians to climate change; and genetic adaptation will depend on the pace and magnitude of climatic change and the generation time of each species (Seebacher et al. 2015). Small body size, low metabolic requirements and behavioral plasticity are also predicted to enhance adaptation and survival to climate change (Lillywhite 2016).

In addition to these threats, epizootic diseases are increasingly affecting wildlife populations, and, to date, it remains poorly understood how the environment shapes most host-pathogen systems. Recently, it has been highlighted that microclimate constraints suppress host thermal behavior that is favorable to disease control. Thus, innate host defenses against these diseases might remain limited in the wild, predisposing to range-wide disease outbreaks and population declines (Beukema et al. 2021). Emerging infectious diseases, such as chytridiomycosis, are associated with population decline and the collapse of amphibian communities, being implicated in the decline of at least 501 species of amphibians worldwide (Scheele et al. 2019). This infectious and fatal disease, caused by a fungus *Batrachochytrium dendrobatidis*, is also related to climate change and amphibians' ecophysiology (Lips et al. 2003, Pounds et al. 2006, Skerratt et al. 2007; see Chapter 5). As global temperatures rise, climatic fluctuations can exceed the limits for certain pathogens, triggering outbreaks and causing some diseases to become more lethal or spread more quickly (Epstein 2001, Harvell et al. 2002). In the specific case of *B. dendrobatidis*, this fungus grows best at moderate temperatures (slightly above 20°C), mainly infecting in winter periods or in environments with mild summers. With the thermal increase due to global warming, these periods may present ideal temperatures to trigger the spread of the fungus on areas that were too cold for it before (Pounds et al. 2006). In addition, during winter periods the amphibian's immune system response tends to be slower, as these individuals depend on thermoregulation to fight certain pathogens (Wright and Cooper 1981, Woodhams et al. 2003, Pounds et al. 2006). The association of these factors raises great concern about the consequences of the increase in temperature in the development of *B. dendrobatidis*. Furthermore, amphibian decline due to disease can lead to cascading effects, as happened in Panama, where frog's collapse led to a decline in the community of snakes (Zipkin et al. 2020). Paradoxically, the fact that thermal tolerance of the fungus is lower than that of many amphibians can be a solution to mitigate its impact in some regions, providing amphibians with warm microhabitats to avoid infection (Hettyey et al. 2019, but see Cohen et al. 2019). This example illustrates the importance and complexity of ecophysiology for animal conservation.

Loss and fragmentation of suitable habitat is an additional threat for amphibian populations. Habitat connectivity appears to play a key role in regional viability of amphibian populations, where prevalent human land-use changes strongly reduce the richness of amphibians (Cordier et al. 2021). Population connectivity is predominantly affected through juvenile dispersal and the short-term impact of habitat loss and fragmentation increases with dispersal ability. Scarcity of trophic resources, alteration of food web dynamics, increasing of diseases or increasing competition due to shifts on species' distributions will also imply unpredictable and potentially drastic impacts on amphibians' populations (Lillywhite 2016). Conservation strategies would benefit from species-specific recommendations and by moving from site-specific actions to implementing conservation plans at multiple scales across broad landscapes (Cushman 2006). Effective wildlife restoration is a critical requirement of many conservation actions. Outcomes of conservation actions can be optimized through knowledge of species' habitat requirements, but, to date, evidence from local research is not usually used to design habitat management. Furthermore, interventions administered externally from the top down, avoiding those developed with multiple stakeholders including land managers and researchers, run the risk of failing to be effective. Overall, the new habitats constructed or restored should be characterized by ecological research to quantify local habitat requirements and working with commercial land managers to ensure equitable benefits prior to designing conservation actions can promote rapid and efficient recovery of wildlife (O'Brien et al. 2021). The availability of shelters (thermal and hydric buffers as burrows and shadowed areas) will have a major role on restoration ecology, to be able to mitigate the impact of climate change on amphibian populations (Lillywhite 2016).

Conclusions

Amphibians live in different types of habitats—aquatic, semi-aquatic, fossorial or terrestrial—in a wide variety of climates and elevations. In addition, some species show diurnal activity while others are nocturnal. This, and fine-grain behavioral and morphological adjustments, lead amphibian species to experience a great diversity of thermal and hydric environments. However, amphibians share two key traits that make their ecophysiology particular among other tetrapod groups: they are ectotherms, and they highly depend on water or moisture. Thermal ecology and water balance have been traditionally considered separately, although for amphibians most published studies assessed or commented in the interactive effect of temperature and water availability.

Both processes, thermal ecology and water balance, would have two main axles: physiological sensitivity and regulation. A recently developed conceptual framework of thermo-hydroregulation integrates thermal and hydric sensitivity (which probably interact on most species) with the shared mechanisms for thermo- and hydroregulation. We think that next steps are to develop this framework for amphibians, to gain a better understanding on how thermal and hydration-state performance curves interact on the different species, and how the model of costs and benefits of thermoregulation can be adapted to assess amphibian thermo-hydroregulation. With all of this in mind, we urge herpetologists, behavioral ecologists, ecophysiologists and conservation biologists to work together to implement this framework on amphibian conservation with the aim of mitigating the effects of habitat loss, emerging diseases and climate change and incorporate the physiological requirements of amphibians for ecological restoration.

Acknowledgements

We thank the editors of this book, Mar Comas and Gregorio Moreno-Rueda, for the invitation to participate and for sharing insightful discussion on amphibians' ecology through many years. We thank Coordenação de Aperfeiçoamento de Pessoal de Nível Superior (CAPES; Brazil) for funding ZO with a postdoctoral grant (ref. PNPD 1694744) and Junta de Andalucía (Spain) and European FEDER funds for supporting ZO and MLR through Research Postdoctoral Contracts.

References

Allen, M., M. Babiker, Y. Chen, H. de Coninck and S. Conors. 2018. Framing and context. *In:* Masson-Delmotte, V., P. Zhai, H.-O. Pörtner, D. Roberts, J. Skea, P.R. Shukla et al. [eds.]. Global Warming of 1.5 C. An IPCC Special Report on the Impacts of Global Warming of 1.5 C above Pre-Industrial Levels and Related Global Greenhouse Gas Emission Pathways, in the Context of Strengthening the Global Response to the Threat of Climate Change, Sustainable Development, and Efforts to Eradicate Poverty. IPCC Special Report.

Anderson, R.C. and D.V. Andrade. 2017. Trading heat and hops for water: Dehydration effects on locomotor performance, thermal limits, and thermoregulatory behavior of a terrestrial toad. Ecol. Evol. 7: 9066–9075.

Angilletta, M.J. 2006. Estimating and comparing thermal performance curves. J. Therm. Biol. 31: 541–545.

Angilletta, M.J. 2009. Thermal adaptation: a theoretical and empirical synthesis. Oxford University Press, Oxford, UK.

Angilletta, M.J., R.B. Huey and M.R. Frazier. 2010. Thermodynamic effects on organismal performance: is hotter better? Physiol. Biochem. Zool. 83: 197–206.

Bancroft, B.A., N.J. Baker, C.L. Searle, T.S. Garcia and A.R. Blaustein. 2008. Larval amphibians seek warm temperatures and do not avoid harmful UVB radiation. Behav. Ecol. 19: 879–886.

Barbeau, T.R. and H.B. Lillywhite. 2005. Body wiping behaviors associated with cutaneous lipids in hylid tree frogs of Florida. J. Exp. Biol. 208: 2147–2156.

Bartelt, P.E., R.W. Klaver and W.P. Porter. 2010. Modeling amphibian energetics, habitat suitability, and movements of western toads, *Anaxyrus* (= *Bufo*) *boreas*, across present and future landscapes. Ecol. Model. 221: 2675–2686.

Beukema, W., F. Pasmans, S. Van Praet, F. Ferri-Yáñez, M. Kelly, A.E. Laking et al. 2021. Microclimate limits thermal behaviour favourable to disease control in a nocturnal amphibian. Ecol. Lett. 24: 27–37.

Bevier, C.R. 2016. Physiological and biochemical correlates of calling behavior in anurans with different calling strategies. pp. 63–79. *In*: Andrade, D.V., C.R. Bevier and J.E. Carvalho [eds.]. Amphibian and Reptile Adaptations to the Environment: Interplay Between Physiology and Behavior. CRC Press Taylor & Francis Group, London, UK.

Blaylock, L.A., R. Ruibal and K. Platt-Aloia. 1976. Skin structure and wiping behavior of phyllomedusine frogs. Copeia 1976: 283–295.

Bodensteiner, B.L., G.A. Agudelo-Cantero, A.A. Arietta, A.R. Gunderson, M.M. Muñoz, J.M. Refsnider et al. 2021. Thermal adaptation revisited: how conserved are thermal traits of reptiles and amphibians? J. Exp. Zool. Part A 335: 173–194.

Bovo, R.P., C.A. Navas, M. Tejedo, S.E. Valença and S.F. Gouveia. 2018. Ecophysiology of amphibians: Information for best mechanistic models. Diversity 10: 118.

Boyer, I., H. Cayuela, R. Bertrand and F. Isselin-Nondedeu. 2021. Improving biological relevance of model projections in response to climate change by considering dispersal amongst lineages in an amphibian. J. Biogeogr. 48: 561–576.

Brannelly, L.A., H.I. McCallum, L.F. Grogan, C.J. Briggs, M.P. Ribas, M. Hollanders et al. 2021. Mechanisms underlying host persistence following amphibian disease emergence determine appropriate management strategies. Ecol. Lett. 24: 130–148.

Brattstrom, B.H. 1963. A preliminary review of the thermal requirements of amphibians. Ecology, 44: 238–255.

Brattstrom, B.H. 1979. Amphibian temperature regulation studies in the field and laboratory. Am. Zool. 19: 345–356.

Brekke, D.R., S.D. Hillyard and R.M. Winokur. 1991. Behavior associated with the water absorption response by the toad, *Bufo punctatus*. Copeia 1991: 393–401.

Brown, P.S., S.A. Hastings and B.E. Frye. 1977. A comparison of the water-balance response in five species of plethodontid salamanders. Physiol. Zool. 50: 203–214.

Canziani, G.A. and M.A. Cannata. 1980. Water balance in *Ceratophrys ornata* from two different environments. Comp. Biochem. Physiol. A 66: 599–603.

Careau, V., P.A. Biro, C. Bonneaud, E.B. Fokam and A. Herrel. 2014. Individual variation in thermal performance curves: swimming burst speed and jumping endurance in wild-caught Tropical Clawed Frogs. Oecologia 175: 471–480.

Carey, C. 1978. Factors affecting body temperatures of toads. Oecologia, 35: 197–219.

Ceballos, G., P.R. Ehrlich and P.H. Raven. 2020. Vertebrates on the brink as indicators of biological annihilation and the sixth mass extinction. Proc. Nat. Acad. Sci. USA 117: 13596–13602.

Cheng, C.S., H. Auld, G. Li, J. Klaassen and Q. Li. 2007. Possible impacts of climate change on freezing rain in south-central Canada using downscaled future climate scenarios. Nat. Hazard Earth Sys. 7: 71–87.

Cohen, J.M., D.J. Civitello, M.D. Venesky, T.A. McMahon and J.R. Rohr. 2019. An interaction between climate change and infectious disease drove widespread amphibian declines. Glob. Change Biol. 25: 927–937.

Cordier, J.M., R. Aguilar, J.N. Lescano, G.C. Leynaud, A. Bonino, D. Miloch et al. 2021. A global assessment of amphibian and reptile responses to land-use changes. Biol. Conserv. 253: 108863.

Costanzo, J.P., M.C.F. do Amaral, A.J. Rosendale and R.E. Lee Jr. 2013. Hibernation physiology, freezing adaptation and extreme freeze tolerance in a northern population of the wood frog. J. Exp. Biol. 216: 3461–3473.

Cushman, S.A. 2006. Effects of habitat loss and fragmentation on amphibians: a review and prospectus. Biol. Conserv. 128: 231–240.

Daszak, P., D.E. Scott, A.M. Kilpatrick, C. Faggioni, J.W. Gibbons and D. Porter. 2005. Amphibian population declines at Savannah River site are linked to climate, not chytridiomycosis. Ecology 86: 3232–3237.

de Andrade, D.V., C.R. Bevier and J.E. De Carvalho [eds.]. 2016. Amphibian and Reptile Adaptations to the Environment: Interplay Between Physiology and Behavior. CRC Press, London, UK.

Diele-Viegas, L.M. and C.F.D. Rocha. 2018. Unraveling the influences of climate change in Lepidosauria (Reptilia). J. Therm. Biol. 78: 401–414.

Duellman, W.E. and L. Trueb. 1994. Biology of Amphibians. Johns Hopkins University Press, Baltimore, USA.

Epstein, P.R. 2001. Climate change and emerging infectious diseases. Microbes Infect. 3: 747–754.

Espinoza, R.E. and S. Quinteros. 2008. A hot knot of toads: aggregation provides thermal benefits to metamorphic Andean toads. J. Therm. Biol. 33: 67–75.

Farallo, V.R., R. Wier and D.B. Miles. 2018. The Bogert effect revisited: Salamander regulatory behaviors are differently constrained by time and space. Ecol. Evol. 8: 11522–11532.

Flannigan, M.D., M.A. Krawchuk, W.J. de Groot, B.M. Wotton and L.M. Gowman 2009. Implications of changing climate for global wildland fire. Int. J. Wildland Fire 18: 483–507.

Freidenburg, L.K. 2017. Environmental drivers of carry-over effects in a pond-breeding amphibian, the Wood Frog (*Rana sylvatica*). Can. J. Zool. 95: 255–262.

Gomez, N.A., M. Acosta, F. Zaidan III and H.B. Lillywhite. 2006. Wiping behavior, skin resistance, and the metabolic response to dehydration in the arboreal frog *Phyllomedusa hypochondrialis*. Physiol. Biochem. Zool. 79: 1058–1068.

Greenberg, D.A. and W.J. Palen. 2021. Hydrothermal physiology and climate vulnerability in amphibians. Proc. R. Soc. B 288: 20202273.

Harvell, C.D., C.E. Mitchell, J.R. Ward, S. Altizer, A.P. Dobson, R.S. Ostfeld et al. 2002. Climate warming and disease risks for terrestrial and marine biota. Science 296: 2158–2162.

Hettyey, A., J. Ujszegi, D. Herczeg, D. Holly, J., Vörös, B.R. Schmidt et al. 2019. Mitigating disease impacts in amphibian populations: capitalizing on the thermal optimum mismatch between a pathogen and its host. Front. Ecol. Evol. 7: 254.

Heyer, W.R., A.S. Rand, C.A.G. Cruz and O.L. Peixoto. 1988. Decimations, extinctions, and colonizations of frog populations in southeast Brazil and their evolutionary implications. Biotropica 20: 230–235.

Hoffmann, E.P., K.L. Cavanough and N.J. Mitchell. 2021. Low desiccation and thermal tolerance constrains a terrestrial amphibian to a rare and disappearing microclimate niche. Conserv. Physiol. 9: coab027.

Huey, R.B. and M. Slatkin. 1976. Cost and benefits of lizard thermoregulation. Q. Rev. Biol. 51: 363–384.

Huey, R.B. and R.D. Stevenson. 1979. Integrating thermal physiology and ecology of ectotherms: a discussion of approaches. Am. Zool. 19: 357–366.

Huey, R.B. 1982. Temperature, physiology, and the ecology of reptiles. pp. 25–74. *In*: Gans, C. and F.H. Pough [eds.]. Biology of the Reptilia, Vol. 12. Academic Press, New York, USA.

Huey, R.B. and D. Berrigan. 2001. Temperature, demography, and ectotherm fitness. Am. Nat. 158: 204–210.

Huey, R.B., P.E. Hertz and B. Sinervo. 2003. Behavioral drive versus behavioral inertia in evolution: a null model approach. Am. Nat. 161: 357–366.

Huey, R.B., M.R. Kearney, A. Krockenberger, J.A. Holtum, M. Jess and S.E. Williams. 2012. Predicting organismal vulnerability to climate warming: roles of behaviour, physiology and adaptation. Phil. Trans. Roy. Soc. B 367: 1665–1679.

IPCC. 2021. Climate Change 2021: The Physical Science Basis. Contribution of Working Group I to the Sixth Assessment Report of the Intergovernmental Panel on Climate Change. V. Masson-Delmotte, P. Zhai, A. Pirani, S.L. Connors, C. Péan, S. Berger et al. [eds.]. Cambridge University Press, Cambridge, UK.

Katzenberger, M., J. Hammond, H. Duarte, M. Tejedo, C. Calabuig and R.A. Relyea. 2014. Swimming with predators and pesticides: how environmental stressors affect the thermal physiology of tadpoles. PLoS One 9: e98265.

Köhler, A., J. Sadowska, J. Olszewska, P. Trzeciak, O. Berger-Tal and C.R. Tracy. 2011. Staying warm or moist? Operative temperature and thermal preferences of common frogs (*Rana temporaria*), and effects on locomotion. Herpetol. J. 21: 17–26.

Lai, J.J., P.C.L. Hou and Y. Steinberger. 2018. Thermal plasticity in the burst swimming of Bufo bankorensis larvae. Ecology and Sustainable Development 1(2): 57–68.

Li, Y., J.M. Cohen and J.R. Rohr. 2013. Review and synthesis of the effects of climate change on amphibians. Integr. Zool. 8: 145–161.

Lillywhite, H.B. 1970. Behavioral temperature regulation in the bullfrog, *Rana catesbeiana*. Copeia 1970: 158–168.

Lillywhite, H.B., P. Licht and P. Chelgren. 1973. The role of behavioral thermoregulation in the growth energetics of the toad, *Bufo boreas*. Ecology 54: 375–383.

Lillywhite, H.B. 2016. Behavior and physiology: an ecological and evolutionary viewpoint on the energy and water relations of ectothermic amphibians and reptiles. pp. 1–40. Andrade, D.V., C.R. Bevier and J.E. Carvalho [eds.]. Amphibian and Reptile Adaptations to the Environment: Interplay Between Physiology and Behavior. CRC Press Taylor & Francis Group, London, UK.

Lips, K.R., D.E. Green and R. Papendick. 2003. Chytridiomycosis in wild frogs from southern Costa Rica. J. Herpetol. 37: 215–218.

Measey, G.J. and S. Van Dongen. 2006. Bergmann's rule and the terrestrial caecilian *Schistometopum thomense* (Amphibia: Gymnophiona: Caeciliidae). Evol. Ecol. Res. 8: 1049–1059.

Mitchell, A. and P.J. Bergmann. 2016. Thermal and moisture habitat preferences do not maximize jumping performance in frogs. Funct. Ecol. 30: 733–742.

Navas, C.A. 1997. Thermal extremes at high elevations in the Andes: physiological ecology of frogs. J. Therm. Biol. 22: 467–477.

Navas, C.A. and L. Otani. 2007. Physiology, environmental change, and anuran conservation. Phyllomedusa 6: 83–103.

Navas, C.A., F.R. Gomes and J.E. Carvalho. 2008. Thermal relationships and exercise physiology in anuran amphibians: integration and evolutionary implications. Comp. Biochem. Physiol. A 151: 344–362.

O'Brien, D., J.E. Hall, A. Miró, K. O'Brien and R. Jehle. 2021. A co-development approach to conservation leads to informed habitat design and rapid establishment of amphibian communities. Ecol. Solut. Evid. 2: e12038.

Ortega, Z., A. Mencía, K. Martins, P. Soares, V.L. Ferreira and L.G. Oliveira-Santos. 2019. Disentangling the role of heat sources on microhabitat selection of two Neotropical lizard species. J. Trop. Ecol. 35: 149–156.

Parmesan, C. 2006. Ecological and evolutionary responses to recent climate change. Annu. Rev. Ecol. Evol. Syst. 37: 637–669.

Pough, F.H., T.L. Taigen, M.M. Stewart and P.F. Brussard. 1983. Behavioral modification of evaporative water loss by a Puerto Rican frog. Ecology 64: 244–252.

Pounds, J.A. and M.L. Crump. 1994. Amphibian declines and climate disturbance: the case of the golden toad and the harlequin frog. Conserv. Biol. 8: 72–85.

Pounds, J.A., M.R. Bustamante, L.A. Coloma, J.A. Consuegra, M.P.L. Fogden, P.N. Foster et al. 2006. Widespread amphibian extinctions from epidemic disease driven by global warming. Nature 439: 161–167.

Richter-Boix, A., M. Katzenberger, H. Duarte, M. Quintela, M., Tejedo and A. Laurila. 2015. Local divergence of thermal reaction norms among amphibian populations is affected by pond temperature variation. Evolution 69: 2210–2226.

Rosset, V. and B. Oertli. 2011. Freshwater biodiversity under climate warming pressure: Identifying the winners and losers in temperate standing waterbodies. Biol. Conserv. 144: 2311–2319.

Rozen-Rechels, D., A. Dupoué, O. Lourdais, S. Chamaillé-Jammes, S. Meylan, J. Clobert et al. 2019. When water interacts with temperature: Ecological and evolutionary implications of thermo-hydroregulation in terrestrial ectotherms. Ecol. Evol. 9: 10029–10043.

Ruibal, R. 1962. The adaptive value of bladder water in the toad, *Bufo cognatus*. Physiol. Zool. 35: 218–223.

Ruibal, R. and S. Hillman. 1981. Cocoon structure and function in the burrowing hylid frog, *Pternohyla fodiens*. J. Herpetol. 15: 403–407.

Scheffers, B.R., D.P. Edwards, A. Diesmos, S.E. Williams and T.A. Evans. 2014. Microhabitats reduce animal's exposure to climate extremes. Glob. Change Biol. 20: 495–503.

Scheele, B., F. Pasmans, L.F. Skerratt, L. Berger, A. Martel, W. Beukema, W. et al. 2019. Amphibian fungal panzootic causes catastrophic and ongoing loss of biodiversity. Science 363: 1459–1463.

Sears, M.W. and M.J. Angilletta. 2015. Costs and benefits of thermoregulation revisited: both the heterogeneity and spatial structure of temperature drive energetic costs. Am. Nat. 185: E94–E102.

Seebacher, F., C.R. White and C.E. Franklin. 2015. Physiological plasticity increases resilience of ectothermic animals to climate change. Nat. Clim. Change 5: 61–66.

Shoemaker, V.H., S.S. Hillman, S.D. Hillyard, D.C. Jackson, L.L. McClanahan, P. Withers et al. 1992. Exchange of water, ions and respiratory gases in terrestrial amphibians. pp. 125–150. *In*: Feder, M.E. and W.W. Burggren [eds.] Environmental Physiology of the Amphibians. The University of Chicago Press, Chicago, USA.

Sinsch, U. 1989. Behavioural thermoregulation of the Andean toad (*Bufo spinulosus*) at high altitudes. Oecologia 80: 32–38.

Sinclair, B.J., K.E. Marshall, M.A. Sewell, D.L. Levesque, C.S. Willett, S. Slotsbo et al. 2016. Can we predict ectotherm responses to climate change using thermal performance curves and body temperatures? Ecol. Lett. 19: 1372–1385.

Skelly, D.K. and L.K. Freidenburg. 2000. Effects of beaver on the thermal biology of an amphibian. Ecol. Lett. 3: 483–486.

Skerratt, L.F., L. Berger, R. Speare, S. Cashins, K.R. McDonald, A.D. Phillott et al. 2007. Spread of chytridiomycosis has caused the rapid global decline and extinction of frogs. EcoHealth 4: 125–134.

Snyder, G.K. and G.A. Hammerson. 1993. Interrelationships between water economy and thermoregulation in the Canyon tree-frog *Hyla arenicolor*. J. Arid Environ. 25: 321–329.

Spotila, J.R. 1972. Role of temperature and water in the ecology of lungless salamanders. Ecol. Monogr. 42: 95–125.

Sun, F., M.L. Roderick and G.D. Farquhar. 2018. Rainfall statistics, stationarity, and climate change. Proc. Nat. Acad. Sci. USA 115: 2305–2310.

Sunday, J.M., A.E. Bates, M.R. Kearney, R.K. Colwell, N.K. Dulvy, J.T. Longino et al. 2014. Thermal safety margins and the necessity of thermoregulatory behavior across latitude and elevation. Proc. Natl Acad. Sci. USA 111: 5610–5615.

Taylor, E.N., L.M. Diele-Viegas, E.J. Gangloff, J.M. Hall, B. Halpern, M.D. Massey et al. 2020. The thermal ecology and physiology of reptiles and amphibians: A user's guide. J. Exp. Zool. A 335: 13–44.

Toledo, R.C. and C. Jared, C. 1993. Cutaneous adaptations to water balance in amphibians. Comp. Biochem. Physiol. A 105: 593–608.

Tracy, C.R. 1976. A model of the dynamic exchanges of water and energy between a terrestrial amphibian and its environment. Ecol. Monogr. 46: 293–326.

Tracy, C.R., K.A. Christian and C.R. Tracy. 2010. Not just small, wet, and cold: effects of body size and skin resistance on thermoregulation and arboreality of frogs. Ecology 91: 1477–1484.

Turriago, J.L., C.A. Parra and M.H. Bernal. 2015. Upper thermal tolerance in anuran embryos and tadpoles at constant and variable peak temperatures. Can. J. Zool. 93: 267–272.

Vázquez, D.P., E. Gianoli, W.F. Morris and F. Bozinovic. 2015. Ecological and evolutionary impacts of changing climatic variability. Biol. Rev. 7: 1–21.

Velasco, J.A., F. Estrada, O. Calderón-Bustamante, D. Swingedouw, C. Ureta, C. Gay et al. 2021. Synergistic impacts of global warming and thermohaline circulation collapse on amphibians. Comm. Biol. 4: 1–7.

Vickers, M., C. Manicom and L. Schwarzkopf. 2011. Extending the cost-benefit model of thermoregulation: high-temperature environments. Am. Nat. 177: 452–461.

Willmer, P., G. Stone and I. Johnston. 2009. Environmental Physiology of Animals. Blackwel, Malden, USA.

Wright, R.K. and E.L. Cooper 1981. Temperature effects on ectotherm immune responses. Develop. Comp. Immunol. 5: 117–122.

Woodhams, D.C., R.A. Alford and G. Marantelli. 2003. Emerging disease of amphibians cured by elevated body temperature. Dis. Aquat. Organ. 55: 65–67.

Zipkin, E.F., G.V. DiRenzo, J.M. Ray, S. Rossman and K.R. Lips. 2020. Tropical snake diversity collapses after widespread amphibian loss. Science 367: 814–816.

CHAPTER 7

The Biogeography of Body Size in Amphibians

Miguel Á. Olalla-Tárraga

INTRODUCTION

Ecogeographical rules describe general trends in the variation of biological traits along geographical gradients (Gaston et al. 2008). They can be defined as "statements that seek to encapsulate responses of the Earth's fauna and flora to the influences of environmental factors in a patterned way" (McDowall 2008). In 1847, the German physiologist Karl Bergmann first noted that body size plays a major role in determining the physiology, ecology and geographic distribution of endothermic vertebrates. He noticed that reduced surface area to volume (A/V) ratios are a selective advantage to better retain body heat and adduced that among closely related species the ones living in cold regions would be larger than those in warmer climates. This was formalized as Bergmann's rule, which is arguably the best-known ecogeographical rule in biology (Blackburn et al. 1999).

The theoretical foundation for Bergmann's rule relies on geometric arguments about heat loss and the importance of A/V ratios for thermoregulation. The larger the A/V ratio, the greater the heat loss through the surface. Bergmann's rule was initially conceived as an ecogeographic pattern for endothermic vertebrate species and originally explained in terms of a physiological adaptive mechanism. However, Ray's (1960) and Lindsey's (1966) pioneering observations that some ectothermic organisms also displayed intra- and interspecific body size clines as a response to environmental gradients fostered the debate around the generality of Bergmann's rule and the validity of traditional explanations (Olalla-Tárraga 2011). Ectotherms do not generate metabolic heat to the extent of mammals and birds and, therefore, mechanisms based on heat conservation processes alone are unsatisfactory for them (Cushman et al. 1993).

Departamento Biología y Geología, Física y Química Inorgánica, Universidad Rey Juan Carlos, 28933, Móstoles (Madrid), Spain.
Email: miguel.olalla@urjc.es

In recent years, there has been a resurgence of interest in studying ecogeographical rules at different levels of the biological organization. Bergmann's rule is nowadays positioned as a central question in ecology for basic and applied reasons. The detection of negative temperature-body size associations over a variety of systems, taxa and spatial and temporal scales involves that body size reductions are generally expected in a context of anthropogenic global warming (Gardner et al. 2011). This has led to the suggestion that, together with the observed changes in the phenology and geographic distributions of species, body-size reductions are a third universal biological response to global warming (see also Daufresne et al. 2009). Gardner et al. (2011) reviewed the existing empirical evidence on body size reductions as a result of recent increasing temperatures, as well as the possible proximate and ultimate underlying mechanisms. These authors identified the importance of conducting broad-scale comparative analyses to assess whether body size reductions are a universal response to changing thermal environments or not. So far, most evidence for this pattern comes from mammals and birds and there is a striking paucity of data for ectotherms. For instance, less than 5% of the species in an extensive literature search of size responses to climate change correspond to terrestrial ectotherms (Sheridan and Bickford 2011).

Tian and Benton (2020) have also made a call for the use of ecogeographical rules as a baseline on which better predictions of organismal responses to global warming can be generated. The potential use of ecogeographic rules as predictive tools for biological responses to climate change is exciting but represents a formidable challenge. The debate on the existence of laws and rules in ecology is an old one and, although perhaps unresolved, most authors would certainly share Lawton's (1999) view on the relevance of historical and evolutionary contingencies. Therefore, the study of the validity of any ecological rule also involves testing it to the point of being able to detect its exceptions. I argued why Bergmann's and other ecogeographical rules should be viewed as correlative laws, in opposition to causal laws, which refer to broad generalizations with no inherent mechanisms attached and subject to the scrutiny of empirical investigation (Olalla-Tárraga 2011). Correlative laws have predictive value but we should be aware that their explanatory value may sometimes be limited. Finding exceptions is indeed a way to gain a better understanding and delimit the domain of applicability of a correlative law (O'Hara 2005). In any case, the first step towards being able to use the predictive potential offered by ecogeographic rules in a climate change context entails reviewing the status and validity of these rules.

Amphibians are ideal organisms to study phenotype-environment interactions. Because of their typical biphasic life cycles, with aquatic eggs and larvae followed by metamorphosis into a terrestrial adult phase, they have to deal with drastic environmental changes across their ontogeny. Indeed, documented responses to anthropogenic climate change by amphibians are numerous as a result of direct effects on physiology (Carey and Alexander 2003). As ectotherms, all their physiological processes are temperature dependent. However, their lack of protective epidermal structures and their permeable skin makes them particularly vulnerable to extreme solar radiations levels. At high ambient temperatures or low humidities, amphibians experience rapid water loss and limit their activity periods (Vitt and Caldwell

2009). Compared to reptiles, higher A/V ratios are associated not only to higher heat exchange rates but also to high rates of water loss, which reduces heat gain via evaporative cooling (Rubalcaba et al. 2019a). Contrarily to other organisms that have evolved surface internalization as an anatomical strategy to alter A/V ratios (Okie 2013), most amphibians fully depend on cutaneous surfaces to maintain homeostasis.

Despite a highly conserved body plan across lineages, amphibians range in length over 250-fold from the 7 mm long frog *Paedophryne amanuensis* to the 33 cm Goliath frog (*Conraua goliath*) or the gigantic 1.8 m Chinese salamander (*Andrias davidianis*) (Levy and Heald 2016). Size variation is widespread across the amphibian phylogeny, with numerous independent transitions to miniature body sizes in all three clades (Orders Anura, Caudata and Gymnophiona). Similarly, albeit less studied, evolutionary transitions to gigantism have occurred across the two-hundred million years of anuran body size evolution (Womack and Bell 2020). Extreme size reductions in the form of miniaturization processes are associated to a number of environmental and physiological challenges. For example, miniature species can be more vulnerable to predators or, given their high A/V ratios, they are more susceptible to desiccation. Hence, the smallest amphibians are typical inhabitants of tropical wet-forest leaf litter or dense, moist moss (Levy and Heald 2016).

Here, I review the current state of knowledge on the interspecific variation of body size in amphibians across large-scale environmental gradients. Although Bergmann's rule was conceived as an interspecific pattern, this version of the rule has received less attention since Rensch (1938) and Mayr (1956) (cited in Blackburn et al. 1999) reformulated it as an intraspecific pattern. The degree of generality of the intraspecific version of the rule for amphibians has been investigated using meta-analytical techniques elsewhere (Ashton 2002, Adams and Church 2008). Therefore, the present chapter focuses on the interspecific level, which can be particularly informative on evolutionary and ecological mechanisms structuring amphibian assemblages. First, I compile the available information on interspecific patterns of body size in amphibians from regional to global scales that have been documented to date, together with the mechanisms that have received most empirical support. In doing so, I bear in mind the need to avoid the conflation between pattern and process if we are to gain a better understanding on the biogeography of body size (Mayr 1963, Olalla-Tárraga 2011). Second, with hindsight, I discuss emerging research perspectives that will help to understand the biogeography of body size in amphibians in the light of recent work on this topic.

Pattern and Process in the Variation of Amphibian Body Size

In 1966, Lindsey documented the existence of latitudinal body size trends in poikilotherm vertebrates by examining maximum sizes of 12,503 species, including marine and freshwater fishes, amphibians and reptiles. The shape of body size frequency distributions for both anurans (657 species) and urodeles (189 species) differed across the three latitudinal bands into which the data were split. Compared to cool and warm temperate regions, the tropical zone showed a pronounced right-skewness (specially for anurans) with a clear dominance of small body sizes. Among the possible hypotheses suggested to explain the observed latitudinal pattern, differential selection pressures in ecological specialisation and reproductive potential between cold regions and the tropics were mentioned, as well as the starvation resistance benefits that larger body sizes confer to better withstand periods of food shortage during winter. Yet, Lindsey (1966) specially highlighted variation in organismal surface-to-volume ratios *via* direct effects on thermoregulation, osmoregulation and respiration as an explanation. Lindsey's study (1966), albeit markedly descriptive in nature, anticipated by almost thirty years the first works that in the 1990s began to document global scale body size-frequency distributions of different animal groups with the statistical rigour of the nascent field of macroecology.

In the origins of macroecology, the massive exploration of body size data to understand the ecological and evolutionary processes underlying the emergence of often right-skewed body size frequency distributions across entire continental faunas mostly focused on birds and mammals (Maurer et al. 1992, Brown 1995, Gaston and Blackburn 2000). The macroecology of herptiles in general, and amphibians in particular, only began to receive increased attention in the mid-2000s. Ashton (2002) was the first one to provide meta-analytical support for the intraspecific version of Bergmann's rule in amphibians, whereas Olalla-Tárraga and collaborators pioneered the exploration of large-scale interspecific size gradients in reptiles (Olalla-Tárraga et al. 2006) and amphibians (Olalla-Tárraga and Rodríguez 2007) (Table 1).

Classical interspecific analyses aimed to study the macroecology of body size adopted a "cross-species" approach by treating each species as an independent datum and using bivariate scatter-plots to examine the covariation of body size and latitude (or occasionally temperature) across species. This method has also been termed the "midpoint approach" (Blackburn and Hawkins 2004) since it involves obtaining a single spatial descriptor to characterize the distribution of each species (usually

Table 1. Studies of the interspecific version of Bergmann's rule in amphibians.

Author	Clade	Number of species	Geographic extent
Lindsey 1966	Anurans and urodeles	846	Global
Olalla-Tárraga and Rodríguez 2007	Anurans and urodeles	112	North America and Europe
Adams and Church 2008	Plethodon salamanders	44	North America
Olalla-Tárraga et al. 2009	Anurans	131	Brazilian Cerrado
Olalla-Tárraga et al. 2010	*Plethodon* salamanders	44	North America
Gouveia et al. 2013	Centrolenidae frogs	148	Neotropics
Gouveia and Correia 2016	Centrolenidae, Phyllomedusinae, *Hypsiboas* and *Plethodon*	229	New World
Rapacciuolo et al. 2017	Amphibians	3281	New World
Pincheira-Donoso et al. 2019	Caecilians	191	Global
Amado et al. 2019a	Anurans	2761	New World
Womack and Bell 2020	Anurans	2434	Global

the latitudinal midpoint of its geographic range) and then plotting these mid-points against species' body sizes. Alternatively, broad-scale body size gradients at the interspecific level have been documented by means of an assemblage-approach. This method, also termed "community approach" (Blackburn and Hawkins 2004), explores geographical patterns within spatial grids covering the study region and combines the species' presences/absences in the cells with their body sizes to obtain cell-average body size values (usually log-transformed geometric means or medians). While the units of analysis in cross-species approaches are single species, in assemblage-based methods they are measures of average body size of all co-occurring species within grid-cells of a particular biogeographic region. The pros and cons of both approaches have been discussed in detail elsewhere (Gaston et al. 2008, Olalla-Tárraga et al. 2010). In essence, cross-species comparisons must be careful to consider possible phylogenetic autocorrelation effects, whereas assemblage approaches are more sensitive to spatial autocorrelation effects that must be controlled for.

After Lindsey's (1966) seminal paper, Olalla-Tárraga and Rodríguez (2007) applied an assemblage-based approach to investigate the existence of geographic body size gradients in amphibian faunas of the Western Palearctic and Nearctic in response to historical and macroclimatic factors. Both anurans and urodeles displayed clear latitudinal size clines, which were consistent across the Holarctic. However, anurans followed Bergmann's rule and urodeles exhibited the reverse trend, with decreasing body size northwards. Beyond a mere pattern description, this study was conceived to generate insight into possible energetic and physiological mechanisms. Instead of predictors such as latitude or elevation, which can lead to spurious interpretations (Hawkins and Diniz-Filho 2004), a set of proxy variables was included for an explicit multiple-hypothesis testing. Specifically, Olalla-Tárraga and Rodríguez (2007) tested a number of hypotheses that had been proposed to account for Bergmann's rule (or its converse): heat balance, migration ability, primary productivity, seasonality, habitat availability, and water availability. As expected from the heat balance hypothesis, potential evapotranspiration (PET, a measure of environmental energy availability, Fisher et al. 2011) was the strongest predictor. The heat balance hypothesis (HBH), suggested by Olalla-Tárraga and Rodríguez (2007) as an extension of Bergmann's original explanation, posits that the rates of heat gain or loss decrease with decreasing A/V ratios in any organism, but notes that the most limiting factor to reach and maintain activity temperatures in cold climates differs between endotherms and ectotherms. While the former produce internal metabolic heat and mostly benefit from reduced A/V ratios for heat conservation, activity options for ectotherms in cold climates critically depend on heat gain. HBH, now a textbook example (Granado Lorencio 2007, Vitt and Caldwell 2009, Meerhoff et al. 2012), considers that many ectothermic vertebrates can behaviorally (and to a lesser extent physiologically) thermoregulate their body temperatures within relatively narrow ranges. For instance, some amphibians can effectively control heat exchange with the environment by means of short-term movements or posturing to maximize heat gain or minimize heat loss (Vitt and Caldwell 2009). In general, while anurans are able to achieve some control over heat exchange with the environment, urodeles are typical thermoconformers and their body temperatures closely parallel variations in the thermal environment (see, e.g., Hutchinson and Dupré 1992). HBH develops different predictions on the relationship between ambient temperature and body size based on the thermoregulatory abilities of species.

For small thermoregulating ectotherms, HBH is coincident with Bergmann's traditional mechanism for endotherms and predicts that large-bodied species (with lower A/V ratios) will be favored in cold climates because of their better heat retention. The largest thermoregulating ectotherms may need some compensatory physiological, morphological and/or behavioral mechanisms to reduce heating times (Olalla-Tárraga et al. 2006). For example, the degree of melanization has been found to increase with decreasing temperature and solar radiation in numerous ectothermic species, including vertebrates (Clusella-Trullas et al. 2007) and invertebrates (Zeuss et al. 2014). This has been termed as the *thermal melanism hypothesis*, and stating that dark (low skin reflectance) individuals are at an advantage in cold climates as they can heat up faster and reach higher equilibrium temperatures than lighter (high skin reflectance) individuals. Compared to non-melanistic forms, melanistic ectotherms achieve higher heating rates and reduce basking times. Both biophysical models and laboratory and field experiments show that heating rates are faster in dark ectotherms than in light-colored ones, which is particularly advantageous for the survival of larger species under low ambient temperature conditions. The adaptive role of melanism in thermoregulation has been documented for a few amphibian species, including *Rana temporaria* (Alho et al. 2011, Vences et al. 2002) and *Bokermannohyla alvarengai* (Tattersall et al. 2006). However, this pattern, together with its genetic and ontogenetic basis, remains largely understudied in amphibians and has received most attention in reptiles (see Clusella-Trullas et al. 2008). Dark pigment variants have been found in several anuran species in the Holarctic such as *Lithobates pipiens*, *Lithobates clamitans*, *Lithobates sylvatica* and *Ptychadena mascareniensis* (Richards and Nace 1983), but the underlying causes for this geographical pattern have not been explored. Early evidence exists that the heat balance and the thermal melanism hypotheses may act concomitantly to generate geographic body size gradients in ectotherms (Reguera et al. 2014, Moreno Azócar et al. 2015). In thermoconformers, whose body temperature fluctuate more closely with ambient temperature, smaller organisms would be favored in cold areas because their greater A/V ratios allow them to have shorter heating times. Under HBH, size increases are expected towards colder areas in anurans and size decreases towards warm territories in urodeles. Still, predictions based on the HBH become complicated in water-limited tropical environments, where lower A/V ratios are also relevant to provide a smaller mass-specific water loss and better desiccation resistance.

Using anuran data from the Brazilian Cerrado, the most extensive woodland-savanna in South America, Olalla-Tárraga et al. (2009) tested the validity of Bergmann's rule and the HBH beyond the Holarctic domains. Interestingly, broad-scale body size patterns also emerged across environmental gradients but not in a latitudinal fashion. The smallest sizes were detected in the south-west bordering the Pantanal, a wetland biome, and the largest ones in the drier areas of the northeast near the limits of the Caatinga semi-arid region. Indeed, water deficit was identified as the best predictor in a relationship that remained consistent even in the presence of strong phylogenetic signal in body size. Species larger than expected by their phylogenetic relatedness tended to more frequently aggregate in drier environments, whereas species smaller than expected by phylogeny were mostly distributed in low-water-deficit areas. These findings suggests a selective advantage for larger anurans under high water deficit conditions in tropical regions. The seasonal climate of the Cerrado is characterized by a six-month period of dry and hot conditions, which imposes serious hydric constraints on its anuran fauna. As wet-skinned ectotherms, amphibians have little resistance to cutaneous evaporative water loss and are particularly sensitive to long droughts. Through a mechanism analogous to Bergmann's heat conservation, A/V ratios in anurans are positively correlated with water loss and can be involved in controlling the water balance. Thus, a simple mechanism to reduce desiccation is decreasing A/V ratios by increasing body size. The benefits of large body size as an adaptation to arid climates in amphibians were already noted by Nevo (1973), who could be considered the proponent of the water availability hypothesis. Subsequent studies in the Neotropical realm have supported both the heat balance hypothesis in Centrolenidae glassfrogs (Gouveia et al. 2013) and the water availability hypothesis in Centrolenidae, Phyllomedusinae leaf frogs and *Hypsiboas* gladiator frogs (Gouveia and Correia 2016). These different investigations have undoubtedly highlighted the importance, and at the same time the difficulty, of taking into account the interactions between thermoregulation and hydroregulation (also see Chapter 6) in order to better understand size variation in amphibians.

In the last five years, the increasing availability of data on the geographic distribution, morphology and phylogenetic relatedness of amphibians has made it possible to extend the geographical and taxonomic scope in the study of large-scale interspecific body size patterns. Across the Americas, Rapacciuolo et al. (2017) have shown that amphibians do not display as pronounced patterns of latitudinal variation in size as birds and mammals. Similarly, these authors only detected a weak influence of human pressure on amphibian body sizes which contrasts with their findings for endotherms. Despite the spatial complexity and the apparent absence of a universal Bergmannian pattern in amphibians, environmental signals in anuran body size variation across the New World are still evident. Amado et al. (2019a) were able to detect an association between body size and PET for a dataset of 2,761 anuran species of the Western Hemisphere. Both cross-species and assemblage-based analyses concurred in identifying environmental constraints on anuran body sizes. This was consistent with previous findings at the regional level and confirmed the relevance of energy and water availability in accounting for interspecific size clines in anurans. Also, the set of spatially-explicit and phylogenetically-informed analyses of Amado et al. (2019a) were able to highlight the role of topography and microhabitat beyond the importance of macroclimatic correlates of body size. The topographical effect is consistent with the observation of Womack and Bell (2020) that, on a global scale, there is a positive correlation between anuran body size and maximum elevation. In terms of microhabitat, aquatic and burrowing anurans responded differently than other lifestyles to environmental conditions (Amado et al. 2019a). For these species, microhabitat preferences are critical to endure high temperatures and evaporative water loss. In this respect, Pincheira-Donoso et al. (2019) have also provided evidence in a global-scale analysis for caecilian amphibians that accounting for microhabitat preferences can increase the explanatory power in macroecological models of body size. As predicted from the *water conservation hypothesis,* environmental humidity levels constraint body size in fossorial but not in aquatic caecilians. Most caecilians occupy underground microhabitats that offer relatively stable thermal environments. Compared to other vertebrates, including amphibians, caecilians display especially high rates of evaporative water loss. The hydroregulatory advantages of larger sizes become evident in the most water-deficient environments along aridity gradients, whereas at the other end of the wetness spectrum selection pressures on body size *via* water-deprivation are relaxed. Because of their elongated and narrow body plan, caecilians are an appropriate group to test heat and water conservation mechanisms linked to A/V ratios. Pincheira-Donoso et al. (2019) argued that their elongated body shape is more prone to water loss and suggested that the interplay between a moisture-demanding morphology and a fossorial lifestyle may have prevented this lineage to successfully colonize extra-tropical environments. Although the interspecific pattern of body size for caecilians has already been described on a global scale (191 species were analysed), there is a striking lack of research at the intraspecific level (but see Measey and van Dongen 2006).

The results of Pincheira-Donoso et al. (2019) for Gymnophiona are also in agreement with findings for above-ground amphibian species. The largest anuran species (*Calyptocephallela gayi* and *Conraua goliath*) are aquatic or semi-aquatic (Womack and Bell 2020) and aquatic-egg-laying species tend to be larger on average than terrestrial-egg-laying species (Gomez-Mestre et al. 2012). In general, most smaller-than-average anurans (< 52 mm) are either arboreal, semi-arboreal or terrestrial (Womack and Bell 2020). While in general most amphibians show evaporative water loss from the skin very similar to the loss of water from a free water surface some species (termed as "atypical" or "waterproof") have cutaneous resistance to evaporative water loss and this ability appears to be connected to lifestyle. Arboreal species tend to have the highest resistance, aquatic species show little or no resistance and terrestrial species tend to have intermediate resistances (Young et al. 2005). Evidence for this pattern is, to my knowledge, so far limited to Australian species and future studies should explore the extent to which lower size limits, microhabitat and cutaneous resistance show correlated evolution.

Bergmann's Rule in the 21st Century: A Research Perspective for Amphibians

All organisms continuously exchange resources (nutrients, oxygen and water) and energy with their environment. A robust theoretical and empirical framework (Okie 2013, Schmidt-Nielsen 1984) shows that these exchanges critically depend on A/V ratios. Decreasing A/V ratios involve lower capacities to passively exchange energy and resources with the environment but a higher potential to conserve them. The profound biological consequences of decreasing A/V ratios, albeit already noted by Galileo, were for the first time discussed and elaborated in the 19th Century (Gould 1966). Despite the relevance of A/V ratios for the heat and water balance of organisms, most studies on Bergmann's rule have not considered body shape and instead use body size metrics (body mass or more frequently body length) that may not respond so strongly to environmental changes. Considering the allometric consequences of volumetric expansion is important because animals do not remain geometrically similar as size increases. Earlier allometric studies in the 19th century detected that body surface area seems a better correlate than body mass for metabolic rates and heat exchanges. In animals such as amphibians, whose transcutaneous fluxes of respiratory gases, osmolytes and water are essential to maintain homeostasis, physiologists have tried to characterize external surface area (A_s) since Rubner's times. However, measurements of A_s are difficult to obtain and involve removing and cutting the skin of euthanized individuals into small pieces (see, e.g., Talbot and Feder 1992 or Klein et al. 2016 and references therein). These data are then used to develop allometric equations at different taxonomic levels and estimate evaporative water loss from body mass data. If possible, alternative non-destructive methods should be used to determine A_s.

In recent years, we have witnessed the arrival of techniques developed at the interface of applied mathematics, computer science and engineering that allow the acquisition, reconstruction, analysis, manipulation, simulation and transmission of complex digital 3D geometries. The methodological and computational advances of digital geometry processing have made possible revolutionary developments in biomedical and anatomical research disciplines, among others. For instance, virtual paleontology is an emerging research field that allows computer-aided visualization and analyses of fossils in 3D (Cunningham et al. 2014). Paleontologists nowadays apply several modern 3D imaging methods to digitally characterize the internal and external morphology of extinct organisms. Along the same lines, biological anthropologists are increasingly acquiring three-dimensional landmark data to study climate-driven adaptive changes in body size and body proportions in modern humans (Betti et al. 2014). Ironically, however, the widespread application of high-resolution 3D digitization techniques in paleontology or anthropology contrasts with a poor state of knowledge of the anatomy in living biotas.

Scanning and data processing technologies have been unaffordable in the past, but cost-effective non-destructive imaging approaches are now available to obtain digital geometric data from 3D surfaces of both extinct and extant species. This includes laser scanning, the most widely used surface-based method in paleontology, and photogrammetry, which has been used to create high-quality 3D duplicates of insect specimens in museum collections (Nguyen et al. 2014) or corals and sponges (Lavy et al. 2015). In amphibians, Wardziak et al. (2014) used magnetic resonance imaging to generate 3D reconstructions of *Lissotriton helveticus* and understand how body surface areas and volumes are related to the water exchange with the environment in these organisms. Although this method allows obtaining 3D meshed geometries that can be used to compute surface area and volumes, it is typically employed to imaging soft tissues in biological specimens and has the disadvantage of requiring long scan times and generating low-resolution images. On the other hand, laser scanning systems may be limited to characterize the fine structures of the smaller species and reflective or iridescent surfaces in amphibians. Iriodophores, a common cromatophore type in the skin of amphibians, are reflective and iridescent pigments which can be, together with the typically moist skin of amphibians, a possible issue to obtain accurate 3D models of this taxon with laser scanners. In contrast, photogrammetric techniques are also relatively cheap, portable and easy to use, which makes them suitable for data gathering in the field and in natural history herpetology collections.

Amado et al. (2019b) have developed surface scanning techniques based on photogrammetry that allow assembling accurate high-resolution 3D duplicates of extant amphibian species from museum specimens. The 3D reconstruction protocol involves three steps (Fig. 1), that include mounting the specimen, obtaining a complete set of 2D images from different orientations and a computationally-demanding process to produce 3D digital models. Then, external surface area and A/V ratios can be computed from 3D models. Using data across the distributional range of the natterjack toad (*Epidalea calamita*), Amado et al. (2019b) compared the performance of "classic" climate-body size models (i.e., with body size measured as snout-to vent length, SVL) with climate-body geometry models (i.e., with A/V ratios as response variable). Interestingly, the explanatory power of SVL models was significantly lower than the one of A/V models that clearly supported the water conservation hypothesis. Although studies based on 3D morphological models are still lacking, the findings of Amado et al. (2019b) suggest that the functional relationship phenotype-climate in amphibians may be stronger than inferred from models that use body length as size metric. Current evidence, although limited, points towards this direction.

Vidal-García et al. (2014) quantified morphological differences in body shape and body size in 129 myobatrachid frog species and found that the former metrics provide a better match with environmental data than the latter ones. Castro et al. (2021) have combined 3D models, geometric morphometric analyses and phylogenetic comparative methods and have found that among the smaller treefrogs more spherical body shapes may be favored as an adaptive response in high water-deficit environments. In the coming years, the development of non-invasive devices and methods to rapidly scan live amphibian and reptile specimens in the laboratory or field opens up a wealth of opportunities to study the ecology and evolution of body size. Research in the laboratory of Professor Duncan Irschick has produced 3D scanning methods based on the Beastcam™ technology that allow obtaining 3D models from small living organisms ranging from 2.5 cm to 35 cm length (Irschick et al. 2020).

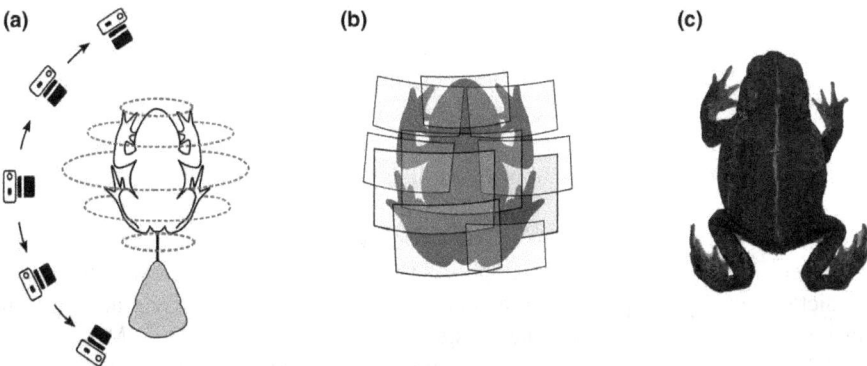

Figure 1. Three-step process to obtain high-resolution natural-colour 3D duplicates of amphibian specimens using photogrammetry. Source: Amado et al. (2019b).

Another methodological shortcoming in the study of Bergmannian patterns is the inability of correlative large-scale multi-species analyses to provide a more comprehensive understanding of the underlying causal forces for the observed body size gradients. Over the last two decades, the rise of macroecology as a biological discipline has allowed the description of numerous broad-scale emergent patterns in the distribution of ecologically relevant characteristics of organisms (such as body size) and their statistical association with environmental variables. Most traditional explorations of Bergmann's rule, either intra- or interspecific, have focused in generating bivariate plots on the relationship between latitude or temperature and body size (see above). Current evidence in support of the heat balance or the water conservation hypotheses comes from large-scale non-manipulative macroecological data for different biogeographic realms but its physiological foundations remain to be further studied. While such statistical non-manipulative approaches are necessary to document biogeographical patterns in the morphology and related traits of organisms, they are limited to disentangle causalities. Lomolino et al. (2006) made a call for a more integrative research approach to ecogeographical rules that goes beyond merely describing patterns and is designed to test causal hypotheses that attempt to explain the primary pattern, its corollaries and exceptions. They argued that such a research agenda involves using phylogenetic comparative approaches and the inclusion of functionally different groups, including other vertebrates aside from the widely studied mammals and birds.

One line of research that is gaining momentum to better understand broad-scale body size gradients involves the use of models based on first biophysical principles and their scaling into macroecological phenomena. Classical species distribution modelling (SDM) approaches are correlative and distribution-based so that they do not explicitly consider the mechanistic link with behavioral, morphological and physiological traits. An alternative set of ecophysiological, trait-based models is emerging to characterize the fundamental Grinnellian niche and predict how these niche dimensions can be reprojected onto the geographical landscape (Kearney and Porter 2009). Mechanistic models are built using balance equations from biophysical ecology principles and characterize how animals transfer energy and mass with their microclimate. While numerous examples exist for correlative SDMs applied to amphibians under different macroclimatic conditions (Araújo et al. 2006), few studies have considered the mechanistic role of microclimate to link the physiology and morphology of amphibians with their environment (but see Kearney et al. 2008, Rubalcaba et al. 2019a, 2019b). Tracy's (1976) thorough study on the mechanistic interactions between water and energy balance of northern leopard frogs (*Lithobates pipiens*) with their environment was pioneer in this respect. Rubalcaba et al. (2019a) derived null model expectations of lizard-like reptiles and anuran-like amphibians from microclimatic variables to explore how biophysical processes related to body temperature (T_b) and cutaneous evaporative water loss (EWL) scale up into interspecific size gradients. Geographical projections of their mechanistic model predictions well corresponded to the empirical patterns of body size that had been documented to emerge across the Western Palaearctic and Nearctic. Mechanistic models require understanding how the thermal and hydric environment experienced by an organism affect their body temperature and evaporative water loss. Perhaps the difficulties involved in characterizing evaporative heat exchange in small wet-skinned ectotherms and the amount of resources and time needed to build and validate these mechanistic models are possible reasons for the low number of studies on biophysical modelling of amphibians. Procedures to mechanistically model climatic constraints have to rely on global fine-resolution datasets (Kearney and Porter 2017) and desirably consider the energy and water balance of different life-cycle stages (eggs, larvae and adult). Also, model sensitivity should be tested whenever possible to different behavioral postures for thermoregulation and hydroregulation.

Conclusions

Ecogeographic rules should play a prominent role in predicting the magnitude and direction of biological responses to climate change (Tian and Benton 2020) such as suggested size reductions (Gardner et al. 2011, Sheridan and Bickford 2011). However, in order to optimize their use as predictive tools, it is particularly important to collect the available information on phenotype-environment relationships for each particular taxonomic group. In the case of amphibians, evidence shows the emergence of large-scale interspecific patterns of body size and the existence of an underlying climatic signal. The disparity of responses among the three orders (Anura, Caudata and Gymnophiona) in terms of geographic variation and degree of environmental influence, with latitudinal patterns only detectable at the regional level, points to the non-universality of Bergmann's rule in amphibians. Overall, at least part of the complexity in documented spatial patterns seems to be associated to the interplay between thermoregulation and hydroregulation mediated by surface area to volume ratios. Both the heat balance and the water conservation hypotheses are identified as the most plausible explanations for the observed patterns. The use of length-based body size metrics (SVL or TL) that do not take into account the importance of A/V ratios for the homeostasis and regulation of transcutaneous fluxes of water, osmolytes and respiratory gases in these organisms could weaken the detection of environmental signals. Therefore, two new avenues of research are suggested, based on the development of 3D morphological models and biophysical models to deepen the mechanistic basis of the observed patterns.

References

Adams, D.C. and J.O. Church. 2008. Amphibians do not follow Bergmann's rule. Evolution 62: 413–420.

Alho, J.S., G. Herczeg, A.T. Laugen, K. Räsänen, A. Laurila and J. Merilä. 2011. Allen's rule revisited: quantitative genetics of extremity length in the common frog along a latitudinal gradient. J. Evol. Biol. 24: 59–70.

Amado, T.F., C.J. Bidau and M.Á. Olalla-Tárraga. 2019a. Geographic variation of body size in New World anurans: energy and water in a balance. Ecography 42: 456–466.

Amado, T.F., M.G. Moreno Pinto and M.Á. Olalla-Tárraga. 2019b. Anuran 3D models reveal the relationship between surface area-to-volume ratio and climate. J. Biogeogr. 46: 1429–1437.

Araújo, M.B., W. Thuiller and R.G. Pearson. 2006. Climate warming and the decline of amphibians and reptiles in Europe. J. Biogeogr. 33: 1712–1728.

Ashton, K.G. 2002. Do amphibians follow Bergmann's rule? Can. J. Zool. 80: 708–716.

Bergmann, C. 1847. Über die verhältnisse der warmeökonomie der thiere zuihrer grosse. Göttinger Studien 1: 595–708.

Betti, L., N. von Cramon-Taubadel, A. Manica and S.J. Lycett. 2014. The interaction of neutral evolutionary processes with climatically-driven adaptive changes in the 3D shape of the human of coxae. J. Human Evol. 73: 64-74.

Blackburn, T.M., K.J. Gaston and N. Loder. 1999. Geographic gradients in body size: a clarification of Bergmann's rule. Divers. Distrib. 5: 165–174.

Blackburn, T.M. and B.A. Hawkins. 2004. Bergmann's rule and the mammal fauna of northern North America. Ecography 27: 715–724.

Brown, J.H. 1995. Macroecology. University of Chicago Press, Chicago, Illinois, USA.

Carey, C. and Alexander M.A. 2003. Climate change and amphibian declines: is there a link? Divers. Distrib. 9: 111–121.

Castro, K.M.S.A., T.F. Amado, M.Á. Olalla-Tárraga, S.F. Gouveia, C.A. Navas and P.A. Martinez. 2021. Water constraints drive allometric patterns in the body shape of tree frogs. Sci. Rep. 11: 1218.

Clusella-Trullas, S., J.H. van Wyk and J.R. Spotila. 2007. Thermal melanism in ectotherms. J. Thermal Biol. 32: 235–245.

Clusella-Trullas, S., J.S. Terblanche, T.M. Blackburn and S.L. Chown. 2008. Testing the thermal melanism hypothesis: a macrophysiological approach. Funct. Ecol. 22: 232–238.

Cunningham, J.A., I.A. Rahman, S. Lautenschlager, E.J. Rayfield and P.C.J. Donoghue. 2014. A virtual world of palaeontology. Trends Ecol. Evol. 29: 347–357.

Cushman, J.H., J.H. Lawton and B.F.J. Manly. 1993. Latitudinal patterns in European ant assemblages: variation in species richness and body size. Oecologia 95: 30–37.

Daufresne, M., K. Lengfellner and U. Sommer. 2009. Global warming benefits the small in aquatic ecosystems. PNAS 106: 12788–12793.

Fisher J.B., R.J. Whittaker and Y. Malhi. 2011. ET come home: potential evapotranspiration in geographical ecology. Global Ecol. Biogogr. 20: 1–18.

Gardner, J.L., A. Peters, M.R. Kearney, L. Joseph and R. Heinsohn. 2011. Declining body size: a third universal response to warming? Trends Ecol. Evol. 26: 285–291.

Gaston, K.J. and T.M. Blackburn. 2000. Pattern and Process in Macroecology. Wiley-Blackwell, London, UK.

Gaston, K.J., S.L. Chown and K.L. Evans. 2008. Ecogeographical rules: elements of a synthesis. J. Biogeogr. 35: 483–500.

Gomez-Mestre, I., R.A. Pyron and J.J. Wiens. 2012. Phylogenetic analyses reveal unexpected patterns in the evolution of reproductive modes in frogs. Evolution 66: 3687–3700.

Gould, S.J. 1966. Allometry and size in ontogeny and phylogeny. Biol. Rev. 41: 587–640.

Gouveia, S.F., R. Dobrovolski, P. Lemes, F.A.S. Cassemiro and J.A.F. Diniz-Filho. 2013. Environmental steepness, tolerance gradient, and ecogeographical rules in glassfrogs (Anura: Centrolenidae). Biol. J. Linn. Soc. 108: 773–783.

Gouveia, S.F. and I. Correia. 2016. Geographical clines of body size in terrestrial amphibians: water conservation hypothesis revisited. J. Biogeogr. 43: 2075–2084.

Granado Lorencio, C. 2007. Avances en Ecología: Hacia un Mejor Conocimiento de la Naturaleza. Serie Ciencias. Editorial Universidad de Sevilla, Spain.

Hawkins B.A. and J.A.F. Diniz-Filho. 2004. 'Latitude' and geographic patterns in species richness. Ecography 27: 268–272.

Hutchinson, V.H. and R.K. Dupré. 1992. Thermoregulation. pp. 206–249. *In:* Feder, M.E. and W.W. Burggren [eds.]. Environmental Physiology of the Amphibians. University of Chicago Press, Chicago, USA.

Irschick, D., Z. Corriveau, T. Mayhan, C. Siler, M. Mandica, T. Gamble et al. 2020. Devices and methods for rapid 3D photo-capture and photogrammetry of small reptiles and amphibians in the laboratory and the field. Herpetol. Rev. 51: 716–725.

Kearney, M., B.L. Phillips, C.R. Tracy, K.A. Christian, G. Betts and W.P. Porter. 2008. Modelling species distributions without using species distributions: The cane toad in Australia under current and future climates. Ecography 31: 423–434.

Kearney, M. and W. Porter. 2009. Mechanistic niche modelling: combining physiological and spatial data to predict species' ranges. Ecol. Lett. 12: 334–350.

Kearney, M.R. and W.P. Porter. 2017. NicheMapR—an R package for biophysical modelling: the microclimate model. Ecography 40: 664–674.

Klein, W., L. Dabés, V.M.G. Bonfim, L. Magrini and M.F. Napoli. 2016. Allometric relationships between cutaneous surface area and body mass in anuran amphibians. Zool. Anz. 263: 45–54.

Lavy, A., G. Eyal, B. Neal, R. Keren, Y. Loya, M. Ilan et al. 2015. A quick, easy and non-intrusive method for underwater volume and surface area evaluation of benthic organisms by 3D computer modelling. Methods Ecol. Evol. 6: 521–531.

Lawton, J.H. 1999. Are there general laws in ecology? Oikos 84: 177–192.

Levy, D.L. and R. Heald. 2016. Biological scaling problems and solutions in amphibians. Cold Spring Harb. Perspect. Biol. 8: a019166.

Lindsey, C.C. 1966. Body sizes of poikilotherm vertebrates at different latitudes. Evolution 20: 456–465.

Lomolino, M.V., D.F. Sax, B.R. Riddle and J.H. Brown. 2006. The island rule and a research agenda for studying ecogeographical patterns. J. Biogeogr. 33: 1503–1510.

Maurer B.A., J.H. Brown and R.D. Rusler. 1992. The micro and macro in body size evolution. Evolution 46: 939–953.

Mayr, E. 1963. Animal Species and Evolution. Belknap Press, London, UK.

McDowall, R.M. 2008. Jordan's and other ecogeographical rules, and the vertebral number in fishes. J. Biogeogr. 35: 501–508.

Measey, G.J. and S. van Dongen. 2006. Bergmann's Rule and the terrestrial caecilian *Schistometopum thomense* (Amphibia: Gymnophiona: Caeciliidae). Evol. Ecol. Res. 8: 1049–1059.

Meerhoff, M., F. Teixeira-de Mello, C. Kruk, C. Alonso, I. González-Bergonzoni, J.P. Pacheco et al. 2012. Environmental warming in shallow lakes: a review of potential changes in community structure as evidenced from space-for-time substitution approaches. Adv. Ecol. Res. 46: 259–349.

Moreno Azócar, D.L., M.G. Perotti, M.F. Bonino, J.A. Schulte, C.S. Abdala and F.B. Cruz. 2015. Variation in body size and degree of melanism within a lizards clade: is it driven by latitudinal and climatic gradients? J. Zool. 295: 243–253.

Nevo, E. 1973. Adaptive variation in size of cricket frogs. Ecology 54: 1271–1281.

Nguyen, C.V., D.R. Lovell, M. Adcock and J. La Salle. 2014. Capturing natural-colour 3D models of insects for species discovery and diagnostics. PLoS ONE 9: e94346.

O'Hara, R.B. 2005. The anarchist's guide to ecological theory. Or, we don't need no stinkin' laws. Oikos 110: 390–393.

Okie, J.G. 2013. General models for the spectra of surface area scaling strategies of cells and organisms: fractality, geometric dissimilitude, and internalization. Am. Nat. 181: 421–439.

Olalla-Tárraga, M.Á., M.Á. Rodríguez and B.A. Hawkins. 2006. Broad-scale patterns of body size in squamate reptiles of Europe and North America. J. Biogeogr. 33: 781–793.

Olalla-Tárraga, M.Á. and M.Á. Rodríguez. 2007. Energy and interspecific body size patterns of amphibian faunas in Europe and North America: Anurans follow Bergmann's rule, urodeles its converse. Global Ecol. Biogeogr. 16: 606–617.

Olalla-Tárraga, M.Á., J.A.F. Diniz-Filho, R.P. Bastos and M.Á. Rodríguez. 2009. Geographic body size gradients in tropical regions: water deficit and anuran body size in the Brazilian Cerrado. Ecography 32: 581–590.

Olalla-Tárraga, M.Á., L.M. Bini, J.A.F. Diniz-Filho and M.Á. Rodríguez. 2010. Cross-species and assemblage-based approaches to Bergmann's rule and the biogeography of body size in *Plethodon* salamanders of eastern North America. Ecography 62: no-no.

Olalla-Tárraga, M.Á. 2011. "Nullius in Bergmann" or the pluralistic approach to ecogeographical rules: a reply to Watt et al. (2010). Oikos 120: 1441–1444.

Pincheira-Donoso, D., S. Meiri, M. Jara, M.Á. Olalla-Tárraga and D.J. Hodgson. 2019. Global patterns of body size evolution are driven by precipitation in legless amphibians. Ecography 42: 1682–1690.

Rapacciuolo, G., J. Marin, G.C. Costa, M.R. Helmus, J.E. Behm, R.M. Brooks et al. 2017. The signature of human pressure history on the biogeography of body mass in tetrapods. Global Ecol. Biogeogr. 26: 1022–1034.

Ray, C. 1960. The application of Bergmann's and Allen's rules to the poikilotherms. J. Morphol. 106: 88–108.

Reguera, S., F.J. Zamora-Camacho and G. Moreno-Rueda. 2014. The lizard *Psammodromus algirus* (Squamata: Lacertidae) is darker at high altitudes. Biol. J. Linn. Soc. 112: 132–141.

Richards, C.M. and G.W. Nace. 1983. Dark pigment variants in anurans: classification, new descriptions, color changes and inheritance. Copeia 1983: 979–990.

Rubalcaba, J.G., S.F. Gouveia and M.Á. Olalla-Tárraga. 2019a. A mechanistic model to scale up biophysical processes into geographical size gradients in ectotherms. Global Ecol. Biogeogr. 28: 793-803.

Rubalcaba, J.G., S.F. Gouveia and M.Á. Olalla-Tárraga. 2019b. Upscaling microclimatic conditions into body temperature distributions of ectotherms. Am. Nat. 193: 677–687.

Sheridan, J.A. and D. Bickford. 2011. Shrinking body size as an ecological response to climate change. Nature Clim. Change 1: 401–406.

Schmidt-Nielsen, K. 1984. Scaling: Why is Animal Size so Important? Cambridge University Press, Cambridge, UK.

Tian, L. and M.J. Benton. 2020. Predicting biotic responses to future climate warming with classic ecogeographic rules. Curr. Biol. 30: R744–R749.

Talbot C.R. and M.E. Feder. 1992. Relationships among cutaneous surface area, cutaneous mass and body mass in frogs: A reappraisal. Physiol. Zool. 65: 1135–1147.

Tattersall, G.J., P.M. Eterovick and D.V. Andrade. 2006. Role of skin colour changes in the regulation of body temperature in basking frogs (*Bokermannohyla alvarengai*). J. Exp. Biol. 209: 1185–1196.

Tracy, C.R. 1976. A model of the dynamic exchange of water and energy between a terrestrial amphibian and its environment. Ecol. Monogr. 46: 293–326.

Vidal-García, M., P.G. Byrne, J.D. Roberts and J.S. Keogh. 2014. The role of phylogeny and ecology in shaping morphology in 21 genera and 127 species of Australo-Papuan myobatrachid frogs. J. Evol. Biol. 27: 181–192.

Vences, M., P. Galán, D.R. Vieites, M. Puente, K. Oetter and S. Wanke. 2002. Field body temperature and heating rates in a montane frog population: the importance of black dorsal pattern for thermoregulation. Ann. Zool. Fenn. 39: 209–220.

Vitt, L.J. and J.P. Caldwell. 2009. Herpetology: An Introductory Biology of Amphibians and Reptiles. Third Edition. Academic Press, San Diego, USA.

Wardziak, T., L. Oxarango, S. Valette, L. Mahieu-Williame and P. Joly. 2014. Modelling skin surface areas involved in water transfer in the Palmate Newt (*Lissotriton helveticus*). Can. J. Zool. 92: 707–714.

Womack, M.C. and R.C. Bell. 2020. Two-hundred million years of anuran body-size evolution in relation to geography, ecology and life history. J. Evol. Biol. 33: 1417–1432.

Young, J.E., K.A. Christian, S. Donnellan, C.R. Tracy and D. Parry. 2005. Comparative analysis of cutaneous evaporative water loss in frogs demonstrates correlation with ecological habits. Physiol. Biochem. Zool. 78: 847–856.

Zeuss, D., R. Brandl, M. Brändle, C. Rahbek and S. Brunzel. 2014. Global warming favours light-coloured insects in Europe. Nat. Commun. 5: 3874.

CHAPTER 8

Evolution and Ecology of Locomotion in Amphibians

Francisco Javier Zamora-Camacho

INTRODUCTION

Evolutionary Aspects of Locomotion in the Animal Kingdom

Movement is inherent to metazoans as we know them today. Even in its most rudimentary expression, the physiology of sessile, primitive sponges depends in the last analysis on the coordinated movement of flagella that regulates the water flow within the organism (Ruppert et al. 2004). However, a vast majority of animals are capable of more sophisticated movements based on the nervous modulation of muscular cells culminating in purposeful locomotion (McMahon 1984). Being heterotrophs, most animals resort to locomotion to reach sources of organic matter they incorporate as food (Budick and O'Malley 2000, Higham 2007). On the flip side, locomotion enables animals to flee from others that attempt to use them as such sources of organic matter by preying on them (Watkins 1996, McGee et al. 2009). Predator-prey interactions are arguably among the primary ecological triggers of the evolution of locomotion (Vermeij 1994, Moore and Biewener 2015). Nonetheless, the consequences of animals' ability to change their location go beyond the eat-or-be-eaten dichotomy. As such, locomotion potentiates disparate aspects of animal ecology. For instance, locomotion enhances social dominance (Garland et al. 1990, Perry et al. 2004), since faster individuals can control more extensive territories (Peterson and Husak 2006). Locomotion also intervenes in sexual selection (Husak and Fox 2008), since faster males are better equipped to defend females from male competitors (Husak et al. 2008), and therefore sire more offspring (Husak et al. 2006).

Given its paramount fitness-enhancing role (Huey et al. 1991, O'Steen et al. 2002, Miles 2004, Meyer-Vernet and Rospars 2016, Lailvaux and Husak 2017,

Departamento de Biodiversidad, Ecología y Evolución, Facultad de Ciencias Biológicas, Universidad Complutense de Madrid, C/José Antonio Novais 12, 28040 Madrid, Spain.
Email: fj.zamcam@gmail.com

Kraskura and Nelson 2018), locomotion is a hardly surprising driving force of animal evolution (Gordon et al. 2017). Locomotion is sustained on the appearance of specialized morphological features (Biewener and Patek 2018). Not only does locomotion entail the development of structures specialized in actual movement, but also of complex perceptual systems for accurate spatial recognition (Gordon et al. 2017, Biewener and Patek 2018). Locomotion involved profound innovations in animal physiology and body plan, and is thus among the major drivers of evolution and diversification in the Metazoa (Gordon et al. 2017). And most interestingly, evidence of locomotion throughout animal evolution can be traced in the fossil record.

Locomotion as a Force in Animal Evolution

Albeit not without debate (Conway Morris 2002, Rasmussen et al. 2002a, Seilacher 2007), the earliest examples of animal locomotion fossil evidence, in the form of movement traces, are claimed to have originated somewhere between 1,000 and 2,500 million years ago (Kauffman and Steidtmann 1981, Seilacher et al. 1998, Rasmussen et al. 2002b, Bengtson et al. 2007, Kauffman et al. 2009). However, more robust evidence dates from somewhat later, around 580–565 million years back (Liu et al. 2010, Pecoits et al. 2012, Chen et al. 2013, Chen et al. 2019, Ivantsov et al. 2019). These latter traces coincide with the appearance of worm-like bilaterians in the Ediacaran Period, their alleged authors.

Bilateral symmetry, a milestone in animal body plan with overriding evolutionary consequences (Genikhovich and Technau 2017, Holló 2017), has long been considered a direct byproduct of locomotion, as bilaterality favors directed movement (Beklemishev 1969, Willmer 1990, Ruppert et al. 2004). Although this assumption has been challenged by the potential alternative or complementary role of internal circulation of food as a trigger of bilaterality, the relevance of locomotion on its radiation and maintenance has not been called into question (Finnerty 2005, Holló and Novák 2012, Genikhovich and Technau 2017).

But locomotion further revolutionized animal evolution by prompting the origin of appendages. The earliest fossil traces compatible with paired locomotive appendages appear in the late Ediacaran Period, 551–541 million years ago (Chen et al. 2018). The subsequent radiation of metazoans diversified appendages in terms or morphology and, supposedly, locomotor functionality (Shu et al. 2014, Mángano and Buatois 2020). From a physiological standpoint, the evolution of different locomotion modes entailed the rearrangement of preexistent neural circuits to govern the structures on which they rely (Jung and Dasen 2015). These events marked the so-called Cambrian explosion, when animal body plans underwent a dramatic radiation leading to today's fauna biodiversity (Finnerty 2003, Budd and Jensen 2017). Nowadays, the overwhelming majority of extant animals, including vertebrates, are bilaterians.

Locomotion is a major trigger of morphological evolution (Botton-Divet et al. 2017, Higham et al. 2017) and a remarkable selection pressure (Calsbeek 2008). According relationships between morphology and locomotor performance have been reported throughout the animal kingdom: locomotor performance improves with larger fins in fish (Li et al. 2016), longer limbs in amphibians (Hudson et al. 2016), reptiles (Zamora-Camacho et al. 2014), and mammals (Day and Jayne 2007), and longer wings in birds (Moreno-Rueda 2003). But not all is about size. For instance, divergent wing shapes suit insects for different locomotory capabilities (Wootton 1992, Johansson et al. 2009). The preeminent effect of locomotion on animal morphological diversity is a direct consequence of its ecological significance.

Costs of Locomotion

However significant is its role in animal ecology , by no means does locomotion come devoid of costs. The mere development and maintenance of appendages requires energy investment (Kooijman 2009, Rombough 2011). Moreover, locomotion involves muscular exertion, which provokes the consumption of a considerable amount of energy and metabolites such as amino acids (Taylor et al. 1980, Brijs et al. 2017). Muscular exertion also prompts the functioning of physiological pathways that may outbalance the production of pro-oxidant chemical species in proportion to that of anti-oxidant defenses (Fisher-Wellman and Bloomer 2009). The outcome of this imbalance can be an insalubrious prevalence of the former, referred to as oxidative stress (Sorci and Faivre 2009). Oxidative stress is capable of damaging a wide array of metabolic processes that are indispensable for homeostasis, resulting in health issues of diverse consideration (Halliwell and Gutteridge 2007).

Besides its metabolic consequences, locomotion may incur ecological costs. For instance, animals that are in motion are more likely to be detected by predators, whereas immobile animals have greater chances to escape notice (Krause and Godin 1995, Geipel et al. 2020). Other costs of locomotion may have a physiological basis and ecological consequences. Such is the case of many ectotherms, which are unsuited for endogenous regulation of body temperature and depend on external heat sources for thermal upkeep and organismal function, which is known as thermoregulation (Hertz et al. 1993, Brown and Au 2009). Due to its thermal dependence, an efficient thermoregulation is essential for ectotherm's locomotion optimization (Paladino 1985, Zamora-Camacho et al. 2015a). Therefore, locomotion in these animals is subjected to the costs of thermoregulation, broadly speaking, time consumption and conspicuousness to predators (Alford and Lutterschmidt 2012).

Consequently, locomotion often appears subjected to trade-offs with other resource-consuming traits, as organisms often have difficulty in satisfying the concurrent needs of several demanding functions (Stearns 1989, Zera and Harshman 2001). From the viewpoint of metabolism, locomotion can compete for resources with the immune system (Zamora-Camacho et al. 2015b), osmoregulation (Li et al. 2019) and detoxification cascades (Handy et al. 1999). From the viewpoint of ecology, locomotion is traded-off with traits as disparate as growth (Berghänel et al. 2015), body resource storage (Gosler et al. 1995) or even gravidity (Lin et al. 2008), the latter two likely mirroring a reduction in locomotor performance caused by an extra weight (Tickle et al. 2010). What is more, diverse aspects of locomotion may trade-off among them within individuals, such as speed versus stamina (Vanhooydonck et al. 2001) or aquatic versus terrestrial locomotion (Shine and Shetty 2001).

Locomotion: a Capability or a Behavior?

From the aforesaid close relationship often found between morphology and locomotor performance, the idea could transpire that the latter is a mere function of the former. But, albeit logic and true to a certain extent, this overly simplistic approach is confronted by the notion that locomotion is subjected to costs. These costs are predicted to be behaviorally regulated according to their ratio with the benefits of locomotion (Biewener and Patek 2018). Indeed, animals do not unvaryingly exploit their locomotor capacity at its full potential (Husak 2006, Wheatley et al. 2018). On the contrary, they may reduce their investment in locomotion, and thus energy expenditure, when the benefits are lower (Irschick et al. 2005), such as in the perceived absence of predators (Zamora-Camacho et al. 2018) or in closed-canopy habitats where they are less conspicuous to predators (Vanhooydonck and Van Damme 2003, Goodman 2009). On the other hand, animals invest more in locomotion when the relative costs are lower, such as in good energy status (Careau and Garland 2015, Lindgren et al. 2018), or when the relative benefits are higher, such as following recent predator attacks (Freymiller et al. 2017). Therefore, locomotion appears simultaneously as a capability that depends on morphology and as a behavior that is regulated so as to maximize its benefits relative to its costs.

A Brief Evolutionary History of Amphibian Locomotion

In evolutionary terms, amphibians were the first vertebrates to shift from aquatic to terrestrial environments, which entailed substantial morphological changes that permitted, among others, land locomotion (Edwards 1977). Amphibians evolved from late-Devonian (circa 380 million years ago) sarcopterygians, also known as lobe-finned fishes. The putative antecessor of tetrapods, *Panderichthys*, had fins with not only bones that were analogous to humerus, radius and ulna of current tetrapods, but also with non-articulated distal radial bones that could be precursors of tetrapod digits (Boisvert et al. 2008). Although without consensus, some authors consider that actual digits were indeed present in *Tiktaalik*, a 375-million-year-old transitional form between sarcopterygians and tetrapods (Laurin 2006). These fins are believed to have allowed their owners to creep through the debris-filled bottom of shallow water bodies, or even to venture into land not far from the shore (Ahlberg and Clack 2006, Clack 2007). But then, about ten million years later, finger-finned *Acanthostega* and *Ichthyostega*, whose relative position in the sequence of tetrapod evolution is disputed (Ahlberg and Milner 1994, Ahlberg and Clack 2006), gave rise to the so-called stem tetrapods, from which all current tetrapods, including modern amphibians, are descendants (Ahlberg and Johanson 1998).

These ungainly steps towards land locomotion also involved greater energy costs than aquatic locomotion, due to the floatability lost, which allegedly resulted in decreased endurance in the early stages of the transition (Bennett 1991). Therefore, further adaptations at the physiologic and metabolic levels were essential for early amphibians to optimize locomotion in their new terrestrial habitats (Garland and Carter 1994). However, the reconstruction of the evolutionary pathways of amphibian physiology is obscured by a broad array of limitations, such as the absence of a record comparable to the fossil evidence of morphological evolution (Bennet and Huey 1990).

Locomotion in Extant Amphibians

Amphibian locomotor performance is closely related to morphology, although with different implications in different taxa (Bennett et al. 1989, Johansson et al. 2010, Hudson et al. 2016). As a consequence of the relevance of morphology on amphibian locomotion, deficiencies in limb anatomy dramatically decrease jumping distance of *Pelobates cultripes* toadlets (Zamora-Camacho and Aragón 2019a). There is also evidence that faster amphibians are more likely to survive predator attacks. Indeed, *Thamnophis sirtalis* snakes capture slower *Pseudacris regilla* frog tadpoles more often (Watkins 1996). However, the fact that *Aeshna cyanea* dragonfly nymphs capture preferentially smaller, but not slower, *Ichthyosaura alpestris* newt larvae may suggest that locomotor performance is not the only nor a universal factor involved in predator avoidance (Gvoždík and Smolinský 2015). Actually, widespread among amphibians are a plethora of antipredator strategies that could make their bearers less dependent on locomotion. These include unpalatability, found in a wide array of larval (Peterson and Blaustein 1991) and adult anurans (O'Donohoe et al. 2019) as well as larval (Gall et al. 2011) and adult urodeles (Marion and Hay 2011). These distasteful, even toxic at times, substances are often accompanied by concomitant aposematic colorations, although cryptic colorations are also common (Rudh and Qvarnström 2013). Noticeably, some amphibians are capable of highly specific antipredator mechanisms, such as protruding sharp ribs in *Pleurodeles waltl* newts (Heiss et al. 2010).

Amphibians are ectotherms, and as such, their locomotor performance is narrowly dependent on environmental temperature, with optima that veer away from cold and hot extremes. Thermal sensitivity of locomotor performance has been detected in aquatic and terrestrial stages of anurans and urodeles (Else and Bennett 1987, Knowles and Weigl 1990, Navas et al. 1999, Herrel and Bonneaud 2012, Baškiera and Gvoždík 2019). Nevertheless, some extent of thermal acclimation is possible. Larval *Xenopus laevis* (Wilson et al. 2000a) and *Limnodynastes peronii* frogs (Wilson and Franklin 1999) accommodate locomotor performance to a given temperature after an acclimation period in laboratory-controlled conditions. However, while adult *X. laevis* retain this capacity (Wilson et al. 2000a), which is also displayed by juvenile and adult *Cynops orientalis* newts (Lu et al. 2016), that is not the case of *L. peronii* (Wilson and Franklin 2000), which suggests some degree of interspecific variability in the capacity of adult amphibians of thermally acclimating locomotor performance. This ability may be of the utmost importance in biological invasions by amphibians. Thermal dependence of locomotor performance has been detected to shift in accordance with the new range occupied by invasive populations of *X. laevis* in France (Araspin et al. 2020). Locomotor performance is indeed an essential trait in amphibian invasion ecology, as evidenced by the fact that individuals from the edge populations in the aforesaid invasion process have greater stamina than those from core populations (Louppe et al. 2017). Similarly, hopping distance of *Rhinella marina* toads at the invasion front in Australia is also greater

(Hudson et al. 2020), and so is their locomotor resistance to suboptimal temperature (Kosmala et al. 2017).

However, there exists another downside of ectothermy for amphibians. Their thin skin is highly permeable to water loss (Toledo and Jared 1993). Consequently, at least in some environments, the same heat acquisition process that improves locomotion may compromise the hydration status of amphibians in their terrestrial stages. Although its particular effects differ among taxa and habitats, dehydration impairs amphibian locomotion in most instances (Moore and Gatten 1989, Titon et al. 2010), with the remarkable exception of some arid-habitat dwelling amphibians (Prates et al. 2013). Therefore, a trade-off between thermoregulation and hydration, which compromises locomotion, has been detected in several amphibian species (Preest and Pough 1989, Köhler et al. 2011, Anderson and Andrade 2017). Again, locomotor resistance to dehydration plays a fundamental role in invasion ecology, as it enables invasive amphibians to colonize dry areas (Roznik et al. 2018). Indeed, locomotion of *R. marina* toads stablished away from their natural range is better suited to resist dehydration at the invasion front than in core populations (Kosmala et al. 2017).

Another side effect of amphibian skin permeability is the uptake of chemicals (Kaufmann and Dohmen 2016). These chemicals may disrupt the normal functioning of amphibian metabolism (Hayes et al. 2010), which can ultimately impair locomotion. Indeed, pesticides may damage locomotor performance of larval anurans, as detected in larvae of *Rana temporaria* exposed to endosulfan (Denoël et al. 2013), *Limnodynastes tasmaniensis* exposed to imidacloprid and copper (Sievers et al. 2018), or *Odontophrynus americanus* exposed to pyriproxyfen (Lajmanovich et al. 2019). However, these effects may vary depending on the chemical composition of the insecticide and the resistance of each particular species to them (Junges et al. 2017). Not only larva locomotion can be affected by pollutants: similar effects have been proven in post-metamorphic *R. temporaria* frogs exposed to fungicides (Adams et al. 2021), *Ambystoma maculatum* salamanders exposed to pesticides (Mitchkash et al. 2014) and adult *Eurycea bislineata* salamanders exposed to mercury (Burke et al. 2010). Even naturally occurring chemicals, such as organic acids, may have a major effect on locomotor performance of *L. peronii* frogs (Barth and Wilson 2010). Salinity may also diminish locomotor performance, as in *Rhinella spinulosa* toads from brackish environments with different degrees of salinity (Sanabria et al. 2018). Unexpectedly, however, jumping distance appears reinforced in roadside *Lithobates sylvaticus* frogs despite being exposed to de-icing salt, likely due to the simultaneous effect of traffic as a selective agent against frogs with the poorest locomotor performance (Brady et al. 2019). Lastly, it is noteworthy that the skin is not the only gate in for pollutants: the intake of microplastics through the digestive system has recently been demonstrated to impair locomotion of *Physalaemus cuvieri* frog polliwogs (Pereira da Costa Araújo and Malafaia 2020).

Locomotion in amphibians can also be subjected to trade-offs. The reason why locomotor performance is impaired in nematode-parasitized *Rhinella horribilis* toadlets (Kelehear et al. 2019) and *Rhinella icterica* adults (Moretti et al. 2014) could be a trade-off between locomotion and the immune system. Nevertheless, locomotion of *P. waltl* newts in their aquatic stage is insensitive to a controlled immune-system activation by lipopolysaccharide of *Escherichia coli* (Zamora-Camacho et al. 2020). Likewise, conflicts between different components of locomotion performance have been detected: endurance and swimming speed are traded-off in *X. laevis* frogs (Wilson et al. 2002).

Locomotion in Larval Amphibians

The ontogeny of early amphibians was gradual (Schoch 2009). Remarkably, following a certain evolutionary stage around the Permo-Carboniferous cline, about 300 million years ago, metamorphosis, to which Alberch (1989) referred as a "condensation of developmental events", became the norm (Schoch 2009). The average current amphibian undergoes a metamorphosis process by which a water-breathing larval stage culminates in an air-breathing adult. The many exceptions to this scheme are secondarily derived from it, and fit in one of two broad categories: neotenic amphibians, whose larval characters are retained into adulthood, or amphibians that skip the free-ranging larval stage, usually spent in body cavities or in nests under the surveillance of the mother (Crump 2010).

Albeit with disparate ecological roles in different groups, amphibian larvae share some traits that are similarly pivotal to their locomotor performance. Most amphibian larvae are suited for aquatic but not for terrestrial locomotion, with few exceptions, like the larvae of some species that deposit their eggs out of the water and can crawl after hatching towards the closest waterbody (Oldham 1977). Amphibian larvae have membranous fins of variable extension on their tails and, to a certain degree, trunks, that enables propulsion in the water (Wassersug 1989). Indeed, locomotor performance of larval urodeles (Fitzpatrick et al. 2003) and anurans (Hoff and Wassersug 2000) is intimately related to the morphology and beat frequency of their tails. Moreover, their bodies are flexible, slender in the case of urodeles and caecilians, and with no bone in anuran tails, so that their locomotion mode is more reliant on maneuverability than it is on sustainable speed (Wassersug 1989).

Larval amphibians are particularly vulnerable to predators, and their prime antipredator strategy is flight (Watkins 1996). Interestingly, the morphology of larval anurans (Wilson et al. 2005) and urodeles (Storfer and White 2004) can be subjected to refined plastic responses to perceived predation risk that can even differ according to the predator species in question. These responses include deeper and more muscular tails, to degrees that can be adjusted for improved speed or maneuverability, tuned to active or sit-and-wait predators respectively (Wilson et al. 2005). However, not in all cases do predator-triggered deeper tails enhance locomotion, but they can also reduce the chances of being attacked in more vulnerable head and body (Van Buskirk et al. 2003). Besides, puzzling interactive effects of predation risk and food shortage on burst speed have been detected in larvae of *R. temporaria* frogs (Lindgren et al. 2018) and *P. waltl* newts (Zamora-Camacho et al. 2018): when at predation risk, larvae under restrictive diets swim faster, maybe as a mechanic consequence of reduced body mass.

Finally, given the relevance of morphology on locomotion, metamorphosis represents in itself a critical stage at which the individual's anatomy is neither optimized for larval nor for post-metamorphic locomotion. During anuran metamorphosis, indeed, both endurance (Wassersug and Sperry 1977) and burst speed (Huey 1980) decrease drastically. Consequently, predator pressure is greater during metamorphosis (Wassersug and Sperry 1977, Arnold and Wassersug 1978, Touchon et al. 2013). Metamorphosis also represents a turning point in urodeles, despite their less drastic change. As a matter of fact, *Triturus cristatus* newts experience a dramatic decrease in swimming speed after metamorphosis (Wilson 2005), and *Ambystoma californiense* salamanders are faster in their larval stage than both in their terrestrial and aquatic post-metamorphic phases (Shaffer et al. 1991). However, locomotor endurance is greater in terrestrial post-metamorphs than in either aquatic phase, which suggests that, at least in some cases, it is the environment rather than the vital stage that determines stamina (Shaffer et al. 1991).

Locomotion in Post-Metamorphic Amphibians

Larval amphibians share relatively comparable body plans that allow underwater life. In that sense, many of the features of larval locomotion are relatable among amphibian clades and even with other taxa. Nevertheless, post-metamorphic amphibians face some particular challenges that are not usually met by their larvae nor by other vertebrates. One of them is derived from potential carryover effects. The environment experienced by tadpoles could have consequences that transcend into the post-metamorphic stages. Some of these carryover effects could directly involve locomotion. First of all, size at metamorphosis is known to improve locomotion and survival, as in *R. marina* toads (Cabrera-Guzmán et al. 2013). On the same note, a faster development in *Rana latastei* tadpoles produces froglets with shorter limbs and consequent poorer jumping performance (Ficetola and de Bernardi 2006). Therefore, factors resulting in smaller metamorphs are also liable to impair their locomotion.

For instance, heat hastens water evaporation, which could compromise the survival of larvae that do not complete metamorphosis before the water body dries. Accordingly, both water temperature and dwindling accelerate metamorphosis: *Bombina variegata* (Sinsch et al. 2020) and *Physalaemus pustulosus* (Charbonnier and Vonesh 2015) froglets resulting from tadpoles artificially raised in ephemeral, warmer ponds emerge earlier, but at the expense of reduced body size, jumping distance and endurance as compared with their counterparts raised in permanent ponds. Noticeably, the ability of *Discoglossus galganoi* tadpoles to metamorphose earlier as a response to desiccating ponds is limited by food availability, which highlights the energy costs that this strategy may entail (Enriquez-Urzelai et al. 2013). On the other hand, temperature alone can reduce time to metamorphosis, maybe as a consequence of enhanced metabolic activity. In laboratory experiments, *D. galganoi* (Álvarez and Nicieza 2002), *Rana lessonae* (Orizaola and Laurila 2009), and *L. sylvaticus* (Gahm et al. 2021) tadpoles reared at higher temperature gave rise to froglets with impaired absolute locomotor performance relative to those reared at lower temperature. These phenomena have been assessed in the wild as well: *L. sylvaticus* froglets naturally raised in closed-canopy ponds tend to be larger and jump farther than their open-canopy counterparts (Boes and Benard 2013). This could be a consequence of closed-canopy ponds being cooler and thus allowing slower development.

Besides physical characteristics, chemical parameters during the larval stage may impact locomotion after metamorphosis. Increased salinity typically reduces metamorph locomotor performance (Kearney et al. 2016, Sanabria et al. 2018). However, this general rule is reversed by species adapted to brackish conditions such as *R. marina*, whose metamorphs, at least in some invasive populations in Australia, have demonstrated to perform worse when reared in low-salinity water as tadpoles (Wijethunga et al. 2016). Surprisingly, whereas the effects of pesticides on locomotion of larval and post-metamorphic stages of amphibians are well known (see above), carryover effects have not been noticed so far, as concluded by a study where *L. sylvaticus* larvae exposed to neonicotinoids originated froglets whose jumping distance was similar to control non-exposed conspecifics (Lee-Jenkins and Robinson 2018). Or what is even more mystifying, *Fejervarya limnocharis* larvae exposed to malathion originated froglets whose jumping distance was greater than non-exposed control group (Kulkarni and Krishnamurthy 2020). The physiological mechanisms involved remain obscure. But these findings, in any case, do not necessarily mean that these substances are not detrimental for survival at this stage, as other aspects of antipredator behavior are indeed affected (Lee-Jenkins and Robinson 2018).

Another remarkable source of carryover effects from larval to post-metamorphic amphibian stages is predation risk. Amphibian larvae under predator pressure tend to metamorphose earlier and at smaller body sizes, although with great variability among species and threats (Relyea 2007). Actually, *Rana ridibunda* tadpoles exposed to dragonfly nymph visual and chemical cues become froglets with shorter, but more muscular hindlimbs, which makes them capable of longer jumps than control conspecifics (Van Buskirk and Saxer 2001). These findings also reveal that other factors besides hindlimb length, namely muscle size, are essential to locomotion in these animals (Moreno-Rueda et al. 2020). More conclusive are the effects of predator pressure when larvae undergo actual damage: partial tail loss—mimicking failed predation—in *P. cultripes* tadpoles resulted in metamorphs with shorter hindlimbs and compromised locomotor performance (Zamora-Camacho and Aragón 2019b).

But the relationships between locomotion of pre- and post-metamorphic amphibians could be under other constraints. Because larvae and adults result from one single genome, both stages of amphibian life cycle could be under an ontogenetic genomic conflict if traits with a positive genetic correlation (via linkage or pleiotropy) are under opposing selection, or if traits with a negative genetic correlation are under consistent selection (Rice 2013). Nonetheless, across-stage genetic correlations tend to be low in amphibians (Phillips 1998), which virtually partitions life-history stages into modules with the ability to response separately to selection (Goedert and Calsbeek 2019). Accordingly, faster *P. regilla* larvae do not metamorph into better jumping performance froglets (Watkins 1997). Similarly, larva and metamorph locomotion are not related in *R. temporaria* (Johansson et al. 2010).

However, potential conflicts that could impact locomotion do not end once metamorphosis is accomplished. Although not few amphibians are strictly aquatic or terrestrial after metamorphosis, the life cycles of vast majority of them alternate both environments. The fact that locomotion of post-metamorphic amphibians should stand disparate pressures in such profoundly dissimilar environments could lead to a design conflict that could compromise optimization in one environment to the detriment of the other. Nevertheless, thus far there is no practical evidence of this phenomenon. Actually, terrestrial and aquatic locomotion do not conflict in females of several salamandrid newt species (Gvoždík and Van Damme 2006). Similarly, jumping and swimming performance appear unrelated in *Rana esculenta* frogs (Nauwelaerts et al. 2007). The evolutionary pressures undergone by simultaneous locomotion in terrestrial and aquatic environments have apparently shaped morphologies that are equally efficient in both.

Locomotion in Post-Metamorphic Urodeles

Urodeles preserve their tails after metamorphosis, and, as a general rule, four limbs of similar size, which makes them the least morphologically divergent from basal amphibians among the three extant lissamphibian orders. At first glance, this general body plan could be suitable for aquatic and terrestrial locomotion. Indeed, speed and endurance improve with body size in *Ambystoma tigrinum* salamanders, although neither longer limbs nor tails appear involved in this relationship (Bennett et al. 1989). Similarly, morphology seems to play a limited role in aquatic locomotion of *P. waltl* newts (Zamora-Camacho et al. 2020). However, other studies have found a positive effect of tail length on swimming speed of female salamandrid newts (Gvoždík and Van Damme 2006). Contradictory findings support both a positive (Lu et al. 2016) and a negative effect of snout-to vent length (SVL) on swimming speed of different salamandrids (Gvoždík and Van Damme 2006). This negative effect of SVL on locomotor performance extends to the terrestrial phase (Gvoždík and Van Damme 2006). This finding is particularly counter-intuitive, as not only limbs are involved in urodele locomotion (Devolvé et al. 1997), but so are axial undulations of the trunk (Cabelguen et al. 2014).

Urodele body plan is not invariable. Indeed, some aquatic clades, such as amphiumids and sirenids, have evolved elongated bodies and reduced limbs, to such an extent that hindlimbs are completely missing. Not only could limb reduction make their bodies more hydrodynamic, but it could as well spare the animals the costs associated with their development (Kooijman 2009, Rombough 2011). Locomotion in these animals is understudied, but sirenids are known to rely on tailbeat, whose period and frequency determine speed while swimming fast (Gillis 1997), whereas they alternate forelimb use with trunk undulatory waves during slow motion (Azizi and Horton 2004). Other urodeles are arboreal, among which the best studied is the genus *Bolitoglossa*, with fully webbed feet that behave as adhesive pads capable of clinging to soft, smooth surfaces (Green and Alberch 1981). Another significant morphological variation within urodeles with repercussions on locomotion is the absence of lungs, characteristic of plethodontids. The absence of lungs has been related to a modest endurance at fast speed (Full 1986), and to oxygen consumption saturation at speeds overwhelmingly lower than those of lunged salamanders, which also take longer to fatigue (Full et al. 1988). Therefore, locomotion performance of lungless salamanders pays the price of reduced oxygen intake.

Locomotion in Post-Metamorphic Anurans

Albeit relatively homogeneous among clades due to conservative and convergent evolution (Moen et al. 2013), anuran characteristic body plan diverges greatly from all other vertebrates. Long deemed an integral adaptation for jumping, reinterpretations of the morphology of the fossil remains of the earliest anuran known, *Triadobatrachus massinoti*, suggest that it primarily moved by means of urodele-like lateral undulations, and that jumping was, at most, only a part of its locomotor repertoire (Sigurdsen et al. 2012, Lires et al. 2016). Long-distance jumping is supposed to have evolved later in the evolution of the clade (Reilly and Jorgensen 2011). Interestingly, the most basal anuran family, leiopelmatids, move by means of quadruped walking and seldom resort to jumping (Reilly et al. 2015).

Regardless of the evolutionary trigger of their body plan, locomotion of most current adult anurans is eminently saltatory, which entailed numerous complex adaptations, such as fused vertebrae, the absence of a tail, and a highly modified pelvis to which particularly elongated hindlimbs appear attached (Handrigan and Wassersug 2007, Buttimer et al. 2020, Petrović et al. 2021). Jumping performance is known to relate positively to hindlimb length in young (Zamora-Camacho and Aragón 2019b) and adult anurans (Johansson et al. 2010, Hudson et al. 2016, Moreno-Rueda et al. 2020), which is a likely consequence of more massive muscles (James et al. 2005). Similarly, everything else being equal, both subadult (Johansson et al. 2010) and adult (Llewelyn et al. 2010, Moreno-Rueda et al. 2020) post-metamorphs with a longer body tend to jump farther, although only up to a certain body length past which locomotion is no longer improved (Moreno-Rueda et al. 2020). However, it is the relationship between hindlimb length and body length that acts as the best predictor of anuran locomotor performance, according to an across-taxa study (Choi et al. 2003). More complex is the role of body mass, as its relation with locomotor performance is positive in young *Pelobates cultripes* (Zamora-Camacho and Aragón 2019b) but negative in young and adult *Pelophylax perezi* (Moreno-Rueda et al. 2020). These apparently contradictory findings could be a consequence of the intricate relationships of body mass with fat storage and muscle mass, and therefore with jumping physics. The complexity of this relationship is highlighted in a study with *L. peronii*, where body mass shows no effect on jumping distance, but a positive one with maximum jumping force and a negative one with mass-specific jumping power (Wilson et al. 2000b).

Nonetheless, there is more to anuran locomotion than jumping. Anurans are by far the group of lissamphibians with a wider array of locomotion modes, likely selected upon by microhabitat use (Citadini et al. 2018). After jumping, the most obvious locomotion mode of anurans is swimming. Obligate and facultative swimming anurans usually have interdigital skin webbings in their hindlimbs, which increase the area for more powerful propulsion (Goldberg and Fabrezi 2008), and is thus positively related to swimming speed (Rebelo and Measey 2019). The extension of this webbing increases with time spent in the water, so that some fully or mostly aquatic frogs, such as *Xenopus* or *Pipa*, have complete fore and hindlimb skin webbings. Besides skin webbings, swimming performance of anurans is positively correlated with body and hindlimb length (Herrel et al. 2012, Rebelo and Measey 2019).

Although jumping and swimming are the most frequent locomotion modes in anurans, some species or clades have evolved particular abilities. Some toads, such as *Epidalea calamita*, are secondarily cursorial, do not jump, and have shorter hindlimbs than their jumping relatives. Even so, their locomotor performance is largely determined by hindlimb length (Zamora-Camacho 2018). Remarkably, in the same way that jumping anurans move by means of leaps that involve frequent stops, *E. calamita* toads also perform intermittent locomotion, which could be useful to distract predators, especially in open habitats (Zamora-Camacho 2018). Other anurans can burrow, with pelobatids and scaphiopodids possessing hard spades in their hindlimbs that allow earth scraping (Savage 1942). Burrowing ability does not appear to be correlated with hindlimb length in *P. cultripes* toads, but is negatively affected by body mass, and sexes display different context-dependent burrowing behavior: whereas females burrow deeper in odorless sand, in the presence of predator kairomones this trend is reverted (Zamora-Camacho et al. 2019). Particularly interesting are the adaptations that allow gliding in *Rhacophorus, Polypedates* (family Racophoridae) and *Ecnomiohyla* (family Hylidae) frogs. Gliding frogs present a set of morphological adaptations such as extra elongate limbs and toes with extensive skin webbings (Emerson and Koehl 1990). These, combined with specific banked and crabbed turning behaviors (McCay 2001), enhance maneuverability during long-distance glides. These adaptations, however, do not seem to have undermined the ability of these frogs to jump, which constitutes plus one example where amphibian morphology is simultaneously optimized for several locomotion modes (McKnight et al. 2020).

Anurans are frequently subjected to sexual dimorphism in locomotor performance. Better locomotor performance in males, apparently related to longer limbs, has been detected, for example, in *Xenopus tropicalis* frogs when swimming (Herrel et al. 2012) and jumping (Herrel et al. 2014), in terrestrial, jumping *R. marina* toads (Llewelyn et al. 2010) and in terrestrial, cursorial *E. calamita* toads (Zamora-Camacho 2018). Stamina is also greater in *X. laevis* males (Louppe et al. 2017). Nonetheless, sexual differences in jumping performance do not appear in *P. perezi* frogs (Moreno-Rueda et al. 2020). Therefore, sexual differences in jumping performance are not universal in anurans, and may differ across species according to factors such as morphology, body size, sexual dimorphism habitat use or reproduction mode. In fact, reproduction may inflict costs on locomotion. The most intuitive factor is body mass, increased in gravid females, which has confirmed negative effects on locomotor performance (Moreno-Rueda et al. 2020). That is the case of *L. sylvatica* frogs, in which females jump shorter distances than males when gravid, but not after oviposition (Brady et al. 2019). Indeed, in an interesting case where the egg burden is transferred to males, these experiences reduce locomotor performance: *Alytes obstetricans* are terrestrial toads whose males carry their clutch on their backs, tangled among their hindlimbs, which shortens their jumping distance (Lange et al. 2021). But it is during mating that anuran locomotion is particularly at stake. These animals reproduce by means of an amplexus of variable duration, in which males grasp females and cling to their backs. The costs that amplexus inflicts on female locomotion have been studied in *R. marina* toads: both terrestrial and aquatic locomotor performance are reduced, in a fashion that is directly related with the body size of the male in terrestrial locomotion (Bowcock et al. 2009).

Locomotion in Post-Metamorphic Caecilians

Caecilians are by a long way the least studied of amphibians, probably as a consequence of their narrow distribution ranges, restricted to the tropics, and their inconspicuous lifestyles (Gower and Wilkinson 2005). Caecilians are in most cases fossorial, which has had an enormous impact on their evolution (Santos et al. 2020). These amphibians are elongated, with compact, ossified, modular skulls that favor burrowing (Bardua et al. 2019). Some caecilians have fenestrated skulls though, which, however, does not impact burrowing ability (Kleinteich et al. 2012). Probably as an adaptation that facilitates subterranean life, limbs are absent in all extant species, although fossil evidence reveals their presence in extinct transitional clades (Jenkins and Walsh 1993). Therefore, all locomotion power relies on axial musculature. While burrowing, caecilians alternate lat-eral undulation with concertina locomotion, in which the whole body is sequentially compressed and expanded in an antero-posterior axis, and with vermiform or internal concertina locomotion, in which the vertebral column and its musculature move independently of the body wall resorting to a hydrostatic mechanism that is unique among vertebrates (O'Reilly et al. 2015). Lateral undulation is primarily used for superficial locomotion, but it is combined with both concertina patterns during burrowing (Herrel and Measey 2010). This combination of movements provides caecilians with a maximum forward force that doubles that of fossorial snakes of comparable sizes (O'Reilly et al. 1997). However, the independence between the vertebrae and the skin is reduced in the most elongated caecilians, which may impose a limitation to body size (Herrel and Measey 2010). In any case, burrowing is more efficient in less compact soils, which are actively selected by caecilians (Ducey et al. 1993). Apparently derived from fossorial relatives, some caecilians are aquatic or semiaquatic, which requires different locomotion modes. Aquatic caecilians use lateral undulations to move, and have lost the ability to use the internal hydrostatic concertina locomotion (Summers and O'Reilly 1997).

References

Adams, E., V. Gerstle and C.A. Brühl. 2021. Dermal fungicide exposure at realistic field rates induces lethal and sublethal effects on juvenile European common frogs (*Rana temporaria*). Environ. Toxicol. (in press).

Ahlberg, P.E. and A.R. Milner. 1994. The origin and early diversification of tetrapods. Nature 368: 507–514.

Ahlberg, P.E. and Z. Johanson. 1998. Osteolepiforms and the ancestry of tetrapods. Nature 395: 792–794.

Ahlberg, P.E. and J.A. Clack. 2006. A firm step from water to land. Nature 440: 748–749.

Alberch, P. 1989. Development and evolution of amphibian metamorphosis. Fortsch. Zool. 35: 163–173.

Alford, J.G. and W.I. Lutterschmidt. 2012. Modeling energetic and theoretical costs of thermoregulatory strategy. J. Biol. Dyn. 6: 63–69.

Álvarez, D. and A.G. Nicieza. 2002. Effects of induced variation in anuran larval development on postmetamorphic energy reserves and locomotion. Oecologia 131: 186–195.

Anderson, R.C.O. and D.V. Andrade. 2017. Trading heat and hops for water: dehydration effects on locomotor performance, thermal limits, and thermoregulatory behavior of a terrestrial toad. Ecol. Evol. 7: 9066–9075.

Araspin, L., A. Serra Martinez, C. Wagener, J. Courant, V. Louppe, P. Padilla et al. 2020. Rapid shifts in the temperature dependence of locomotor performance in an invasive frog, *Xenopus laevis*, implications for conservation. Integr. Comp. Biol. 60: 456–466.

Arnold, S.J. and R.J. Wassersug. 1978. Differential predation on metamorphic anurans by gartner snakes (*Thamnophis*): social behavior as a possible defense. Ecology 59: 1014–1022.

Azizi, E. and J.M. Horton. 2004. Patterns of axial and appendicular movements during aquatic walking in the salamander *Siren lacertina*. Zoology 107: 111–120.

Bardua, C., M. Wilkinson, D.J. Gower, E. Sherratt and A. Goswami. 2019. Morphological evolution and modularity of the caecilian skull. BMC Evol. Biol. 19: 30.

Barth, B.J. and R.S. Wilson. 2010. Life in acid: interactive effects of pH and natural organic acids on growth, development and locomotor performance of larval striped marsh frogs (*Limnodynastes peronii*). J. Exp. Biol. 213: 1293–1300.

Baškiera, S. and L. Gvoždík. 2019. Repeatability of thermal reaction norms for spontaneous locomotor activity in juvenile newts. J. Therm. Biol. 80: 126–132.

Beklemishev, W.N. 1969. Principles of Comparative Anatomy of Invertebrates. The University of Chicago Press, Chicago, USA.

Bengtson, S., B. Rasmussen and B. Krapež. 2007. The Paleoproterozoic megabiotic Stirling biota. Palaeobiology 33: 351–381.

Bennett, A.F., T. Garland and P.L. Else. 1989. Individual correlation of morphology, muscle mechanics, and locomotion in a salamander. Am. J. Physiol. 256: R1200–R1208.

Bennett, A.F. and R.B. Huey. 1990. Studying the evolution of physiological performance. pp. 251–284. In: Futuyma, D.J. and J. Antonovics [eds.]. Oxford Surveys in Evolutionary Biology. Oxford University Press, New York, USA.

Bennett, A.F. 1991. The evolution of activity capacity. J. Exp. Biol. 160: 1–23.

Berghänel, A., O. Schülke and J. Ostner. 2015. Locomotor play drives motor skill acquisition at the expense of growth: a life history trade-off. Sci. Adv. 1: e1500451.

Biewener, A.A. and S.N. Patek. 2018. Animal locomotion. Second Edition. Oxford University Press, New York, USA.

Boes, M.W. and M.F. Benard. 2013. Carry-over effects in nature: effects of canopy cover and individual pond on size, shape, and locomotor performance of metamorphosing wood frogs. Copeia 4: 717–722.

Boisvert, C.A., E. Mark-Kurik and P.E. Ahlberg. 2008. The pectoral fin of *Panderichthys* and the origin of digits. Nature 456: 636–638.

Botton-Divet, L., R. Cornette, A. Houssaye, A.C. Fabre and A. Herrel. 2017. Swimming and running: a study of the convergence in long bone morphology among semi-aquatic mustelids (Carnivora: Mustelidae). Biol. J. Linn. Soc. 121: 38–49.

Bowcock, H., G.P. Brown and R. Shine. 2009. Beastly bondage: the costs of amplexus in cane toads (*Bufo marinus*). Copeia 2009: 29–36.

Brady, S.P., F.J. Zamora-Camacho, D. Goedert, M. Comas and R. Calsbeek. 2019. Fitter frogs from polluted ponds: the complex impacts of human-altered environments. Evol. Appl. 12: 1360–1370.

Brijs, J., E. Sandblom, H. Sundh, A. Gräns, J. Hinchcliffe, A. Ekström, K. Sundell, C. Olsson, M. Axelsson and N. Pichaud. 2017. Increased mitochondrial coupling and anaerobic capacity minimizes aerobic costs of trout in the sea. Sci. Rep. 7: 45778.

Brown, R.P. and T. Au. 2009. The influence of metabolic heat on body temperature of a small lizard, *Anolis carolinensis*. Comp. Biochem. Physiol. A 153: 181–184.

Budd, G.E. and S. Jensen. 2017. The origin of the animals and a 'Savannah' hypothesis for early bilaterian evolution. Biol. Rev. 92: 446–473.

Budick, S.A. and D.M. O'Malley. 2000. Locomotor repertoire of the larval zebrafish: swimming, turning and prey capture. J. Exp. Biol. 203: 2565–2579.

Burke, J.N., C.M. Bergeron, B.D. Todd and W.A. Hopkins. 2010. Effects of mercury on behavior and performance of norther two-lined salamanders (*Euricea bislineata*). Environ. Pollut. 158: 3546–3551.

Buttimer, S.M., N. Stepanova and M.C. Womack. 2020. Evolution of the unique anuran pelvic and hind limb skeleton in relation to microhabitat, locomotor mode, and jump performance. Integr. Comp. Biol. 60: 1330–1345.

Cabelguen, J.M., V. Charrier and A. Mathou. 2014. Modular functional organisation of the axial locomotor system of salamanders. Zoology 117: 57–63.

Cabrera-Guzmán, E., M.R. Crossland, G.P. Brown and R. Shine. 2013. Larger body size at metamorphosis enhances survival, growth and performance of young cane toads (*Rhinella marina*). PLoS ONE 8: e70121.

Calsbeek, R. 2008. An ecological twist on the morphology-performance-fitness axis. Evol. Ecol. Res. 10: 197–212.

Careau, V. and T. Garland. 2015. Energetics and behavior: many paths to understanding. Trend. Ecol. Evol. 30: 365–366.

Charbonnier, J.F. and J.R. Vonesh. 2015. Consequences of life history switch point plasticity for juvenile morphology and locomotion in the Túngara frog. PeerJ 3: e1268.

Chen, Z., C. Zhou, M. Meyer, K. Xiang, J.D. Schiffbauer, X. Yuan et al. 2013. Trace fossil evidence for Ediacaran bilaterian animals with complex behaviors. Precambrian Res. 224: 690–701.

Chen, Z., X. Chen, C. Zhou, X. Yuan and S. Xiao. 2018. Late Ediacaran trackways produced by bilaterian animals with paired appendages. Sci. Adv. 4: eaao6691.

Chen, Z., C. Zhou, X. Yuan and S. Xiao. 2019. Death march of a segmented and trilobate bilaterian elucidates early animal evolution. Nature 573: 412–415.

Choi, I., J.H. Shim and R.E. Ricklefs. 2003. Morphometric relationships of take-off speed in anuran amphibians. J. Exp. Zool. 299A: 99–102.

Citadini, J.M., R. Brandt, C.R. Williams and F.R. Gomes. 2018. Evolution of morphology and locomotor performance in anurans: relationships with microhabitat diversification. J. Evol. Biol. 31: 371–381.

Clack, J.A. 2007. Devonian climate change, breathing, and the origin of the tetrapod stem group. Integr. Comp. Biol. 47: 510–523.

Conway Morris, S. 2002. Ancient animals or something else entirely? Science 298: 57–58.

Crump, M.L. 2010. Amphibian diversity and life history. pp. 1–19. *In*: Dodd, K.C. [ed.]. Amphibian ecology and conservation: A handbook of techniques. Oxford University Press, Oxford, UK.

Day, L.M. and B.C. Jayne. 2007. Interspecific scaling of the morphology and posture of the limbs during locomotion in cats (Felidae). J. Exp. Biol. 210: 642–654.

Denoël, M., S. Libon, P. Kestemont, C. Brasseur, J.F. Focant and E. De Pauw. 2013. Effects of a sublethal pesticide exposure on locomotor behavior: a video-tracking analysis in larval amphibians. Chemosphere 90: 945–951.

Devolvé, I., T. Bem and J.M. Cabelguen. 1997. Epaxial and limb muscle activity during swimming and terrestrial stepping in the adult newt, *Pleurodeles waltl*. J. Neurophysiol. 78: 638–650.

Ducey, P.K., D.R. Formanowicz, L. Boyet, J. Mailloux and R.A. Nussbaum. 1993. Experimental examination of burrowing behavior in caecilians (Amphibia: Gymnophiona): effects of soil compaction on burrowing ability of four species. Herpetologica 49: 450–457.

Edwards, J.L. 1977. The evolution of terrestrial locomotion. pp. 553–577. *In*: Hecht, M.K., P.C. Goody and B.M. Hecht [eds.]. Major patters in vertebrate evolution. Plenum, New York. USA.

Else, P.L. and A.F. Bennett. 1987. The thermal dependence of locomotor performance and muscle contractile function in the salamander *Ambystoma tigrinum nebulosum*. J. Exp. Biol. 128: 219–233.

Emerson, S.B. and M.A.R. Koehl. 1990. The interaction of behavioral and morphological change in the evolution of a novel locomotion type: "flying frogs". Evolution 44: 1931–1946.

Enriquez-Urzelai, U., O. San Sebastián, N. Garriga and G.A. Llorente. 2013. Food availability determines the response to pond desiccation in anuran tadpoles. Oecologia 173: 117–127.

Ficetola, G.F. and F. de Bernardi. 2006. Trade-off between larval development and post-metamorphic traits in the frog *Rana latastei*. Evol. Ecol. 20: 143–158.

Finnerty, J.R. 2003. The origins of axial patterning in the metazoa: how old is bilateral symmetry? Int. J. Dev. Biol. 47: 523–529.

Finnerty, J.R. 2005. Did internal transport, rather than directed locomotion, favor the evolution of bilateral symmetry in animals? BioEssays 27: 1174–1180.

Fisher-Wellman, K. and R.J. Bloomer. 2009. Acute exercise and oxidative stress: a 30 year history. Dyn. Med. 8: 1.

Fitzpatrick, B.M., M.F. Benard and J.A. Fordyce. 2003. Morphology and escape performance of tiger salamander larvae (*Ambystoma tigrinum mavortium*). J. Exp. Zool. 297A: 147–159.

Freymiller, G.A., M.D. Whitford, T.E. Higham and R.W. Clark. 2017. Recent interactions with snakes enhance escape performance of desert kangaroo rats (Rodentia: Heteromyidae) during simulated attacks. Biol. J. Linn. Soc. 122: 651–660.

Full, R.J. 1986. Locomotion without lungs: energetics and performance of a lungless salamander. Regul. Integr. Comp. Physiol. 251: R775–R780.

Full, R.J., B.D. Anderson, C.M. Finnerty and M.E. Feder. 1988. Exercising with and without lungs: the effects of metabolic cost, maximal oxygen transport and body size on terrestrial locomotion in salamander species. J. Exp. Biol. 138: 471–485.

Gahm, K., A.Z.A. Arietta and D.K. Skelly. 2021. Temperature-mediated trade-off between development and performance in larval wood frogs (*Rana sylvatica*). J. Exp. Zool. 335: 146–157.

Gall, B.G., A.N. Stokes, S.S. French, E.A. Schlepphorst, E.D. Brodie and E.D. Brodie. 2011. Tetrodoxin levels in larval and metamorphosed newts (*Taricha granulosa*) and palatability to predator dragonflies. Toxicon. 57: 978–983.

Garland, T., E. Hankins and R.B. Huey. 1990. Locomotor capacity and social dominance in male lizards. Funct. Ecol. 4: 243–250.

Garland, T. and P.A. Carter. 1994. Evolutionary physiology. Annu. Rev. Physiol. 56: 579–621.

Geipel, I., C.E. Kernan, A.S. Litterer, G.G. Carter, R.A. Page and H.M. ter Hofstede. 2020. Predation risks of signalling and searching: bats prefer moving katydids. Biol. Lett. 16: 20190837.

Genikhovich, G. and U. Technau. 2017. On the evolution of bilaterality. Development 144: 3392–3404.

Gillis, G.B. 1997. Anguilliform locomotion in an elongate salamander (*Siren intermedia*): effects of speed on axial undulatory movements. J. Exp. Biol. 200: 767–784.

Goedert, D. and R. Calsbeek. 2019. Experimental evidence that metamorphosis alleviates genomic conflict. Am. Nat. 194: 356–366.

Goldberg, J. and M. Fabrezi. 2008. Development and variation of the anuran webbed feet (Amphibia, Anura). Zool. J. Linn. Soc. 152: 39–58.

Goodman, B.A. 2009. Nowhere to run: the role of habitat openness and refuge use in defining patterns of morphological and performance evolution in tropical lizards. J. Evol. Biol. 22: 1535–1544.

Gordon, M.S., R. Blickhan, J.O. Dabiri and J.J. Videler. 2017. Animal Locomotion: Physical Principles and Adaptations. CRC Press, Boca Raton, USA.

Gosler, A.G., J.J.D. Greenwood and C. Perrins. 1995. Predation risk and the cost of being fat. Nature 377: 621–623.

Gower, D.J. and M. Wilkinson. 2005. Conservation biology of caecilian amphibians. Conserv. Biol. 19: 45–55.

Green, D.M. and P. Alberch. 1981. Interdigital webbing and skin morphology in the neotropical salamander genus *Bolitoglossa* (Amphibia; Plethodontidae). J. Morphol. 170: 273–282.

Gvoždík, L. and R. Van Damme. 2006. *Triturus* newts defy the running-swimming dilemma. Evolution 60: 2110–2121.

Gvoždík, L. and R. Smolinský. 2015. Body size, swimming speed, or thermal sensitivity? Predator-imposed selection on amphibian larvae. BMC Evol. Biol. 15: 238.

Halliwell, B. and J.M.C. Gutteridge. 2007. Free radicals in biology and medicine. Oxford University Press, Oxford, UK.

Handrigan, G.R. and R.J. Wassersug. 2007. The anuran Bauplan: a review of the adaptive, developmental, and genetic underpinnings of frog and tadpole morphology. Biol. Rev. 82: 1–25.

Handy, R.D., D.W. Sims, A. Giles, H.A. Campbell and M.M. Musonda. 1999. Metabolic trade-off between locomotion and detoxification for maintenance of blood chemistry and growth parameters by rainbow trout (*Oncorrhynchus mykiss*) during chronic dietary exposure to copper. Aquat. Toxicol. 47: 23–41.

Hayes T.B., P. Falso, S. Gallipeau and M. Stice. 2010. The cause of global amphibian declines: a developmental endocrinologist's perspective. J. Exp. Biol. 213: 921–933.

Heiss, E., N. Natchev, D. Salaberger, M. Gumpenberger, A. Rabanser and J. Weisgram. 2010. Hurt yourself to hurt your enemy: new insights on the function of the bizarre antipredator mechanism in the salamandrid *Pleurodeles waltl*. J. Zool. 280: 156–162.

Herrel, A. and G.J. Measey. 2010. The kinematics of locomotion in caecilians: effects of substrate and body shape. J. Exp. Zool. 313A: 301–309.

Herrel, A. and C. Bonneaud. 2012. Temperature dependence of locomotor performance in the tropical clawed frog, *Xenopus tropicalis*. J. Exp. Biol. 215: 2465–2470.

Herrel, A., L.N. Gonwouo, E.B. Fokam, W.I. Ngundu and C. Bonneaud. 2012. Intersexual differences in body shape and locomotor performance in the aquatic frog, *Xenopus tropicalis*. J. Zool. 287: 311–316.

Herrel, A., M. Vasilopoulou-Kampitsi and C. Bonneaud. 2014. Jumping performance in the highly aquatic frog, *Xenopus tropicalis*: sex-specific relationships between morphology and performance. PeerJ 2: e661.

Hertz, P.E., R.B. Huey and R.D. Stevenson. 1993. Evaluating temperature regulation by field-active ectotherms: the fallacy of the inappropriate question. Am. Nat. 142: 796–818

Higham, T.E. 2007. The integration of locomotion and prey capture in vertebrates: morphology, behavior and performance. Integr. Comp. Biol. 47: 82–95.

Higham, T.E., T. Gamble and A.P. Russell. 2017. On the origin of frictional adhesion in geckos: small morphological changes lead to a major biomechanical transition in the genus *Gonatodes*. Biol. J. Linn. Soc. 120: 503–517.

Hoff, K.S. and R.J. Wassersug. 2000. Tadpole locomotion: axial movement and tail functions in a largely vertebraeless vertebrate. Am. Zool. 40: 62–76.

Holló, G. and M. Novák. 2012. The manoeuvrability hypothesis to explain the maintenance of bilateral symmetry in animal evolution. Biol. Direct 7: 22.

Holló, G. 2017. Demystification of animal symmetry: symmetry is a response to mechanical forces. Biol. Direct 12: 11.

Hudson, C.M., G.P. Brown and R. Shine. 2016. Athletic anurans: the impact of morphology, ecology and evolution on climbing ability in invasive cane toads. Biol. J. Linn. Soc. 119: 992–999.

Hudson, C.M., M. Vidal-García, T.G. Murray and R. Shine. 2020. The accelerating anuran: evolution of locomotor performance in cane toads (*Rhinella marina*, Bufonidae) at an invasion front. Proc. R. Soc. B 287: 20201964.

Huey, R.B. 1980. Sprint velocity of tadpoles (*Bufo boreas*) through metamorphosis. Copeia 1980: 537–540.

Huey, R.B., A.E. Dunham and K.L. Overall. 1991. Variation in locomotor performance in demographically known populations of the lizard *Sceloporus merriami*. Physiol. Zool. 63: 845–872.

Husak, J.F. 2006. Does survival depend on how fast you can run or how fast you do run? Funct. Ecol. 20: 1080–1086.

Husak, J.F., S.F. Fox, M.B. Lovern and R.A. Van Den Bussche. 2006. Faster lizards sire more offspring: sexual selection on whole-animal performance. Evolution 60: 2122–2130.

Husak, J.F. and S.F. Fox. 2008. Sexual selection on locomotor performance. Evol. Ecol. Res. 10: 213–228.

Husak, J.F., S.F. Fox and R.A. Van Den Bussche. 2008. Faster male lizards are better defenders not sneakers. Anim. Behav. 75: 1725–1730.

Irschick, D.J., A. Herrel, B. Vanhooydonck, K. Huyghe and R. Van Damme. 2005. Locomotor compensation creates a mismatch between laboratory and field estimates of escape speed in lizards: a cautionary tale for performance-to-fitness studies. Evolution 59: 1579–1587.

Ivantsov, A., A. Nagovitsyn and M. Zakrevskaya. 2019. Traces of locomotion of Ediacaran macroorganisms. Geosciences 9: 395.

James, R.S., R.S. Wilson, J.E. de Carvalho, T. Kohlsdorf, F.R. Gomes and C.A. Navas. 2005. Interindividual differences in leg muscle mass and pyruvate kinase activity correlate with interindividual differences in jumping performance of *Hyla multilineata*. Physiol. Biochem. Zool. 78: 857–867.

Jenkins, F.A. and D.M. Walsh. 1993. An Early Jurassic caecilian with limbs. Nature 365: 246–250.

Johansson, F., M. Söderquist and F. Bokma. 2009. Insect wing shape evolution: independent effects of migratory and mate guarding flight on dragonfly wings. Biol. J. Linn. Soc. 97: 362–372.

Johansson, F., B. Lederer and M.I. Lind. 2010. Trait performance correlations across life stages under environmental stress conditions in the common frog, *Rana temporaria*. PLoS ONE 5: e11680.

Jung, H. and J.S. Dasen. 2015. Evolution of patterning systems and circuit elements for locomotion. Dev. Cell 32: 408–422.

Junges, C.M., M.I. Maglianese, R.C. Lajmanovich, P.M. Peltzer and A.M. Attademo. 2017. Acute toxicity and etho-toxicity of three insecticides used for mosquito control on amphibian tadpoles. Water Air Soil Pollut. 228: 143.

Kaufmann, K. and P. Dohmen. 2016. Adaption of a dermal in vitro method to investigate the uptake of chemicals across amphibian skin. Environ. Sci. Eur. 28: 10.

Kauffman, E.G. and J.R. Steidtmann. 1981. Are these the oldest metazoan trace fossils? J. Paleontol. 55: 923–947.

Kauffman, E.G., E.R. Elswick, C.C. Johnson and K. Chamberlain. 2009. The first diversification of metazoan life: biogeochemistry and comparative morphology of 1.9–2.5 billion year old trace fossils to Phanerozoic counterparts. *In*: 9th North American Paleontological Convention Proceedings: Cincinnati Museum Center Scientific Contributions 3: 62.

Kearney, B.D., P.G. Byrne and R.D. Reina. 2016. Short- and long-term consequences of developmental saline stress: impacts on anuran respiration and behaviour. R. Soc. Open Sci. 3: 150640.

Kelehear, C., K. Saltonstall and M. Torchin. 2019. Negative effects of parasitic lung nematodes on the fitness of a neotropical toad (*Rhinella horribilis*). Parasitology 146: 928–936.

Kleinteich, T., H.C. Maddin, J. Herzen, F. Beckmann and A.P. Summers. 2012. Is solid always best? Cranial performance in solid and fenestrated caecilian skulls. J. Exp. Biol. 215: 833–844.

Knowles, T.W. and P.D. Weigl. 1990. Thermal dependence of anuran burst locomotor performance. Copeia 1990: 796–802.

Köhler, A., J. Sadowska, J. Olszewska, P. Trzeciak, O. Berger-Tal and C.R. Tracy. 2011. Staying warm or moist? Operative temperature and thermal preferences of common frogs (*Rana temporaria*), and effects on locomotion. Herpetol. J. 21: 17–26.

Kooijman, S.A.L.M. 2009. Dynamic energy budget theory for metabolic organisation, 3rd ed. Cambridge University Press, New York, USA.

Kosmala, G., K. Christian, G. Brown and R. Shine. 2017. Locomotor performance of cane toads differs between native-range and invasive populations. R. Soc. Open Sci. 4: 170517.

Kraskura, K. and J.A. Nelson. 2018. Hypoxia and sprint swimming performance of juvenile striped bass, *Morone saxatilis*. Physiol. Biochem. Zool. 91: 682–690.

Krause, J. and J.G.J. Godin. 1995. Predator preferences for attacking particular prey group sizes: consequences for predator hunting success and prey predation risk. Anim. Behav. 50: 465–473.

Kulkarni, K. and S.V. Krishnamurthy. 2020. Performance studies on jumping behaviour in froglets exposed to commercial grade malathion. J. Environ. Biol. 41: 1450–1454.

Lailvaux, S.P. and J.F. Husak. 2017. Predicting life-history trade-offs with whole-organism performance. Integr. Comp. Biol. 57: 325–332.

Lajmanovich, R.C., P.M. Peltzer, C.S. Martinuzzi, A.M. Attademo, A. Bassó and C.L. Colussi. 2019. Insecticide pyriproxyfen (Dragón®) damage biotransformation, thyroid hormones, heart rate, and swimming performance of *Odontophrynus americanus* tadpoles. Chemosphere 220: 714–722.

Lange, L., L. Bégué, F. Brischoux and O. Lourdais. 2021. The costs of being a good dad: egg-carrying and clutch size impair locomotor performance in male midwife toads (*Alytes obstetricans*). Biol. J. Linn. Soc. 132: 270–282.

Laurin, M. 2006. Scanty evidence and changing opinions about evolving appendages. Zool. Scr. 35: 667–668.

Lee-Jenkins, S.S.Y. and S.A. Robinson. 2018. Effects of neonicotinoids on putative escape behavior of juvenile wood frogs (*Lithobates sylvaticus*) chronically exposed as tadpoles. Environ. Toxicol. Chem. 37: 3115–3123.

Li, J., X. Lin, C. Zhou, P. Zeng, Z. Xu and J. Sun. 2016. Sexual dimorphism and its relationship with swimming performance in *Tanichthys albonubes* under laboratory conditions. Chin. J. Appl. Ecol. 27: 1639–1646.

Li, J., X. Xu, W. Li and X. Zhang. 2019. Linking energy metabolism and locomotor variation to osmoregulation in Chinese shrimp *Fenneropenaeus chinensis*. Comp. Biochem. Physiol. B 234: 58–67.

Lin, C.X., L. Zhang and X. Ji. 2008. Influence of pregnancy on locomotor and feeding performances of the skink, *Mabuya multifasciata*: why do females shift thermal preferences when pregnant? Zoology 111: 188–195.

Lindgren, B., G. Orizaola and A. Laurila. 2018. Interacting effects of predation risk and resource level on escape speed of amphibian larvae along a latitudinal gradient. J. Evol. Biol. 31: 1216–1226.

Lires, A.I., I.M. Soto and R.O. Gómez. 2016. Walk before you jump: new insights on early frog locomotion from the oldest known salientian. Paleobiology 42: 612–623.

Liu, A.G., D. McIlroy and M.D. Brasier. 2010. First evidence for locomotion in the Ediacara biota from the 565 Ma Mistaken Point Formation, Newfoundland. Geology 38: 123–126.

Llewelyn, J., B.L. Phillips, R.A. Alford, L. Schwarzkopf and R. Shine. 2010. Locomotor performance in an invasive species: cane toads from the invasion front have greater endurance, but not speed, compared to conspecifics from a long-colonised area. Behav. Ecol. 162: 343–348.

Louppe, V., J. Courant and H. Herrel. 2017. Differences in mobility at the range edge of an expanding invasive population of *Xenopus laevis* in the west of France. J. Exp. Biol. 220: 278–283.

Lu, H.L., Q. Wu, J. Geng and W. Dang. 2016. Swimming performance and thermal resistance of juvenile and adult newts acclimated to different temperatures. Acta Herpetol. 11: 189–195.

McCay, M.G. 2001. Aerodynamic stability and maneuverability of the gliding frog *Polypedates dennysi*. J. Exp. Biol. 204: 2817–2826.

McKnight, D.T., J. Nordine, B. Jerrett, M. Murray, P. Murray, R. Moss et al. 2020. Do morphological adaptations for gliding in frogs influence clinging and jumping? J. Zool. 310: 55–63.

Mángano, M.G. and L.A. Buatois. 2020. The rise and early evolution of animals: where do we stand from a trace-fossil perspective? Interface Focus 10: 20190103.

Marion, Z.H. and M.E. Hay. 2011. Chemical defense of the eastern newt (*Notophthalmus viridescens*): variation in efficiency against different consumers and in different habitats. PLoS ONE 6: e27581.

McGee, M.R., M.L. Julius, A.M. Vajda, D.O. Norris, L.B. Barber and H.L. Schoenfuss. 2009. Predator avoidance performance of larval fathead minnows (*Pimephales promelas*) following short-term exposure to estrogen mixtures. Aquat. Toxicol. 91: 355–361.

McMahon, T.A. 1984. Muscles, reflexes, and locomotion. Princeton University Press, Princeton, USA.

Meyer-Vernet, N. and J.P. Rospars. 2016. Maximum relative speeds of living organisms: why do bacteria perform as fast as ostriches? Phys. Biol. 13: 066006.

Miles, D.B. 2004. The race goes to the swift: fitness consequences of variation in sprint performance in juvenile lizards. Evol. Ecol. Res. 6: 63–75.

Mitchkash, M.G., T. McPeek and M.D. Boone. 2014. The effects of 24-h exposure to carbaryl or atrazine on locomotor performance and overwinter growth and survival of juvenile spotted salamanders (*Ambystoma maculatum*). Environ. Toxicol. Chem. 33: 548–552.

Moen, D.S., D.J. Irschick and J.J. Wiens. 2013. Evolutionary conservatism and convergence both lead to striking similarity in ecology, morphology and performance across continents in frogs. Proc. R. Soc. B 280: 20132156.

Moore, F.R. and R.E. Gatten. 1989. Locomotor performance of hydrated, dehydrated, and osmotically stressed anuran amphibians. Herpetologica 45: 101–110.

Moore, T.Y. and A.A. Biewener. 2015. Outrun or outmaneuver: predator-prey interactions as a model system for integrating biomechanical studies in a broader ecological and evolutionary context. Integr. Comp. Biol. 55: 1188–1197.

Moreno-Rueda, G. 2003. The capacity to escape from predators in *Passer domesticus*: an experimental study. J. Ornithol. 144: 438–444.

Moreno-Rueda, G., A. Requena-Blanco, F.J. Zamora-Camacho, M. Comas and G. Pascual. 2020. Morphological determinants of jumping performance in the Iberian green frog. Curr. Zool. 66: 417–424.

Moretti, E.H., C.B. Madelaire, R.J. Silva, M.T. Mendonça and F.R. Gomes. 2014. The relationships between parasite intensity, locomotor performance, and body condition in adult toads (*Rhinella icterica*) from the wild. J. Herpetol. 48: 277–283.

Nauwelaerts, S., J. Ramsay and P. Aerts. 2007. Morphological correlates of aquatic and terrestrial locomotion in a semi-aquatic frog, *Rana esculenta*: no evidence for a design conflict. J. Anat. 210: 304–317.

Navas, C.A., R.S. James, J.M. Wakeling, K.M. Kemp and I.A. Johnston. 1999. An integrative study of the temperature dependence of whole animal and muscle performance during jumping and swimming in the frog *Rana temporaria*. J. Comp. Physiol. B 169: 588–596.

O'Donohoe, M.E.A., M.C. Luna, E. Regueira, A.E. Brunetti, N.G. Basso, J.D. Lynch et al. 2019. Diversity and evolution of the parotoid macrogland in true toads (Anura: Bufonidae). Zool. J. Linn. Soc. 187: 453–478.

Oldham, R.S. 1977. Terrestrial locomotion in two species of amphibian larva. J. Zool. 181: 285–295.

O'Reilly, J.C., D.A. Ritter and D.R. Carrier. 1997. Hydrostatic locomotion in a limbless tetrapod. Nature 386: 269–272.

O'Reilly, J.C., A.P. Summers and D.A. Ritter. 2015. The evolution of the functional role of trunk muscles during locomotion in adult amphibians. Am. Zool. 40: 123–135.

Orizaola, G. and A. Laurila. 2009. Microgeographic variation in the effects of larval temperature environment on juvenile morphology and locomotion in the pool frog. J. Zool. 277: 267–274.

O'Steen, S., A.J. Cullum and A.F. Bennett. 2002. Rapid evolution of escape ability in Trinidadian guppies (*Poecilia reticulata*). Evolution 56: 776–784.

Paladino, F.V. 1985. Temperature effects on locomotion and activity bioenergetics of amphibians, reptiles and birds. Am. Zool. 25: 965–972.

Pecoits, E., K.O. Konhauser, N.R. Aubet, L.M. Heaman, G. Veroslavsky, R.A. Stern et al. 2012. Bilaterian burrows and grazing behavior at >585 million years ago. Science 336: 1693–1696.

Pereira da Costa Araújo, A. and G. Malafaia. 2020. Can short exposure to polyethylene microplastics change tadpoles' behavior? A study conducted with neotropical tadpole species belonging to order anura (*Physalaemus cuvieri*). J. Hazard. Mater. 391: 122214.

Perry, G., K. Levering, I. Girard and T. Garland. 2004. Locomotor performance and social dominance in male *Anolis cristatellus*. Anim. Behav. 67: 37–47.

Peterson, C.C. and J.F. Husak. 2006. Locomotor performance and sexual selection: individual variation in sprint speed of collared lizards (*Crotaphytus collaris*). Copeia 2006: 216–224.

Peterson, J.A. and A.R. Blaustein. 1991. Unpalatability in anuran larvae as a defense against natural salamander predators. Ethol. Ecol. Evol. 3: 63–72.

Petrović, T.G., T. Vukov and N. Tomašević Kolarov. 2021. Patterns of correlations and locomotor specialization in anuran limbs: association with phylogeny and ecology. Zoology 144: 125864.

Phillips, P.C. 1998. Genetic constraints at the metamorphic boundary: morphological development in the wood frog, *Rana sylvatica*. J. Evol. Biol. 11: 453–463.

Prates, I., M.J. Angilleta, R.S. Wilson, A.C. Niehaus and C.A. Navas. 2013. Dehydration hardly slows hopping toads (*Rhinella granulosa*) from xeric and mesic environments. Physiol. Biochem. Zool. 86: 451–457.

Preest, M.R. and F.H. Pough. 1989. Interaction of temperature and hydration on locomotion of toads. Funct. Ecol. 3: 693–699.

Rasmussen, B., S. Bengtson, I.R. Fletcher and N.J. McNaughton. 2002a. Ancient animals or something else entirely? Response. Science 298: 58–59.

Rasmussen, B., S. Bengtson, I.R. Fletcher and N.J. McNaughton. 2002b. Discoidal impressions and trace-like fossils more than 1200 million years old. Science 296: 1112–1115.

Rebelo, A.D. and J. Measey. 2019. Locomotor performance constrained by morphology and habitat in a diverse clade of African frogs (Anura: Pyxicephalidae). Biol. J. Linn. Soc. 127: 310–323.

Reilly, S. and M.E. Jorgensen. 2011. The evolution of jumping in frogs: morphological evidence for the basal anuran locomotor condition and the radiation of locomotor system in crown group anurans. J. Morphol. 272: 149–168.

Reilly, S., R. Essner, S. Wren, L. Easton and P.J. Bishop. 2015. Movement patterns in leiopelmatid frogs: insights into the locomotor performance of basal anurans. Behav. Process. 121: 43–53.

Relyea, R.A. 2007. Getting out alive: how predators affect the decision to metamorphose. Oecologia 152: 389–400.

Rice, W.R. 2013. Nothing in genetics makes sense except in light of genomic conflict. Ann. Rev. Ecol. Evol. Syst. 44: 217–237.

Rombough, P. 2011. The energetic of embryonic growth. Resp. Physiol. Neurobiol. 178: 22–29.

Roznik, E.A., C.A. Rodriguez-Barbosa and S.A. Johnson. 2018. Hydric balance and locomotor performance of native and invasive frogs. Front. Ecol. Evol. 6: 159.

Rudh, A. and A. Qvarnström. 2013. Adaptive colourations in amphibians. Sem. Cell Dev. Biol. 24: 553–561.

Ruppert, E.E., R.S. Fox and R.D. Barnes. 2004. Invertebrate Zoology. A Functional Evolutionary Approach. Brooks/Cole-ThompsonLearning, Belmont, USA.

Sanabria, E., L. Quiroga, C. Vergara, M. Banchig, C. Rodriguez and E. Ontivero. 2018. Effect of salinity on locomotor performance and thermal extremes of metamorphic Andean toads (*Rhinella spinulosa*) from Monte Desert, Argentina. J. Therm. Biol. 74: 195–200.

Santos, R.O., M. Laurin and H. Zaher. 2020. A review of the fossil record of caecilians (Lissamphibia: Gymnophionomorpha) with comments on its use to calibrate molecular timetrees. Biol. J. Linn. Soc. 131: 737–755.

Savage, R.M. 1942. The burrowing and emergence of the spade-foot toad, *Pelobates fuscus fuscus*. Proc. Zool. Soc. Lond. 112: 21–35.

Schoch, R.R. 2009. Evolution of life cycles in early amphibians. Ann. Rev. Earth Plant Sci. 37: 135–162.

Seilacher, A. 2007. Trace Fossil Analysis. Springer, Berlin, Germany.

Seilacher, A., P.K. Bose and F. Pflüger. 1998. Triploblastic animals more than 1 billion years ago: trace fossil evidence from India. Science 282: 80–83.

Shaffer, H.B., C.C. Austin and R.B. Huey. 1991. The consequences of metamorphosis on salamander (Ambystoma) locomotor performance. Physiol. Zool. 64: 212–231.

Shine, R. and S. Shetti. 2001. Moving in two worlds: aquatic and terrestrial locomotion in sea snakes (*Laticauda colubrina*, Laticaudidae). J. Evol. Biol. 14: 338–346.

Shu, D., Y. Isozaki, X. Zhang, J. Han and S. Maruyama. 2014. Birth and early evolution of metazoans. Gondwana Res. 25: 884–895.

Sievers, M., R. Hale, S.E. Swearer and K.M. Parris. 2018. Contaminant mixtures interact to impair predator-avoidance behaviours and survival in a larval amphibian. Ecotoxicol. Environ. Safe. 161: 482–488.

Sigurdsen, T., D.M. Green and P.J. Bishop. 2012. Did *Triadobatrachus* jump? Morphology and evolution of the anuran forelimb in relation to locomotion in early salientians. Fieldiana Life Earth Sci. 2012: 77–89.

Sinsch, U., F. Leus, M. Sonntag and A.M. Hantzschmann. 2020. Carry-over effects of the larval environment on the post-metamorphic performance of *Bombina variegata* (Amphibia, Anura). Herpetol. J. 30: 126–134.

Sorci, G. and B. Faivre. 2009. Inflammation and oxidative stress in vertebrate host–parasite systems. Philos. Trans. R. Soc. B 364: 71–83.

Stearns, S.C. 1989. Trade-offs in life-history evolution. Funct. Ecol. 3: 259–268.

Storfer, A. and C. White. 2004. Phenotypically plastic responses of larval tiger salamanders, *Ambystoma tigrinum*, to different predators. J. Herpetol. 38: 612–615.

Summers, A.P. and J.C. O'Reilly. 1997. A comparative study of locomotion in the caecilians *Dermophis mexicanum* and *Typhlonectes natans* (Amphibia: Gymnophiona). Zool. J. Linn. Soc. 121: 65–76.

Taylor, C.R., N.C. Heglund, T.A. McMahon and T.R. Looney. 1980. Energetic cost of generating muscular force during running. A comparison of large and small animals. J. Exp. Biol. 86: 9–18.

Tickle, P.G., M.F. Richardson and J.R. Codd. 2010. Load carrying during locomotion in the barnacle goose (*Branta leucopsis*) the effect of load placement and size. Comp. Biochem. Physiol. A 156: 309–317.

Titon, B., C.A. Navas, J. Jim and F. Ribeiro Gomes. 2010. Water balance and locomotor performance in three species of neotropical toads that differ in geographical distribution. Comp. Biochem. Physiol. A 156: 129–135.

Toledo, R.C. and C. Jared. 1993. Cutaneous adaptations to water balance in amphibians. Comp. Biochem. Physiol. A 105: 593–608.

Touchon, J.C., R.R. Jiménez, S.H. Abinette, J.R. Vonesh and K.M. Warkentin. 2013. Behavioral plasticity mitigates risk across environments and predators during anuran metamorphosis. Oecologia 173: 801–811.

Van Buskirk, J. and G. Saxer. 2001. Delayed costs of an induced defense in tadpoles? Morphology, hopping, and development rate at metamorphosis. Evolution 55: 821–829.

Van Buskirk, J., P. Anderwald, S. Lüpold, L. Reinhardt and H. Schuler. 2003. The lure effect, tadpole tail shape, and the target of dragonfly strikes. J. Herpetol. 37: 420–424.

Vanhooydonck, B., R. Van Damme and P. Aerts. 2001. Speed and stamina trade-off in lacertid lizards. Evolution 55: 1040–1048.

Vanhooydonck, B. and R. Van Damme. 2003. Relationships between locomotor performance, microhabitat use and antipredator behaviour in lacertid lizards. Funct. Ecol. 17: 160–169.

Vermeij, G.J. 1994. The evolutionary interaction among species: selection, escalation, and coevolution. Annu. Rev. Ecol. Syst. 25: 219–236.

Wassersug, R.J. and D.G. Sperry. 1977. The relationships of locomotion to differential predation on *Pseudacris triseriata* (Anura: Hylidae). Ecology 58: 830–839.

Wassersug, R.J. 1989. Locomotion in amphibian larvae (or "why aren't tadpoles built like fishes"?). Am. Zool. 29: 65–84.

Watkins, T.B. 1996. Predator-mediated selection on burst swimming performance in tadpoles of the Pacific tree frog, *Pseudacris regilla*. Physiol. Biochem. Zool. 69: 154–167.

Watkins, T.B. 1997. The effect of metamorphosis on the repeatability of maximal locomotor performance in the pacific tree frog *Hyla regilla*. J. Exp. Biol. 200: 2663–2668.

Wheatley, R., A.C. Niehaus, D.O. Fisher and R.S. Wilson. 2018. Ecological context and the probability of mistakes underlie speed choice. Funct. Ecol. 32: 990–1000.

Wijethunga, U., M. Greenlees and R. Shine. 2016. Living up to its name? The effect of salinity on development, growth, and phenotype of the "marine" toad (*Rhinella marina*). J. Comp. Physiol. B 186: 205–213.

Willmer, P. 1990. Invertebrate Relationships. Patterns in Animal Evolution. Cambridge University Press, Cambridge, UK.

Wilson, R.S and C.E. Franklin. 1999. Thermal acclimation of locomotor performance in tadpoles of the frog *Limnodynastes peronii*. J. Comp. Physiol. B 169: 445–451.

Wilson, R.S. and C.E. Franklin. 2000. Inability of adult *Limnodynastes peronii* (Amphibia: Anura) to thermally acclimate locomotor performance. Comp. Biochem. Physiol. A 127: 21–28.

Wilson, R.S., R.S. James and I.A. Johnston. 2000a. Thermal acclimation of locomotor performance in tadpoles and adults of the aquatic frog *Xenopus laevis*. J. Comp. Physiol. B 170: 117–124.

Wilson, R.S., C.E. Franklin and R.S. James. 2000b. Allometric scaling relationships of jumping performance in the striped marsh frog, *Limnodynastes peronii*. J. Exp. Biol. 203: 1937–1946.

Wilson, R.S., R.S. James and R. Van Damme. 2002. Trade-offs between speed and endurance in the frog *Xenopus laevis*: a multi-level approach. J. Exp. Biol. 205: 1145–1152.

Wilson, R.S. 2005. Consequences of metamorphosis for the locomotor performance and thermal physiology of the newt *Triturus cristatus*. Physiol. Biochem. Zool. 78: 967–975.

Wilson, R.S., P.G. Kraft and R. Van Damme. 2005. Predator-specific changes in the morphology and swimming performance of larval *Rana lessonae*. Funct. Ecol. 19: 238–244.

Wootton, R.J. 1992. Functional morphology of insect wings. Ann. Rev. Entomol. 37: 113–140.

Zamora-Camacho, F.J., S. Reguera, M.V. Rubiño-Hispán and G. Moreno-Rueda. 2014. Effects of limb length, body mass, gender, gravidity, and elevation on escape speed in the lizard *Psammodromus algirus*. Evol. Biol. 41: 509–517.

Zamora-Camacho, F.J., M.V. Rubiño-Hispán, S. Reguera and G. Moreno-Rueda. 2015a. Thermal dependence of sprint performance in the lizard *Psammodromus algirus* along a 2200-meter elevational gradient: cold-habitat lizards do not perform better at low temperatures. J. Therm. Biol. 52: 90–96.

Zamora-Camacho, F.J., S. Reguera, M.V. Rubiño-Hispán and G. Moreno-Rueda. 2015b. Eliciting an immune response reduces sprint speed in a lizard. Behav. Ecol. 26: 115–120.

Zamora-Camacho, F.J. 2018. Locomotor performance in a running toad: roles of morphology, sex, and agrosystem versus natural habitat. Biol. J. Linn. Soc. 123: 411–421.

Zamora-Camacho, F.J., J. García-Astilleros and P. Aragón. 2018. Does predation risk outweigh the costs of lost feeding opportunities or does it generate a behavioural trade-off? A case study with Iberian ribbed newt larvae. Biol. J. Linn. Soc. 125: 741–749.

Zamora-Camacho, F.J., L. Medina-Gálvez and S. Zambrano-Fernández. 2019. The roles of sex and morphology in burrowing depth of Iberian spadefoot toads in different biotic and abiotic environments. J. Zool. 309: 224–230.

Zamora-Camacho, F.J and P. Aragón. 2019a. Hindlimb abnormality reduces locomotor performance in *Pelobates cultripes* metamorphs but is not predicted by larval morphometrics. Herpetozoa 32: 125–131.

Zamora-Camacho, F.J. and P. Aragón. 2019b. Failed predator attacks have detrimental effects on antipredatory capabilities through developmental plasticity in *Pelobates cultripes* toads. Funct. Ecol. 33: 846–854.

Zamora-Camacho, F.J., M. Comas and G. Moreno-Rueda. 2020. Immune challenge does not impair short-distance escape speed in a newt. Anim. Behav. 167: 101–109.

Zera, A.J. and L.G. Harshman. 2001. The physiology of life history trade-offs in animals. Ann. Rev. Ecol. Syst. 32: 95–126.

CHAPTER 9

Diversity of Reproductive Strategies in the Amphibia

Balázs Vági[1],* and *Tamás Székely*[1,2]

INTRODUCTION

The reproductive strategies in amphibians are the most diverse among tetrapods, and only rivalled by that of the bony fish among the vertebrates. While amphibians are often considered as an ancient vertebrate lineage that first conquered terrestrial habitats, in fact, extant amphibians (Lissamphibia) are a rather modern clade which reached most of its current diversity after the Cretaceous-Paleogene extinction event (approx. 66 MYA, Jetz and Pyron 2018; Chapter 2). Their contemporary species numbers are comparable to other major tetrapod lineages, like Squamata, Aves or Mammalia. While we might think about amphibians as animals which are bound to aquatic habitats at least during their reproduction and early developmental stages, amphibian life cycles and reproductive strategies are much more diverse, especially in tropical and subtropical areas. This complexity involves variation in fertilisation modes, in sites for egg deposition and larval development, alternative ontogenetic pathways like biphasic versus direct development or paedomorphosis, and tremendous variation in parental care. Although the occupation of terrestrial habitats was first achieved much earlier and independently by the first tetrapods, gaining more independence from water is still one of the major driving forces of the reproductive diversity of modern amphibians.

Most reviews on the diversity of amphibian reproductive strategies were published over a decade (e.g., Salthe 1969, Salthe and Duellman 1973, Crump 1995, 1996, Haddad and Prado 2005, Wells 2007), and since then, several new

---

[1] Department of Evolutionary Zoology & Human Biology, University of Debrecen, Egyetemtér 1, H-4032, Hungary.

[2] Milner Centre for Evolution, University of Bath, Bath BA2 7AY, UK.
  Email: T.Szekely@bath.ac.uk

* Corresponding author: bi.vagi@gmail.com

publications arose on different aspects of the reproductive biology. Therefore, it is worth revisiting this field and summarise recent findings and hypotheses. Lissamphibia is monophyletic and all of its three orders, frogs (Anura), salamanders (Urodela or Caudata) and caecilians (Gymnophiona) made remarkable steps towards terrestriality and invented various forms of parental care. However, the three orders have characteristically different body plan, life styles and reproductive biology, and all three have considerably diverse reproduction in their unique way (Table 1). Below we discuss the reproductive strategies of anurans, urodeles and caecilians in separate sections.

Diversity of Reproductive Strategies in Anurans

In the order Anura (tailless amphibians: frogs and toads) the reproductive modes show immense variation (Gomez-Mestre et al. 2012), which can be partially explained by their greatest phylogenetic diversity within the Lissamphibia (Anura: approx. 7000 frog and toad species in 55 families; Lissamphibia in total: approx. 7800 species in 75 families—Amphibiaweb 2021).

Mate Acquisition and Fertilisation

Despite the predominantly external fertilisation which does not necessarily suppose a physical contact between the male and the female, anuran pairs are formed often long before egg-laying. Males guard the females by embracing them in a firm hug called amplexus (Arak 1983). Amplexus have some distinct forms (Table 1): it is usually inguinal (the male grabs the female in the waist region) in more ancient lineages (e.g., spadefoot toads, fire-bellied toads) and axillary (the male grabs the females just behind the forelimbs) in more derived evolutionary lineages (e.g., treefrogs, true frogs and toads). There are some other specialised forms, as cephalic amplexus in poison arrow frogs (Dendrobatidae). The males of the desert rain frogs (*Breviceps*) that could not grab the stocky gravid females glue themselves onto the females' back using sticky skin secretions. In other frogs the amplexus is altogether missing and eggs are fertilised without a direct contact between the pair (Wells 2007).

The time frame for breeding is mostly governed by climatic conditions, and this is a starting point of reproductive diversity. Species living in more seasonal climates, like many from the temperate region or tropical species which experience a well-defined wet season have an explosive breeding system, where the whole population breeds in a few days (Arak 1983). In the usually male-biased breeding aggregations of frogs, males involve in scramble competition for mating opportunities. The reproductive success is usually body size-dependent: larger, stronger males are more successful finding a mate (Davies and Halliday 1978, Roberts et al. 1999, Vági and Hettyey 2016). Further males may join the breeding pair, and by releasing sperm the resulting egg clutch exhibit multiple paternity (D'Orgeix 1996, Byrne and Roberts 1999, Lodé and Lesbarrères 2004, Sztatecsny et al. 2006, Rausch et al. 2014). In extremely male-biased operational sex ratios (i.e., many more males are sexually active than females) the breeding males form a mating ball with the females inside (Verrell and McCabe 1986), which can have serious consequences, as females may get drowned during the wrestling. The foam nest builder rhacophorids are

Table 1. Comparison between the reproductive diversity in frogs and toads (Anura), newts and salamanders (Caudata) and caecilians (Gymnophiona). Based on Wells (2007), San Mauro et al. (2014), Vági et al. (2019, 2022).

	Anura	Caudata	Gymnophiona
Amplexus	mostly present in various forms (see text), but can be absent	present, e.g., in *Pleurodeles waltl* absent, e.g., in *Triturus* spp.	absent
Fertilisation	external, rarely internal	internal (spermatophore), rarely external	internal (phallodeum)
Developmental mode			
Paedomorphosis	absent	- obligate; e.g., Proteidae - facultative; e.g., *Lissotriton vulgaris*	absent
Metamorphosis	common	common in most families	present (e.g., *Grandisonia*)
Direct development	present in many families (e.g., Microhylidae, Craugastoridae)	common in Plethodontidae	present (e.g., *Typhlonectes*)
Parental care (caring sex)			
No care	common (80–90% of species)	common (30–50% of species)	absent
Male-only care	present	present in families with external fertilisation	absent
Female-only care	present	present in families with internal fertilisation, common in Plethodontidae	common (e.g., *Siphonops*)
Biparental care	rarely present (e.g., *Ranitomeya imitator*)	absent	absent
Parental care (forms and functions)			
Nest building	foam, bubble and other types of nests (e.g., *Rhacophorus*, *Engystomops*)	absent, but egg concealing occurs (e.g., *Triturus*, *Lissotriton*)	absent
Attendance	male and female egg attendance are the most widespread care (10–20% of species); coevolved with terrestriality	male attendance in aquatic environments (*Andrias*, *Siren*); female attendance in terrestrial environments (Plethodontidae)	female attendance is ubiquitous in oviparous species (e.g., *Ichthyophis*)
Transportation	Rare. Eggs (*Flectonotus*, *Alytes*), tadpoles (Dendrobatidae) and juveniles (*Sphenophryne*) can be transported	absent	absent
Internal brooding	rarely occurs (e.g., *Assa*, *Gastrotheca*, *Pipa*, *Rheobatrachus*)	absent	absent
Viviparity	rarely occurs (*Nectophrynoides*, *Limnonectes larvaepartus*)	rarely occurs (*Salamandra*, *Lyciasalamandra*)	present (e.g., *Typhlonectes*)
Feeding	Rare. Trophic egg feeding (e.g., *Oophaga*), secretions (e.g., *Rhinoderma*, *Ecnomyohyla*)	absent	dermatophagy (e.g., *Siphonops*) and uterophagy (e.g., *Herpele*)

characterised by the most male-biased matings among vertebrates (Kusano et al. 1991, Byrne et al. 2002, Byrne and Whiting 2011). It is likely that in these species the males are forced to cooperate with each other and with the female, since any bodily fight would destroy the nest which is under construction. Instead of direct wrestling, the males' competition for the fertilisation of the eggs is decided by sperm competition. For better performance in the latter, these frogs have disproportionately large testes (Kusano et al. 1991, Vági et al. 2020). On the other hand, species with more prolonged breeding may defend territories or other resources, and the role of female choice is more important (e.g., Ryan 1980, Summers et al. 1999). In addition to explosive breeders and resource-based breeding systems there are also frog species which gather in mating leks, defending a small, symbolic mating ground (e.g., Bourne 1992, Friedl and Klump 2005).

The length of the amplexus prior to and during the egg-laying is variable. The time needed for the egg-laying depends much on the form of the created egg-mass. True toads (Bufonidae) which lay long, thin egg-strings only release a small number of eggs at once, and the whole process of egg-laying can last for hours (Davies and Halliday 1978). Obviously, the males need to synchronise their gamete release to that of the females, which can be a time- and energy-consuming process. This can explain why male toads get exhausted after two or three breeding events and refute further mating opportunities or achieve a decreased success in fertilisation (Hettyey et al. 2009). In contrast, males of, e.g., most of true frogs and related families (Ranoidea) release all or a large portion of their eggs, rapidly forming a more-or-less spherical egg mass. Most anuran eggs are involved in a gelatinous outer layer, which protects the eggs from predation and desiccation. Large amount of this gelatinous envelope can insulate the eggs and help creating more favourable thermal conditions for embryonic development. This can be especially useful in the large mats of aggregated egg clutches of some species which reproduce in the cold early spring in cool temperate regions (e.g., *Rana temporaria, Lithobates sylvaticus*).

Although the fertilisation of the eggs is predominantly external in anurans (Beck 1998), there are some notable exceptions. Both species of North American tailed frogs (Ascaphiidae), one of the most ancient lineages of frogs, have internal fertilisation with the help of a tail-like appendage of the males. Other internal fertilising frogs do not have any copulatory organ, the sperm is transmitted by direct contact of the cloaca. Obviously, all ovoviviparous and viviparous lineages have internal fertilisation, including the small African toadlets *Nimbaphrynoides* and *Nectophrynoides* (Liedtke et al. 2017), and the recently described *Limnonectes larvaepartus*, which give birth to tadpoles (Iskandar et al. 2014).

Environments for Egg-deposition and Larval Development

Anurans eggs and larvae can develop in diverse environments. The first categorisation of anuran reproductive modes was proposed by Boulenger (1886) with 10 categories, then Salthe and Duelmann (1973) revisited the topic nearly a century later, which induced numerous research in the field. The most detailed classification was made by Duellman and Trueb (1986), later supplemented by Haddad and Prado (2005) and Schulte et al. (2020). The latest review on amphibian reproductive modes

Table 2. Traits used for the classification of amphibian reproductive modes, following the study of Nunes-de-Almeida et al. (2021). Based on the combinations of these traits, they could identify 74 different reproductive modes in amphibians: 71 in anurans, 16 in urodelans and 7 in caecilians. In this classification they concentrated on life history differences and environments for egg deposition, and ignored behavioural diversity, i.e., amplexus, breeding systems or active parental care.

(1) Reproduction type				
Oviparity		Viviparity (incl. ovoviviparity without maternal provisioning)		
(2) Oviposition macrohabitat:				
Environment		Animal (oviduct, pouch, vocal sac, stomach)		
(3) Spawning type				
Froth (foam and bubble nests)		Non-froth		
(4) Oviposition substrate				
Aquatic	Non-aquatic			In/on animal
(5) Medium surrounding the eggs				
Lentic	Lotic	Terrestrial	Non-oviductal	Oviductal/uterine
(6) Nest construction				
Constructed nest	Adopted nest		No nest	
(7) Oviposition microhabitat				
Floating	Ground	Subaquatic ground	Depression	Burrow
Subaquatic chamber	Insect mound	Rock	Wall	Plant leaf
Plant branch	Plant root	Subaquatic plant	Water-filled reservoir	Reservoir without water
(8) Embryonic development				
Indirect		Direct		
(9) Embryonic nutrition				
Lecitotrophic		Matrotrophic		
(10) Larval and newborn nutrition				
Exotrophic (with and without parental feeding)		Endotrophic		
(11) Place of larval development				
Lentic	Lotic	Terrestrial		Internal/on animal

(Nunes-de-Almeida et al. 2021) established a classification based on the combination of 11 individual reproductive traits (Table 2).

Duellman and Trueb's (1986) original categorisation was mostly based on one biogeographic region, the Brasilian Atlantic rainforests, but developed strategies can arise in other tropical regions as well, like Sub-Saharan Africa (Liedtke et al. 2017, Lion et al. 2019). Some new reproductive mode, e.g., mud-packing on the eggs (Gururaja et al. 2014) or giving birth to larvae, was described from Southeast Asia (Kusrini et al. 2015), increasing the number of reproductive modes over 40. The classification based on multiple aspects of egg and larval development. One of the most important dichotomy is a basic division between aquatic and terrestrial egg-laying (Gomez-Mestre et al. 2012, Lion et al. 2019). Aquatic environments can be further divided into lentic (standing water) and lotic (flowing water) types, and small, often ephemeral water bodies, e.g., phytotelms, wooden holes, or natural or constructed basins. Therefore, anurans may use arboreal and subterranean sites for egg-laying and during their larval development, vastly expanding the available terrestrial niches.

The development of the larvae can be classified into three basic types. Exotrophic larvae feed on external food sources like algae, detritus or other animals (including their siblings), or sometimes fed by the parents via trophic eggs or secretions. Endotrophic larvae usually stay in protected nest sites and feed exclusively on the yolk content of their egg; while direct-developing species simply skip the larval phase and complete their development inside the egg (or in the body of the mother, Gomez-Mestre et al. 2012). Interestingly, while direct development is a key innovation and thus, ubiquitous in some frog lineage (like the clade of Central and South American rainfrogs of the families Craugastoridae, Eleutherodactylidae and Strabomantidae), in other groups it is a variable trait; we find species both with free-living larvae and with direct development in the genera of *Pipa* (Pipidae) and *Gastrotheca* (Hemiphractidae), both having highly specialised internal brooding strategies.

Parental Care

The diversity of parental care forms in frogs and toads is unparalleled among tetrapods (i.e., terrestrial vertebrates: amphibians, reptiles, birds and mammals). Interestingly, only 10–20% of the species shows any form of parental care, however, caring ranges from relatively simple forms as egg attendance to complex solutions for the internal development and cooperation between the parents (Crump 1996, Furness and Capellini 2019, Vági et al. 2019). The main functions of care provision are protection of the eggs, tadpoles and juveniles; and nourishing the offspring. All of these can be provided by both male and female frogs (Vági et al. 2019, 2020).

The prevalence of male and female parental care is comparable among frogs: around 20% of species cares by the male only and around 10% by the female only, whereas 5% of species exhibit biparental care (incl. cooperative nest building) of the clutch and/or eggs (n = 1044 species, Vági et al. 2019; see also Furness and Capellini 2019). The evolutionary drivers of care, however, have remained an unresolved issue. First, the mode of fertilisation is long considered an important predictor for the care-providing sex, however, this link was failed to be demonstrated in frogs, possibly due to the relatively low number of internally fertilised lineages (Beck 1998). This means that female care is surprisingly widespread in lineages with external fertilisation. Second, territorial males often attend egg clutches (Vági et al. 2020), but there are also many instances of female attendance and this behaviour proved to be plastic both within a species (Ringler et al. 2015) and labile in evolutionary terms (Delia et al. 2017, Furness and Capellini 2019). Last, in biparental nest building the parents cooperate in similar roles, while in other species the parents fulfil complementary roles (Furness and Capellini 2019), where usually feeding is the females' task (Brown et al. 2010).

Nest building is often considered a form of parental care, albeit it precedes fertilisation. Some anurans build nests as a passive protection for their eggs. In aquatic breeders there are submerged nests and constructed basins which provide a predator-free environment for the developing tadpoles (Martins et al. 1998, Schäfer et al. 2019). In some frog species basin construction is a variable trait, and its presence is depending on the actual environmental conditions (Martins et al. 1998). In species where the males are the main nest builders the constructed structures may represent quality of the builder, and can be the basis for female choice. In other species foam nests are constructed in a cooperative fashion by the male(s) and the female. Nests provide efficient shelter from predators and from adverse climatic conditions. However, we found no trace that nest building is linked to the climatic environment (Vági et al. 2020), suggesting that the nests themselves provide some independence from climate and enabled the colonisation of less suitable habitats. Foam nests were indeed key innovation in the radiation of at least two successful lineages, the South-East Asian and African arboreal rhacophorids (Meegaskumbura et al. 2015) and the leptodactylids of South America (Heyer 1969, Pereira et al. 2017).

Attendance of the eggs and sometimes the larvae and juveniles are widespread, yet these seem to be the most labile forms of anuran parental care (Vági et al. 2019, Furness and Capellini 2019). Due to the lack of detailed behavioural observation attending and actual active protection of the egg clutch can be hardly discerned. However, there is a growing number of evidence which confirms the importance of the parent's presence near the clutch and specify functions of this care form. Amphibian eggs developing in a terrestrial environment are at the risk of desiccation, and attending parents can minimise this risk by hydric brooding: urinating on the eggs or moisture them and reduce evaporation by the physical contact with their own body (Bickford 2004, Poo and Bickford 2013). Many anurans also actively protect the clutch against various predators, like wasps or katydids, which can cause substantial mortality of the eggs and larvae (Chuang et al. 2017, Delia et al. 2017). Fungal or bacterial infections may also contribute to the mortality of unattended eggs (Simon 1983).

Transportation of eggs, tadpoles or juveniles evolved multiple times among anurans (Furness and Capellini 2019, Vági et al. 2020). Eggs may be placed onto the females back and remain exposed throughout their development, like in *Flectonotus* species in the Hemiphractidae family. Other hemiphractids enclose eggs in variously developed dorsal pouches. In the European midwife toads (*Alytes*) the males wrap the egg strings around their hind limb. These long-term forms of transportation last until the hatching of the tadpoles or, in species with direct development, the juveniles, like in some small microhylids from New Guinea (Köhler and Günther 2008). In contrast, poison arrow frogs (Dendrobatidae) deposit their clutch on the ground, and after hatching, they carry tadpoles to streams or small arboreal basins like phytotelmata of the bromeliads. This single, short term transportation mostly carried out by the male, but in one genus, *Oophaga*, it is performed by the females. Some behavioural plasticity also occurs, as the female may take over transportation when the male would not provide it (Ringler et al. 2015). By delivering tadpoles to these small water bodies the parents may reduce competition and predation risk (McKeon and Summers 2013). The parents may also assess the quality of the phytotelms to some degree based on the nutrients available for the tadpoles (Poelman et al. 2013) and the presence of previously deposited larvae (Poelman and Dicke 2007, Rojas 2014).

Feeding of young is a care type which is usually provided by female frogs, except in very rare cases where the males provide nutrients through external (skin) or internal secretions (Goycoechea et al. 1986, Hansen 2012). In contrast, the females usually provide food for the tadpoles by producing infertile trophic eggs. This behaviour typically occurs as uniparental female care, but in some rare case it is a part of a cooperation between parents, where the male fulfils the role of a guard for the new generation and a guide for the female (Tumulty et al. 2014). An adaptive significance of feeding is that the species may occupy food-scarce environments. Typically, feeding occurs where smaller water bodies are used for breeding, like tree holes or phytotelmata. The species with the most complex, biparental breeding system, *Ranitomeya imitator* can use *Heliconia* phytotelms which contain only a few millilitres of water (Brown et al. 2010). The main advantage of these small water bodies is that they lack predators and competitors. Another potential benefit is that poison arrow frog females may transfer alkaloids to the tadpoles via trophic egg feeding, which can contribute to the chemical defence of the tadpoles (Saporito et al. 2019).

Active feeding of the tadpoles only provides nutrients until metamorphosis, and the presence of endotrophy and direct development should be also considered a special case of offspring nourishment (Vági et al. 2019). In these forms of development all nutrient needed until the completion of the metamorphosis is provided by the egg's yolk content. However, an important difference is that all these nutrients should be provided at once at the time of the egg formation. Moreover, species with endotrophic tadpoles and direct development do not need to evolve complex and intricate behaviours for parental care.

An interesting aspect of anuran parental care is the various modes of viviparity and internal brooding. In one of the only truly viviparous lineages, *Nectophrynoides* and *Nimbaphrynoides* toadlets from Africa both lecitotrophy (or ovoviviparity) and matrotrophy (viviparity in the strict sense) occur. Other species may brood the eggs and/or tadpoles in variously developed dorsal pouches of the females (*Gastrotheca*, *Flectonotus*), embedded within the dorsal skin (*Pipa*) or even in the stomach (in the extinct *Rheobatrachus* species) of the females, or in paired lateral pouches (*Assa*) or in the vocal sac (*Rhinoderma*) of the males.

Evolutionary Associations between Ecology, Life History and Parental Care

Several ecological, life history and climatic variables have been identified having an association with parenting, although the jury is still out on which of these variables (or suits of variables) may have driven the evolution of caring. For instance, an important predictor of parental care forms is terrestrial reproduction (Gomez-Mestre et al. 2012, Vági et al. 2019). Parental care is associated with basic life history measures, like egg size or clutch volume. A link between parental care and larger eggs is long suggested (Shine 1978), and it was verified by phylogenetic studies in frogs (Summers et al. 2006, Gomez-Mestre et al. 2012, Furness et al. 2022). However, we should also consider that terrestrial egg laying is both a predictor of larger eggs (Rollinson and Rowe 2018) and parental care (Vági et al. 2019), thus, it can mediate the effect of parental care on egg size. The larger size of terrestrial eggs can be explained by their longer development in direct-developing species (many terrestrial reproducing species complete their metamorphosis within the egg membrane). Another possible explanation is a smaller surface-to-mass ratio which can prevent desiccation and, in contrast to aquatic environments, does not cause gas exchange difficulties (Rollinson and Rowe 2018). In direct-developing species, the anatomical, physiological and ecological constraints on body size should be also taken into account, which are probably different for tadpoles and metamorphosed froglets (small tadpoles can scrape or filtrate tiny organisms or food particles, while juvenile frogs have to cope with small terrestrial invertebrates—see also this argument in urodelans).

Diversity of Reproductive Strategies in Urodelans

Urodela (newts and salamanders) is the sister-clade of Anura within the Lissamphibia group, however, their diversity is more limited both in terms of species numbers and reproductive modes. However, they still show considerable variation, and also have a tendency to reproduce independently from water (Table 1, Fig. 1). In addition, they show a number of different lifestyles, for instance, paedomorphic urodelans complete their whole life cycle in water (Fig. 1), whereas others spent substantial part or all of their life in terrestrial habitats (e.g., many salamander).

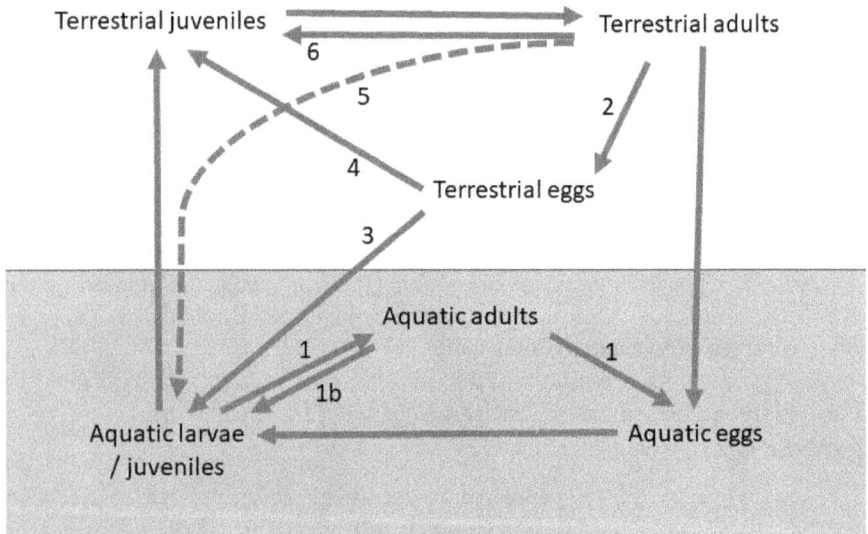

Figure 1. The diversity of amphibian life cycles. The outermost arrows represent the most typical life cycle with aquatic eggs and larvae, and terrestrial juveniles and adults (e.g., *Rana temporaria, Triturus cristatus*). (1) Fully aquatic amphibians (e.g., *Pipa, Andrias*), incl. paedomorphic urodelans (e.g., *Necturus*) skip the terrestrial phase. (1b) Some fully aquatic caecilians (*Typhlonectes*) give birth to juveniles, also skipping the aquatic egg phase. (2) Many terrestrial anurans, salamanders and caecilians lay terrestrial eggs. These can develop into (3) aquatic larvae (e.g., *Agalychnis; Desmognathus*) or (4) to terrestrial juveniles by direct development (e.g., *Speleomantes; Eleutherodactylus*). Instead of egg laying terrestrial adults may (5) give birth to aquatic larvae (*Salamandra salamandra; Limnonectes larvaepartus*); or (6) to fully metamorphosed terrestrial juveniles (*Lyciasalamandra; Nectophrynoides*). Note that by (4) and (6) the life cycle is fully terrestrial.

Pairing and Fertilisation

Fertilisation shows higher variance among urodeles than in anurans (Table 1), however, this variation can be retracted to a single evolutionary switch from external to internal fertilisation (Vági et al. 2022). In basal clades of urodelans (such as the Asian salamanders Hynobiidae, the giant salamanders Cryptobranchidae and sirens Sirenidae) the fertilisation is external (Reinhard et al. 2013), like in many fish. These families reproduce exclusively in water. In more developed clades a spermatophore evolved which is laid on the ground by the male and sucked in by the female's cloaca. The spermatophore transfer can occur both in aquatic and terrestrial environments. Usually it is preceded by a courtship during which visual, chemical and tactile channels for communication are extensively used (Houck and Arnold 2003). Amplexus can precede mating in some of the species, but it has been lost in many urodelans (Table 1).

Habitats for Egg Laying and for the Offspring

Salamanders may use either aquatic or terrestrial habitats for reproduction. Aquatic breeding habitats can be classified into lentic and lotic types that refer to standing and flowing water bodies, respectively (Salthe 1969, Nussbaum 1985). Some lotic breeders lay eggs in semiaquatic conditions, above the water, under rocks and logs next to the stream, or in seepage areas. These species have an aquatic larval stage. Species with lentic larvae may also lay eggs in a terrestrial environment. They usually lay eggs before the onset of the rainy season, before the egg-laying sites become filled with water. Other species switched to a completely terrestrial reproduction, and lay direct-developing eggs in terrestrial environment. The latter reproductive mode is typical among lungless salamanders (Plethodontidae), and presumably had an important role in the diversification of this group. Terrestrial reproduction paved the way for occupying completely new environment for urodelans, namely the arboreal habitats, which are exclusively used by the direct-developing plethodontids. However, the high number of species in Plethodontidae, nearly 500 species—two third of total species richness in Caudata—may also be explained by the dispersal abilities of completely terrestrial species are restricted, thus, their complete switch to terrestriality likely favoured genetic isolation and enhanced speciation processes.

It has been suggested that both lotic and terrestrial breeder salamanders produce proportionately larger eggs than lentic breeders. This discrepancy is explained by a multitude of potential factors: (i) according to Shine (1978) parental care increased developmental time by creating a "safe harbour" for the embryos inside the protected eggs; (ii) in contrast, Nussbaum argued that the size of prey available in different habitats was the main driving force as the hatchlings of stream and terrestrial breeders need to cope with larger food items (Nussbaum 1985, 1987); (iii) while Rollinson and Rowe (2018) implied that lower oxygen availability in aquatic environments limited egg size, as in larger eggs the oxygen diffusion is not as efficient. As no study compared the predictions of these three main hypotheses, more research is needed to clarify the evolution of salamander egg size.

Paedomorphosis

Several salamanders retain their larval external gills while they reach sexual maturity and never leave the water. This phenomenon is called paedomophosis and its different forms occur in around 10% of urodelans (Denoël et al. 2005). While in some groups (mudpuppies and olms, sirens, amphiumas) paedomorphosis is obligate and all individuals retain larval morphology, in other taxa the occurrence of paedomorphosis shows considerable plasticity. Within the genus *Ambystoma* both paedomorphic and biphasic species (i.e., which undergo metamorphosis) occur, and the metamorphosis of some paedomorphic forms (i.e., the well-known Mexican axolotl, *A. mexicanum*) can be induced by hormonal treatment. In contrast, facultative paedomorphosis may occur among normally biphasic species, like many Eurasian newts (e.g., Ceacero et al. 2010).

In urodelans, the main difference between larvae and adults is the breathing apparatus and related functions (i.e., feeding mechanism), but otherwise they move and live in a similar fashion. This can explain why paedomorphosis is never documented among Anurans: presumably the phenotypic divergence and the discrepancy in lifestyle between larvae and adults is more significant in frogs and toads. Facultative paedomorphosis creates phenotypic plasticity which can be explained by a complex cost-benefit framework (Denoël et al. 2005, Lejeune et al. 2018). However, obligate paedomorphosis can create evolutionary plasticity, as it potentially deliberates the ontogenetic pathways from the constraints that bind together the adult and larval phenotype in the species with biphasic development (Bonett and Blair 2017).

Parental Care

In salamanders, parental care diversity does not approach the level of frogs (Table 1). Male care occurs only in the more ancient families which use external fertilisation, while female care occurs only in internal fertilisers (Reinhard et al. 2013, Vági et al. 2022). Biparental care has not been reported from urodelans. They do not build elaborate nests, however, the females of some aquatic-breeding newt species may conceal the eggs. Lotic breeders can hide the eggs under stones and other underwater objects (Nussbaum 1985, 1987), while lentic breeding newts in Salamandridae deposit them separately to the leaves of aquatic vegetation and often wrap the leaf around the egg (Tóth et al. 2011). This extended egg-laying period only occurs among internal fertilisers, where fertilisation may be separated from egg-laying in space and time. Therefore, it created an opportunity for the elongation of the egg-laying process with only the participation of the female. The individual placing of the eggs not only reduces predation risk, but also decreases the potential competition between the siblings.

In the Caudata, the only widespread form of parental care is egg attendance, however, its presence can be variable. Males in externally fertilising species may care for egg clutches of multiple females at the same time, so paternal care coevolved with polygyny and nest site defence (Browne et al. 2014). In species with internal fertilisation terrestrial egg laying is associated with maternal care (Vági et al. 2022). In some species mothers attend the clutch after the hatching of the juveniles (Oneto et al. 2010), however, this should not last long as the hatchling start hunting for live prey on their own soon after hatching. Like in frogs, egg attendance can be an effective way to reduce the risks of predation, pathogens and desiccation (Croshaw and Scott 2005).

Viviparity (including ovoviviparity) likely evolved twice among salamanders, as there is a completely ovoviviparous and viviparous clade within Salamandridae, and a single species within the direct-developing lungless lineage which is presumably ovoviviparous. Ovoviviparous salamandrids give birth to larvae, while viviparous species to fully terrestrial juveniles. Like in some frog genera (e.g., *Pipa, Gastrotheca*), the presence of larval and direct development can be plastic within a species (*Salamandra algira,* García-París et al. 2003) and shows interspecific variance within the *Salamandra* genus. Interestingly, direct development appears to be basal in viviparous *Salamandra*. The embryos of the viviparous salamanders initially consume the unfertilised eggs in the uterus of the mother. Later they switch to feed on their developing siblings (adelphophagy or intrauterine cannibalism, in viviparous subspecies of *Salamandra salamandra*; Dopazo and Korenblum 2000) or to feeding on epithelial uterine cells (in *Salamandra atra* and *S. lanzai*; Guex and Chen 1986, Greven 1998).

In contrast to frogs, the evolutionary drivers of parental care in urodeles are little investigated. While fertilisation mode determines the care-providing sex, it seems likely that male and female parental care had different predictors. In males, territoriality and the fertilisation of multiple clutches may have been the driver of care in aquatic egg-laying sites, while internal fertilisation and female attendance likely opened the avenue for invading terrestrial niches (Vági et al. 2022).

Diversity of Reproductive Strategies in Caecilians

Caecilians (order: Gymnophiona) are the least-known of the three lissamphibian orders. This is mainly caused by the secretive (underground) life style and the fact that their distribution restricted to tropical and subtropical areas. While their anatomy and phylogenetic relations become better known in recent decades, behavioural observations on their life history are still scarce. Most observations came from captive specimens, and details of the reproduction of most of the species were never observed in the field. Despite these difficulties, several fundamental details of their reproductive biology, behaviour and evolution were published in recent years, and the main phylogenetic transitions in their reproductive biology can be traced back using comparative methods.

In all caecilian the fertilisation is internal and performed by a modified section of the caecum; the so-called phallodeum, an eversible part of the cloaca which is situated at the very end of the body (Wells 2007). The most ancient lineages of caecilians have aquatic larvae, however, the eggs are not placed directly into water, rather concealed in burrows and under logs and rocks next to water bodies, similarly to some salamander and many frog species (Wells 2007). Some caecilians return to aquatic life as adults. More developed lineages switched to direct development skipping the free-living larval phase, but there are also examples for a return to aquatic larvae from direct development (San Mauro et al. 2014). Viviparity evolved multiple times during caecilian evolution. The maternal attendance by coiling around the clutch is ubiquitous in caecilians (at least in species with known reproductive behaviour), and the mother often remains with the hatchlings or newborn juveniles for some time. In some caecilians the juveniles are precocial and ready for an independent life, in other species they are altricial and need to stay with the mother in the early their life. Caecilian embryos do not feed on trophic eggs, however, in some species they scrape off hypertrophic epithelium of the oviduct (uterus) using their specialised, vernal teeth (San Mauro et al. 2014). Another extraordinary way of feeding offspring is dermatophagy, when the juveniles scrape off hypertrophied layers of the mother's outer epidermis, using the same vernal dentition (Kupfer et al. 2006). According to phylogenetic reconstructions, dermatophagy evolved from uterophagy (Kupfer et al. 2016). By providing alternative pathways for offspring nourishment via uterotrophy and dermatotrophy the females may invest less yolk into the eggs (Kupfer et al. 2016).

Conclusions

Amphibians provide a unique opportunity to understand the evolution of breeding strategies, because many aspects of their reproductive behaviour show incredible diversity and sometimes plasticity. Moreover, the three extant amphibian orders found different solutions for the challenges of reproduction. General pattern shows that anurans, urodelans and caecilians all invented internal fertilisation, increased protection of the progeny and nutrition for the offspring, and many of these strategies paved their way towards a more terrestrial lifestyle.

As the reproductive behaviour of amphibians is still relatively unexplored compared to other groups, such as birds or mammals, there are still a lot of possible research directions. In a recent review, Schulte et al. (2020) recommended the use of novel techniques from individual tracking to genomic, transcriptomic and hormonal analyses to investigate unexplored connections of amphibian reproductive biology. These can help understanding connections which are largely unexplored to date, like proximate causes of intricate reproductive behaviour or the role of relatedness between individuals. These can substantially influence many aspects of complex reproduction, from communication between individuals to distribution in the habitat. However, along advanced and multidisciplinary studies basic natural history observations are still warranted, especially about salamanders and caecilians (Schulte et al. 2020).

It is also important to understand how these complex adaptations influence the persistence of threatened amphibian species. It is possible that complex reproductive adaptations secure higher survival rates for the eggs and the offspring, which would make these species less vulnerable to anthropogenic threats. On the other hand, it is also plausible that species with intricate reproduction evolved under stable environmental conditions and respond badly to perturbations—at least, the high number of threatened species with advanced reproductive strategies is alarming. A better understanding of the ecological and behavioural drivers of amphibian reproductive diversity is strongly warranted to enhance the conservation of this vertebrate class with exceptional evolutionary and ecological importance.

Acknowledgements

BV and TS were supported by the National Research, Development and Innovation Office of Hungary (PD 132819 to BV; ÉLVONAL KKP-126949, K-116310 to TS). TS was also funded by The Royal Society (Wolfson Merit Award WM170050, APEX APX\R1\191045).

References

Amphibiaweb. 2021. University of California, Berkeley, USA. Available at https://www.amphibiaweb. org.

Arak, A. 1983. Male-male competition and mate choice in anuran amphibians. pp. 181–205. *In*: Bateson, P. [ed.]. Mate choice. Cambridge University Press, Cambridge, UK.

Beck, C.W. 1998. Mode of fertilization and parental care in anurans. Anim. Behav. 55: 439–449.

Bickford, D.P. 2004. Differential parental care behaviors of arboreal and terrestrial microhylid frogs from Papua New Guinea. Behav. Ecol. Sociobiol. 55: 402–409.

Bonett, R.M. and A.M. Blair. 2017. Evidence of complex life cycle constraints on salamander body form diversification. Proc. Natl. Acad. Sci. USA 114: 9936–9941.

Boulenger, G.A. 1886. Remarks in connection with the preceding note. The Annals and Magazine of Natural History 17: 463–464.

Bourne G.R. 1992. Lekking behavior in the neotropical frog *Ololygon rubra*. Behav. Ecol. Sociobiol. 31: 173–180.

Brown, J.L., V. Morales and K. Summers. 2010. A key ecological trait drove the evolution of biparental care and monogamy in an amphibian. Am. Nat. 175: 436–446.

Browne, R.K., H. Li, Z. Wang, S. Okada, P. Hime, A. MacMillan et al. 2014. The giant salamanders (Cryptobranchidae): Part B. Biogeography, ecology and reproduction. Amph. Rept. Cons. 5: 30–50.

Byrne, P.G. and J.D. Roberts. 1999. Simultaneous mating with multiple males reduces fertilization success in the myobatrachid frog *Crinia georgiana*. Proc. R. Soc. Lond. B 266: 717–721.

Byrne, P.G., J.D. Roberts and L.W. Simmons. 2002. Sperm competition selects for increased testes mass in Australian frogs. J. Evol. Biol. 15: 347–355.

Byrne, P.G. and M.J. Whiting. 2011. Effect of simultaneous polyandry on offspring fitness in an African tree frog. Behav. Ecol. 22: 385–391.

Ceacero, F., D. Donaire-Barroso, E. García-Muñoz, J.F. Beltrán and M. Tejedo. 2010. On the occurrence of facultative paedomorphosis in three newt species in Southern Iberian Peninsula. Amphibia-Reptilia 31: 571–575.

Chuang, M.-F., W.-H. Lee, J.-S. Sun, C.-H. You, Y.-C. Kam and S. Poo. 2017. Predation risk and breeding site value determine behavior and indirectly affect survivorship of their offspring. Behav. Ecol. Sociobiol. 71: 122.

Croshaw, A. and D.E. Scott. 2005. Experimental evidence that nest attendance benefits female marbled salamanders (*Ambystoma opacum*) by reducing egg mortality. Am. Midl. Nat. 154: 398–411.

Crump, M.L. 1995. Parental care. pp. 518–567. *In*: Heathvole, H. and B.K. Sullivan [eds.]. Amphibian Biology, volume 2: Social Behaviour. Surrey Beatty & Sons, Chipping Norton, UK.

Crump, M.L. 1996. Parental care among the Amphibia. Adv. Stud. Behav. 25: 109–144.

Davies, N.B. and T.R. Halliday. 1978. Deep croaks and fighting assessment in the toad *Bufo bufo*. Nature 274: 683–685.

Delia, J., L. Bravo-Valencia and K.M. Warkentin. 2017. Patterns of parental care in neotropical glassfrogs: fieldwork alters hypotheses of sex-role evolution. J. Evol. Biol. 30: 898–914.

Denoël, M., P. Joly and H.H. Whiteman. 2005. Evolutionary ecology of facultative paedomorphosis in newts and salamanders. Biol. Rev. 80: 663–671.

Dopazo, H.J. and M. Korenblum. 2000. Viviparity in *Salamandra salamandra* (Amphibia: Salamandridae): Adaptation or exaptation? Herpetologica 56: 144–152.

D'Orgeix, C.A. 1996. Multiple paternity and the breeding biology of the red-eyed treefrog, *Agalychnis callidryas*. PhD thesis, Virginia Tech, Blacksburg.

Duellman, W.E. and L. Trueb. 1986. Biology of amphibians. McGraw-Hill, New York, USA.

Friedl, T.P.W. and G.M. Klump. 2005. Sexual selection in the lek-breeding European treefrog: body size, chorus attendance, random mating and good genes. Anim. Behav. 70: 1141–1154.

Furness, A.I. and I. Capellini. 2019. The evolution of parental care diversity in amphibians. Nat. Commun. 10: 4709.

Furness, A.I., C. Venditti and I. Capellini. 2022. Terrestrial reproduction and parental care drive rapid evolution in the trade-off between offspring size and numbers across amphibians. PLoS Biology 20: e3001495.

García-París, M., M. Alcobendas, D. Buckley and D.B. Wake. 2003. Dispersal of viviparity across contact zones in Iberian populations of fire salamanders (*Salamandra*) inferred from discordance of genetic and morphological traits. Evolution 57: 129–143.

Gomez-Mestre, I., R.A. Pyron and J.J. Wiens. 2012. Phylogenetic analyses reveal unexpected patterns in the evolution of reproductive modes in frogs. Evolution 66: 3687–3700.

Goycoechea, O., O. Garrido and B. Jorquera. 1986. Evidence for a trophic paternal-larval relationship in the frog *Rhinoderma darwinii*. J. Herpetol. 20: 168–178.

Greven, H. 1998. Survey of the oviduct of salamandrids with special reference to the viviparous species. J. Exp. Zool. 282: 507–525.

Guex, G.-D. and P.S. Chen. 1986. Epitheliophagy: Intrauterine cell nourishment in the viviparous alpine salamander, *Salamandra atra* (Laur.). Experientia 42: 1205–1218.

Gururaja, K.V., K.P. Dinesh, H. Priti and G. Ravikanth. 2014. Mud-packing frog: a novel breeding behaviour and parental care in a stream dwelling new species of *Nyctibatrachus*. Zootaxa 3796: 33–61.

Haddad, C.F.B. and C.P.A. Prado. 2005. Reproductive modes in frogs and their unexpected diversity in the Atlantic Forest of Brazil. BioScience 55: 208–217.

Hansen, R.W. 2012. About our cover: *Ecnomiohyla rabborum*. Herpetol. Rev. 42: 3.

Hettyey, A., B. Vági, G. Hévizi and J. Török. 2009. Changes in sperm stores, ejaculate size, fertilization success, and sexual motivation over repeated matings in the common toad, *Bufo bufo* (Anura: Bufonidae). Biol. J. Linn. Soc. 96: 361–371.

Heyer, H.R. 1969. The adaptive ecology of the genus *Leptodactylus* (Amphibia, Leptodactylidae). Evolution 23: 421–28.

Houck, L.D. and S.J. Arnold. 2003. Courtship and mating behaviour. *In*: Jamieson, B.G.M. [ed.]. Reproductive Biology and Phylogeny of Urodela. CRC Press, Boca Raton, USA.

Iskandar, D.T., B.J. Evans and A.J. McGuire. 2014. A novel reproductive mode in frogs: a new species of fanged frog with internal fertilization and birth of tadpoles. PLoS ONE 9: e115884.

Jetz, W. and R.A. Pyron. 2018. The interplay of past diversification and evolutionary isolation with present imperilment across the amphibian tree of life. Nat. Ecol. Evol. 2: 850–858

Köhler, F. and R. Günther. 2008. The radiation of microhylid frogs (Amphibia: Anura) on New Guinea: A mitochondrial phylogeny reveals parallel evolution of morphological and life history traits and disproves the current morphology-based classifications. Mol. Phylogenet. Evol. 47: 353–365.

Kupfer, A., H. Müller, A.M. Antoniazzi, C. Jared, H. Greven, R.A. Nussbaum et al. 2006. Parental investment by skin feeding in a caecilian amphibian. Nature 440: 926–929.

Kupfer, A., E. Maxwell, S. Reinhard and S. Kuehnel. 2016. The evolution of parental investment in caecilian amphibians: a comparative approach. Biol. J. Linn. Soc. 119: 4–14.

Kusano, T., M. Toda and K. Fukuyama. 1991. Testes size and breeding systems in Japanese anurans with special reference to large testes in the treefrog, *Rhacophorus arboreus*. Behav. Ecol. Sociobiol. 29: 27–31.

Kusrini, M.D., J.J.L. Rowley, L.R. Khairunnisa, G.M. Shea and R. Altig. 2015. The reproductive biology and larvae of the first tadpole-bearing frog, *Limnonectes larvaepartus*. PLoS One 10: e116154.

Lejeune, B., N. Sturaro, G. Lepoint and M. Denoël. 2018. Facultative paedomorphosis as a mechanism promoting intraspecific niche differentiation. Oikos 127: 427–439.

Liedtke, H.C., H. Müller, J. Hafner, J. Penner, D.J. Gower, T. Mazuch et al. 2017. Terrestrial reproduction as an adaptation to steep terrain in African toads. Proc. R. Soc. Lond. B 284: 20162598.

Lion, M.B., G.G. Mazzochini, A.A. Garda, M.T. Lee, D. Bickford, G.C. Costa et al. 2019. Global patterns of terrestriality in amphibian reproduction. Glob. Ecol. Biogeogr. 28: 744–756.

Lodé, T. and D. Lesbarrères. 2004. Multiple paternity in *Rana dalmatina*, a monogamous territorial breeding anuran. Naturwissenschaften 91: 44–47.

Martins, M., J.P. Pombal and C.F.B. Haddad. 1998. Escalated aggressive behaviour and facultative parental care in the nest building gladiator frog, *Hyla faber*. Amphibia-Reptilia 19: 65–73.

McKeon, C.S. and K. Summers. 2013. Predator driven reproductive behavior in a tropical frog. Evol. Ecol. 27: 725–737.

Meegaskumbura, M., G. Senevirathne, S.D. Biju, S. Garg, S. Meegaskumbura, R. Pethijagoda et al. 2015. Patterns of reproductive mode evolution in Old World tree frogs (Anura, Rhacophoridae). Zool. Scr. 44: 509–522.

Nunes-de-Almeida, C.H.L., C.F.B. Haddad and L.F. Toledo 2021. A revised classification of amphibian reproductive modes. Salamandra 57: 413–427.

Nussbaum, R.A. 1985. The evolution of parental care in salamanders. University of Michigan Press, Ann Arbor, USA.

Nussbaum, R.A. 1987. Parental care and egg size in salamanders: An examination of the safe harbor hypothesis. Res. Popul. Ecol. 29: 27–44.

Oneto, F., D. Ottonello, M.V. Pastorino and S. Salvidio, S. 2010. Posthatching parental care in salamanders revealed by infrared video surveillance. J. Herpetol. 44: 649–653.

Pereira, E.B., P.N. Pinto-Ledesma, C. Gomez de Freitas, F. Villalobos, R. Garcia Colevatti and N. Medeiros Marciel. 2017. Evolution of anuran foam nest: trait conservatism and lineage diversification. Biol. J. Linn. Soc. 122: 814–823.

Poelman, E.H. and M. Dicke. 2007. Offering offspring as food to cannibals: oviposition strategies of Amazonian poison frogs (*Dendrobates ventrimaculatus*). Evol. Ecol. 21: 215–227.

Poelman, E.H., R.P.A. van Wijngaarden and C.E. Raaijmakers. 2013. Amazon poison frogs (*Ranitomeya amazonica*) use different phytothelm characteristics to determine their suitability for egg and tadpole deposition. Evol. Ecol. 27: 661–647.

Poo, S. and D.P. Bickford. 2013. The adaptive significance of egg attendance in a South-East Asian tree frog. Ethology 119: 671–679.

Rausch, A.M., M. Sztatecsny, R. Jehle, E. Ringler and W. Hödl. 2014. Male body size and parental relatedness but not nuptial colouration influence paternity success during scramble competition in *Rana arvalis*. Behaviour 151: 1869–1884.

Reinhard, S., S. Voitel and A. Kupfer. 2013. External fertilisation and paternal care in the paedomorphic salamander *Siren intermedia* Barnes, 1826. Zool. Anz. 253: 1–5.

Ringler, E., A. Pašukonis, W. Tecumseh Fitch, L. Huber, L., W. Hödl and M. Ringler. 2015. Flexible compensation of uniparental care: female poison frogs take over when males disappear. Behav. Ecol. 26: 1219–1225.

Roberts, J.D., R.J. Standish, P.G. Byrne and P. Doughty. 1999. Synchronous polyandry and multiple paternity in the frog *Crinia georgiana* (Anura: Myobatrachidae). Anim. Behav. 57: 721–726.

Rojas, B. 2014. Strange parental decisions: fathers of the dyeing poison frog deposit their tadpoles in pools occupied by large cannibals. Behav. Ecol. Sociobiol. 68: 551–559.

Rollinson, N. and L. Rowe. 2018. Oxygen limitation at the larval stage and the evolution of maternal investment per offspring in aquatic environments. Am. Nat. 191: 604–619.

Ryan, M.J. 1980. Female mate choice in a neotropical frog. Science 209: 523–525.

Salthe, S.N. 1969. Reproductive modes and the number and sizes of ova in Urodeles. Am. Midl. Nat. 81: 467–490.

Salthe, S.N. and W.E. Duellman 1973. Quantitative constraints associated with reproductive mode in anurans. pp. 229–249. *In:* Vial, J.L. [ed.]. Evolutionary Biology of the Anurans, University of Missouri Press, Columbia, USA.

San Mauro, D., D.J. Gower, H. Müller, S.P. Loader, R. Zardoya, R.A. Nussbaum et al. 2014. Life-history evolution and mitogenomic phylogeny of caecilian amphibians. Mol. Phylogenet. Evol. 73: 177–184.

Saporito, R.A., M.W. Russell, C.L. Richards-Zavacki and M.B. Dugas. 2019. Experimental evidence for maternal provisioning of alkaloid defences in a dendrobatid frog. Toxicon 161: 40–43.

Schäfer, M., S.J. Tsekané, F.A.M. Tchassem, S. Drakulić, M. Kameni, N.L. Gonwouo et al. 2019. Goliath frogs build nests for spawning – the reason for their gigantism? J. Nat. Hist. 53: 1263–1276.

Schulte, L.M., E. Ringler, B. Rojas and J.L. Stynoski. 2020. Developments in amphibian parental care research: history, present advances and future perspectives. Herpetol. Monogr. 34: 71–97.

Shine, R. 1978. Propagule size and parental care: the "safe harbour" hypothesis. J. Theor. Biol. 75: 417–424.

Simon, M.P. 1983. The ecology of parental care in a terrestrial breeding frog from New Guinea. Behav. Ecol. Sociobiol. 14: 61–67.

Summers K., R. Symula, M. Clough and T. Cronin. 1999. Visual mate choice in poison frogs. Proc. R. Soc. Lond. B 266: 2141–2145.

Summers, K., C.S. McKeon and H. Heying. 2006. The evolution of parental care and egg size: a comparative analysis in frogs. Proc. R. Soc. Lond. B 273: 687–692.

Sztatecsny, M., R. Jehle, T. Burke and W. Hödl. 2006. Female polyandry under male harassment: the case of the common toad (*Bufo bufo*). J. Zool. 270: 517–522.

Tóth, Z., H. Hoi and A. Hettyey. 2011. Intraspecific variation in the egg-wrapping behaviour of female smooth newts, Lissotriton vulgaris. Amphibia-Reptilia 32: 77–82.

Tumulty, J., V. Morales and K. Summers. 2014. The biparental care hypothesis for the evolution of monogamy: experimental evidence in an amphibian. Behav. Ecol. 25: 262–270.

Vági, B. and A. Hettyey. 2016. Intraspecific and interspecific competition for mates: *Rana temporaria* males are effective satyrs of *Rana dalmatina* females. Behav. Ecol. Sociobiol. 70: 1477–1484.

Vági, B., D. Marsh, G. Katona, Z. Végvári, R.P. Freckleton, A. Liker and T. Székely. 2022. The evolution of parental care in salamanders. Sci. Rep. 12: 16655.

Vági, B., Z. Végvári, A. Liker, R.P. Freckleton and T. Székely. 2019. Terrestriality and the evolution of parental care in frogs. Proc. R. Soc. Lond. B 286: 20182737.

Vági, B., Z. Végvári, A. Liker, R.P. Freckleton and T. Székely. 2020. Climate and mating systems as drivers of global diversity of parental care in frogs. Glob. Ecol. Biogeogr. 29: 1373–1386.

Verrell, P.A. and N. McCabe. 1986. Mating balls in the common toad, *Bufo bufo*. British Herpetol. Soc. Bull. 16: 28–29.

Wells, K.D. 2007. The Ecology and Behaviour of Amphibians. University of Chicago Press, Chicago, USA and London, UK.

CHAPTER 10

Chemical Communication and Deterrence in Amphibians

Maider Iglesias-Carrasco[1],* and *Bob B.M. Wong*[2]

INTRODUCTION

Chemical communication is omnipresent in the life of amphibians. Considered to be the most widespread and most ancient form of communication, some of the earliest identified chemicals involved in communication in amphibians are believed to have originated some 300 million years ago. Indeed, in both aquatic and terrestrial environments, chemical signals and cues have played a central role in the success of this very diverse vertebrate group – from mediating interactions between conspecifics to predator detection and deterrence.

This chapter synthesises the importance of chemical communication in amphibians, the functioning of the olfactory system, and how the environments in which amphibians inhabit have shaped the types of molecules used. The chapter also provides an extensive review of the social interactions that are mediated by chemical communication, as well as the myriad of different contexts in which amphibians use chemical compounds to interact with other species, and how chemical communication is integrated with other sensory channels. Lastly, the chapter explains how different human activities can potentially impinge on chemical communication, how its disruption can have detrimental conservation and evolutionary consequences, and how a better understanding of chemical communication in amphibians can be harnessed to address key challenges in a changed and changing world.

[1] Doñana Biological Station – CSIC, Sevilla, Spain.
[2] School of Biological Sciences, Monash University, Australia.
* Corresponding author: miglesias15@gmail.com

What Do Amphibians Use Chemical Communication For?

Sources and Types of Chemical Signals and Their Receptive Channels

Chemical communication plays a role in almost every facet of an amphibian's life—from facilitating social interactions to helping individuals to detect food, predators, as well as suitable habitats, shelter and oviposition sites. Amphibians are unique among vertebrates because most of them have a life style that covers both terrestrial and aquatic realms. Indeed, in taxa with life cycles that bridge these very different environments, such as many anurans, complex olfactory systems need to be completely reconfigured during metamorphosis (Belanger and Corkum 2009, Weiss et al. 2021). Since water and air differ in their density and viscosity, diffusion of odours through each medium poses distinct challenges that, in turn, determines the types of molecules used in each of the two environments. For example, in aquatic environments, amphibians typically rely on hydrophilic molecules—such as small organic molecules or large proteins—that are easily dissolved in water. By contrast, in terrestrial habitats, olfaction is based on either highly volatile molecules that are transmitted through the air or non-volatile molecules that need to be deposited on surfaces, such as soil, rocks or even directly onto other individuals (reviewed in Weiss et al. 2021).

Differences in the environment and type of molecules used during chemical communication often require different anatomies of the olfactory system between larvae and adults, and also between different amphibian groups. In general, however, the olfactory system in amphibians is quite similar to other tetrapods (Moreno et al. 2008), as this sensory channel is highly phylogenetically conserved. In this regard, chemical compounds are usually detected by the sensory neurons located in the nasal cavity (Wyatt 2017), although, in some salamanders (e.g., many of the plethodontids), rather than entering the nasal cavity, some of the chemical signals used during mating are delivered through the skin, which seems to be the ancestral condition in the group (Woodley 2010). Caecilians also have a paired tentacle in the head, a specialised organ connected to the olfactory system via the tentacular duct, that is likely to capture chemical cues present in the environment (Schmidt and Wake 1990). The anatomy and functioning of the olfactory system in amphibians have already been detailed in several previous reviews (Woodley 2010, Weiss et al. 2021, Woodley and Staub 2021), but briefly, the amphibian olfactory system comprises a main olfactory system (MOS) and an accessory olfactory system composed of the vomeronasal organ (VNO, see Fig. 1). These two components of the olfactory system appear to serve a complementary purpose in the detection of a range of different chemical compounds (Baum and Cherry 2015). Nevertheless, it is still not well understood which receptors are involved or how these receptors bind to different chemical molecules to process the chemical information received (see Woodley and Staub 2021 for a review in this topic).

In this chapter we use the term *signal* to refer to the chemical compounds released intentionally by an individual to transfer certain information that alters the behaviour of the receiver in an advantageous way for the sender. These signals are typically used in the case of communication between individuals of the same species, and are often referred to as pheromones (Wyatt 2017). In contrast, we use the term *cue*

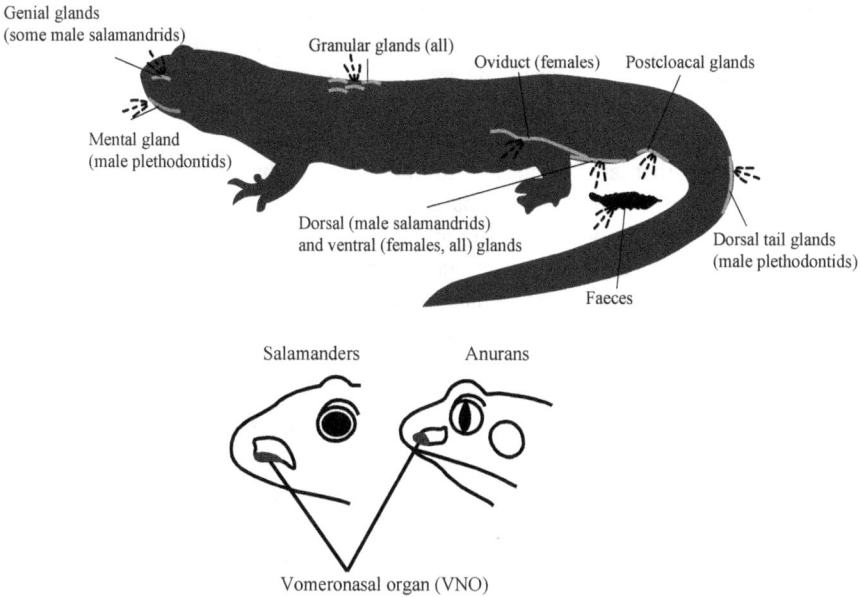

Figure 1. Chemical compounds in amphibians are produced and secreted from glands, the cloaca, the oviduct (in females), or may even be present in faeces. While some glands are specific to certain taxonomic groups, others are more widespread. The main sensory channel is the nasal cavity where the olfactory system is situated. This system is comprised of the main olfactory system and the accessory vomeronasal organ. In some salamanders, chemical compounds are also transferred transdermally.

to refer to a chemical compound that is not released intentionally by the sender, but that it is used by the receiver as a guide to modify their behaviour to their advantage.

In amphibians, chemical compounds are typically released from the skin, cloaca, oviduct and/or via the faeces. The skin of amphibians is composed of multiple glands. While some of these glands are scattered throughout the body, others are more localised with specific functions that, in some cases, are limited to specific taxonomic groups. For example, the shape and the distribution of granular or poison glands (Fig. 1) along the body differs between species: they can be clustered in specific regions of the body vulnerable to predator attack (e.g., head), or sparsely distributed across the body (Toledo and Jared 1995). Specialised skin glands with specific functions can also be important during interactions between conspecifics. For example, in some salamanders, territorial demarcation typically involves the postcloacal gland, the contents of which are released when individuals press the ventral part of their tail onto the substrate (Largen and Woodley 2008). Perhaps the best studied glands involved in chemical communication are those that play a critical role in breeding. Specifically, the development of glands involved in mating is often associated with the breeding season and are mediated by sexual hormones, such as androgens and prolactin (Woodley 1994). For instance, male plethodontid salamanders develop a mental gland, a specialised gland comprising a cluster of mucous glands in the chin, that males use to transfer pheromones to females during courtship to increase their receptivity (Houck 2009). In addition, some plethodontids also develop specialised dorsal tail glands with a similar purpose to the mental gland. The transfer of these courtship pheromones is likely to be critical for male reproductive success. In anurans, males of some species produce chemical compounds through specialised femoral glands that develop during the breeding season on the ventral flanks (Vences et al. 2007).

In addition to skin glands, some amphibians have cloacal glands comprising exocrine glands that are connected to the cloacal chamber (Sever 2003). In certain amphibian groups, such glands are also associated with the production of chemical compounds important during mating and courtship. For example, in some salamandrids, such as aquatic breeding newts, males develop enlarged cloaca during the breeding season, which releases pheromones directly towards females during the tail fanning dance performed by males as part of their courtship repertoire (see Fig. 2 for representation of the tail fanning behaviour). Cloacal glands seem to be important for female chemical signal production too, as suggested by the fact that female ventral glands only produce secretions during the breeding season (Sever 1988). In females, the oviduct might also play a role in the secretion of important chemical signals. For instance, males of red-bellied newts are known to be attracted to water that has previously contained females, but not when the oviduct of the females have been experimentally removed (Toyoda et al. 1994). Last, faecal pellets in some species may also contain chemical compounds that provide important information about the identity of individuals or the condition of potential mates (Jaeger and Wise 1991, Lee and Waldman 2002).

Recent evidence suggests that symbiotic skin bacteria might also contribute to the production, not only of chemicals with antibacterial and antifungal properties

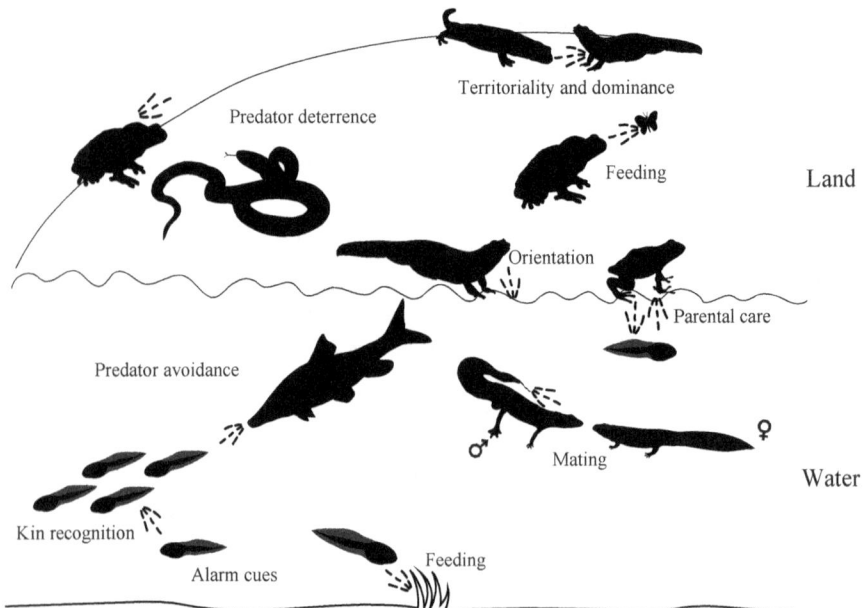

Figure 2. Amphibians use chemical signals to communicate with each other and to learn about their surroundings. Both adults and larvae make use of chemical communication in a myriad of contexts, including in social interactions with conspecifics, in the detection of predators, and to find food. Chemical compounds are important both in terrestrial and aquatic environments. Modified from Weiss et al. (2021).

important for defence against emergent diseases (see Chapters 4 and 5), but also chemical signals involved in intraspecific communication. For example, in the tree frog *Boana prasina*, the skin microbiota produces a sex-specific volatile compound that is likely to be involved in social signalling (Brunetti et al. 2019). However, understanding the ecological role of host-microbiome interactions during chemical communication requires further research.

Different sources and glands secrete different kinds of chemical compounds with specific functions (see Woodley 2014, Woodley and Staub 2021 for more comprehensive reviews of the topic). Briefly, while molecules secreted by poison glands include peptides, amines, alkaloids and steroids (Perry 2014, Sousa-Filho et al. 2016, Lüddecke et al. 2018), most pheromones involved in mating are peptides and proteins. For instance, the group of proteins belonging to the sodefrin precursor-like factor (SPF) family has been identified in several groups of salamanders and some anurans, and is known to influence female behaviour towards males, as well as female interest in mating (VanBocxlaer et al. 2015, Bossuyt et al. 2019). Within the SPF family, sodefrin was the first amphibian pheromone identified, which is secreted by the cloacal gland of the Japanese firebelly newt *Cynops pyrrhogaster* (Kikuyama et al. 1995). However, SPFs are not the only chemical compounds involved in reproduction. Indeed, in some cases, SPFs are just one of the complex cocktail of pheromones that are released during mating. For example, in some newts, a protein called persuasin is released by the male's cloacal gland during courtship which, in combination with SPF proteins, increases female receptivity (Maex et al. 2018). Although some other pheromones involved in reproduction have also been identified in amphibians, most of these come from only a handful of species. However, it is clear that different types of chemical compounds are involved in the diverse social behaviour of amphibians. Therefore, to gain a better understanding of the evolution of chemical communication in amphibians, we need to learn more about the nature of the compounds involved, whether they are specific to certain groups, and whether the same compounds can have different functions.

Intraspecific Chemical Communication: The Complex Social Life of Amphibians

Chemical signals play a vital role in the varied and complex social behaviours of amphibians, including sexual interactions, parental care, territoriality, aggression, and dominance (See Fig. 2).

Chemical Communication in Amphibian Larvae

Chemical communication mediates a range of ecologically important behaviours in the early life stage of amphibians that are critical to survival. This is particularly true in the context of grouping or social aggregations, which can confer fitness benefits by diluting the risk of predation and/or by improving foraging efficiency. The tadpoles of many anuran amphibians, for example, are known to exhibit schooling behaviour, with evidence that tadpoles preferentially school with conspecifics, and, more specifically, with siblings (Blaustein and Waldman 1992). Experiments suggest that waterborne odour cues are important, especially in the identification of kin (Waldman 1985, Eluvathingal et al. 2009), which is likely to have a genetic basis that is mediated by self-referent major histocompatibility complex (MHC) matching (Villinger and Waldman 2008). Recognition of kin is likely to also provide advantages other than those associated with group formation, including heightened inclusive fitness by facilitating the survival of siblings. For example, in polymorphic species comprising of both carnivorous and herbivorous individuals, the former are known to preferentially prey upon non-siblings (Pfennig et al. 1993).

Chemical signals are important in facilitating the interactions between amphibian larvae and their parents. An interesting use of chemical signals in the context of parental care is evidenced in the egg-feeding behaviour of some anurans. A good example of this comes from dart frogs and several other anuran species that breed in the small water-filled cavities of plants (i.e., phytotelmata). Although breeding in such cavities reduces the risk of predation on offspring, the microhabitats themselves are often quite depauperate when it comes to food availability. When a female approaches the phytotelmata where her tadpoles live, the offspring aggregate around her and induce the female to lay non-fertilised eggs to feed them (Kam et al. 1996). In an experiment in the Taiwanese tree-frog *Kurixalus eiffingeri*, researchers found that tadpoles responded to female odour, rather than visual cues (Kam and Yang 2002), suggesting that chemical detection of the female triggers the egg-begging behaviour of tadpoles.

Chemical Communication in Adults

Chemically mediated interactions are also widespread in adult amphibians. As in the case of larvae, some adult amphibians also use chemical cues and signals to aggregate. For example, some caecilian species form aggregations in shelters that are marked with chemicals to attract conspecifics (Warbeck et al. 1996). Aggregations can be particularly important in stressful environments where solitary living can be challenging. This is the case in Luschan's salamanders *Lyciasalamanda luschani* (formerly *Martensiella* genus), where social information, encoded in chemical signals deposited by conspecifics on the ground, is critical in allowing others to locate and identify adequate shelter sites (Gautier et al. 2006), which is essential to survival by reducing the risk of desiccation.

Within aggregations, chemical communication can also play a key role in influencing social interactions, including the establishment of dominance rankings and hierarchies. In extreme cases, proximity with conspecifics can lead to competition and monopolization of resources, such as food or mates, via chemically-mediated territorial and dominance behaviours. In this context, scent markings used to demarcate a territory can serve multiple purposes, including the repelling of intruders, or the identification of neighbours (i.e., dear enemy effect) and competitors. This might be especially important during male-male interactions. Specifically, in many species, males compete vehemently with one another for access to females, and chemical signals can be important in mediating the extent and nature of the interactions that can take place among competing rivals. The use of chemical signals during male-male interactions has been shown in laboratory experiments where, in some cases, scents released by reproductively active males have been found to repel competitors

(Park and Propper 2001), or result in the escalation of aggressive behaviours and courtship investment by the signal receiver (King et al. 2005, Aragón 2009).

Demarcation of territories with chemical signals can also be used to attract potential mates during the breeding season, with evidence that amphibians can use such chemical signals to determine the sex of conspecifics, to locate prospective suitors, and to determine the reproductive status of individuals (Jaeger 1981, Marco et al. 1998, Dantzer and Jaeger 2007). The use of chemical signals for recognition of potential mates has been well studied in salamanders and, to a lesser extent, caecilians (Warbeck and Parzefall 2001). For example, plethodontid salamanders are known to locate and recognise potential mates via the non-volatile pheromones present on the ground by engaging in nose tapping behaviour (a type of chemo-investigation). While anurans are perhaps best known for the use of acoustic signals in mate attraction, we now know that chemical signals can also play a crucial role. Specifically, while frogs and toads typically use vocalisations to locate mates from a distance, it has been suggested that chemical signals might also be important, especially when mates are nearby (see, for example, Byrne and Keogh 2007).

In the context of reproduction, chemical signals can also be important in courtship, mate choice and successful insemination. The use of chemicals during reproduction is perhaps most important in salamanders because they rarely vocalise, are mainly nocturnal, and show complex courtship behaviours. In plethodontids with terrestrial copulation, for example, during courtship, the male rubs his mental gland on the female skin to transfer a complex mix of pheromones. These courtship pheromones alter female behaviour by increasing her receptivity and, if successful in his efforts, induces the typical tail-straddling behaviour of plethodontids, which involves the female straddling the tail of the male to walk in tandem. In salamanders that copulate in water, such as newts, males use a tail fanning dance (Fig. 2) to waft courtship pheromones towards the female. In such instances, transfer of chemical signals during courtship is critical for female mate choice since chemical signals produced by the male can reveal important information to choosy females about the quality of the signaller as a potential mate (Gosling and Roberts 2001, Johansson and Jones 2007, Chouinard 2012). Similarly, males can also use female chemical signals to gain valuable information about the female's reproductive status and condition (Marco et al. 1998). It is therefore likely that the chemical signals exchanged during these elaborate and ritualised courtship displays—and the corresponding quality assessment of potential mates—will determine whether or not insemination occurs. Importantly, chemical signals even play a critical role in the synchronization of sperm transfer and insemination. For instance, during the tandem walk of male and female plethodontids, the former releases pheromones through the skin of the female to coordinate the deposition of the spermatophore (a sperm cap with a gelatinous base) by the male, and its collection and internal insemination by the female. In salamandrids, if a male is successful in his courtship efforts, the breeding ritual ends with the female following the male and touching his tail with her snout to induce the male to deposit his spermatophore on the ground so that it can be lodged into her cloaca where internal fertilization takes place.

After mating, amphibian adults use chemical communication to orientate towards oviposition or nesting sites to lay their eggs or, in the case of live-bearing species, to deposit their larvae. Some amphibian species are philopatric and often return to their home pond for reproduction (e.g., Smith and Green 2005, Gamble et al. 2007). Although orientation involves many sensory channels, evidence suggests that individuals with impaired olfaction have difficulties finding their natal pond and individuals are also known to prefer the odour of their home ponds over other ponds (Forester and Wisnieski 1991, Ishii et al. 1995, Sinsch 2006). Such recognition of home ponds is likely to be a consequence of imprinting during larval stages (Ogurtsov and Bastakov 2001), suggesting that the environment experienced during aquatic life stages might play a vital role in the behaviour of adults. Another intriguing example of the use of chemical signals to identify adequate oviposition sites comes from the American poison frogs. Some poison frog species transport their individual tadpoles from egg deposition sites to phytotelmata (Summers and McKeon 2004, Wells 2007). However, some phytotelmata are better than others. This is because the phytotelmata may already be occupied by a predatory species or by older tadpoles that could benefit from simultaneously feeding on younger siblings and eliminating competitors (Brown et al. 2009). As a consequence, before depositing new tadpoles, some poison frog parents are known to actively avoid already occupied cavities (Poelman and Dicke 2007, Brown et al. 2008, Stynoski 2009) and do so by examining the water for chemical cues of potential predators (Schulte et al. 2011). In salamanders, chemical signals may also play a key role in helping parents to locate their nests, which obviously can have a direct bearing on egg survival (Forester 1986).

Interspecific Chemical Communication: Foraging, Predator Avoidance and Deterrence

Chemical cues and signals also mediate interactions between species, especially in the context of foraging, predation risk assessment, and predator deterrence.

Amphibians, both as larvae and as adults, use chemical cues to detect food. Most adult amphibians are predators and feed on insects or other vertebrates, so visual motion detection has always been thought to be the main sensory channel used by individuals to locate their prey. However, olfaction has been shown to be important in modulating visual signals (Michaels et al. 2018). For example, studies have found that some terrestrial species, such as the leopard frog *Rana pipiens*, use olfactory cues to guide their approach to prey (Shinn and Dole 2018). The detection of prey via chemical cues is likely to be even more important for fossorial species, such as caecilians, or species that inhabit murky waters, including many tadpoles that often live in habitats with low visibility (Himstedt and Simon 1995, South et al. 2020). For example, in the bronze frog *Rana temporalis*, tadpoles responded to chemical, but not visual, food cues (Veeranagoudar et al. 2004), suggesting that at least in some species, foraging is based predominantly on chemical cues.

Detection of chemical cues released by other species is not only important for foraging, but can also play a crucial role in avoiding predators. Amphibians are preyed upon by a myriad of predators, including snakes, fish, mammals, birds, insect larvae, and even other amphibians. Being able to effectively detect and respond to chemical cues released by potential predators is likely to be critical to survival. The first reaction of animals to the presence of a predator is avoidance of a potential attack, and amphibian larvae, adults, and even embryos, use chemical cues to avoid predators. For example, adult amphibians are known to rely on olfaction to steer clear of ponds containing predators, such as snakes or fish (Gonzalo et al. 2006, Hamer et al. 2011). As already discussed, the avoidance of ponds with predators can also enable amphibians to choose a more suitable and safer habitat for their offspring (reviewed in Buxton and Sperry 2017). We have already seen that poison frogs can smell and avoid potential cannibalistic tadpoles in phytotelmata, but similar habitat selection based on chemical cues is also common in species that breed and oviposit in large ponds. For instance, experiments have shown that adult amphibians preferentially avoid ovipositing in water taken from ponds containing the odour cues of predatory fish (e.g., Binckley and Resetarits 2003, Rieger et al. 2004, Kloskowski 2020). In addition, embryos of some amphibian species can respond to chemical cues released by predators or injured conspecifics to modulate the timing of their hatching to reduce their risk of predation (Chivers et al. 2001, Touchon et al. 2006).

However, even despite the best efforts of parents to choose safe habitat for offspring development, the larvae of many amphibian species may still encounter a range of predators during development. Not surprisingly perhaps, like adults, amphibian larvae also use chemical cues released by predators to assess potential risk and to adjust their behaviour accordingly to reduce the chances of being eaten. Some of the best examples of this come from studies of anuran tadpoles. In general, tadpoles react to predators by reducing their activity and increasing refuge-seeking behaviour. Evidence suggests that such behavioural shifts appear to be linked to the detection of predator-specific scent (i.e., kairomones), cues released by conspecifics when attacked, eaten or digested (i.e., alarm cues), or a combination of both predator and conspecific cues (Schoeppner and Relyea 2009). However, since a reduction in activity levels usually trades-off with other activities, such as feeding (Watson et al. 2004), an appropriate risk-assessment is necessary, and, in many cases, such assessment can depend on, for instance, the type of chemical cue that is detected, and the type of predator involved (Chivers and Smith 1998, Ferrari et al. 2007, 2010). For example, tadpoles are more likely to show stronger reactions to the presence of a predator when the predator has previously fed on other tadpoles of the same species (Schoeppner and Relyea 2009), likely because the combination and presence of both predatory and tadpole conspecific cues may convey more imminent danger. Interestingly, evidence suggests that tadpoles may not react to the odour cues of novel predators or species that they do not recognise (Nunes et al. 2013), probably because they cannot recognise or properly assess the risk associated with such predators. This lack of recognition of novel predators can be especially problematic with the introduction of invasive predatory species that can threaten native tadpole populations. Interestingly, the high evolutionary pressure to respond to predators could lead to some species learning to not only recognise the alarm cues produced by the tadpoles of other species (i.e., interspecific recognition, Chivers et al. 2002), but also the kairomones of unknown novel predators, such as invasive crayfish (Polo-Cavia and Gomez-Mestre 2014). This learning process is usually based on the association of another individual's alarm cues with the kairomones released by the novel predator. Long-term exposure to predator chemical cues during larval development can also lead to plastic phenotypic responses and the development of morphologically inducible defences, such as caudal crests that can help prey to escape from predators (Relyea and Auld 2005, Schoeppner and Relyea 2005).

From an antipredatory perspective, apart from using chemicals as cues in predator detection and avoidance, chemicals can also be deployed in predator deterrence. In all anurans, salamanders and caecilians, predator deterrence is often achieved using toxic chemical compounds present in the skin glands of adults, larvae, or even eggs. The type and amount of toxins are known to vary between life stages (Hayes et al. 2009) and between populations of the same species that differ in predation pressure (Mailho-Fontana et al. 2019). In addition, there is variation among amphibian groups in the type of toxins used against predators, with tetrodotoxins in some salamanders, and bufodienolids and bufotoxins in bufonids (i.e., true toads), being arguably the most widespread. Amphibian toxins are used as unpalatable compounds to avoid immediate and, in some cases, future ingestion via learning processes by predators (i.e., 'associative learning'). For instance, predatory larvae of the diving beetle *Dytiscus verticalis* avoided prey newts after learning to associate the unpalatability of such prey with the odour obtained by dipping cotton swabs in newt extract (e.g., Brodie and Formanowicz 1981). Similarly, many amphibians combine chemical deterrents with other sensory modalities to advertise their unpalatability. In tadpoles, such conspicuousness can be achieved through the formation of large aggregations (Peterson and Blaustein 1991), while, in many adults, unpalatability may be advertised through aposematic colouration (see Chapter 12) and conspicuous defensive behaviours (Ferreira et al. 2019). Therefore, multimodal communication, where animals combine chemical signals with other signalling modalities, is likely to be key during interactions between amphibians and their potential predators.

Integration of Chemical Communication with Other Sensory Channels

Chemical communication is only one of a multitude of modalities that amphibians can use to understand their surroundings and to communicate with each other (see Chapters 11 and 12). Multimodal communication involves the simultaneous use of signals that are received through more than one sensory channel (reviewed in Partan and Marler 2005). In general, organisms may use multimodal signalling to improve the detection and discrimination of other individuals, reduce the ambiguity and enhance the processing of messages sent using only one channel, or, in the context of sexual selection, to better detect potential mates and/or evaluate the quality of the signaller as a potential suitor (Candolin 2003, Elias et al. 2003, Higham and Hebets 2013, Halfwerk et al. 2014).

Multimodal communication in a sexual context has been documented in several amphibian species. In Hyperoliid frogs, males have a prominent gular patch that releases volatile compounds when they are singing (Starnberger et al. 2013). The use of chemical compounds by the male, in combination with his acoustic display, is likely to play an important role in mate detection and mate assessment by choosy females. Similarly, mate attraction in many species of newts also involve the use of chemical and visual signals (Halliday 1975). For example, males of the palmate newt *Lissotriton helveticus* develop duck-like webbing in their hind limbs, a caudal crest, and a caudal filament at the tip of the tail (Secondi et al. 2009, Iglesias-Carrasco et al. 2016), that they show to the females while simultaneously releasing pheromones during the mating dance. In this regard, both chemical and visual signals could play a reinforcing role in mate assessment or convey different aspects of mate quality to choosy females (Candolin 2003).

The evolution of multimodal signalling will depend on the cost and benefits of using multiple channels to convey important information, and these benefits might depend on the environmental sensory conditions in which animals send and receive different signals. For example, multiple signalling is likely to evolve in fluctuating environments or in environments where the use of a single sensory channel might be ineffective (Bro-Jørgensen 2010). However, as human activities disrupt different communication channels (e.g., through noise pollution, release of chemicals compounds, etc.), the costs of using specific signals are likely to increase. Here, an important challenge for future research will be to determine whether individuals can flexibly adjust their reliance on different kinds of signals across different sensory modalities, whether the costs of human-mediated disruptions to sensory channels are too high to maintain the value of the signal, and whether the disruption of specific channels will shape the evolution of novel communication channels (Heuschele and Candolin 2007, Halfwerk and Slabbekoorn 2015, Candolin and Wong 2019).

Anthropogenic Impacts on Chemical Communication

Amphibians are among the most threatened vertebrates on Earth, with populations undergoing precipitous declines worldwide as human activities destroy and alter their habitats and ecosystems. Particularly insidious are pollutants that can alter chemical communication and behaviours associated with such communication (reviewed in Lürling and Scheffer 2007).

Pollution can affect the effectiveness of chemical cues and signals, and the ability of receivers to detect and perceive them. For example, tadpoles of the Cuban tree frog *Osteopilus septentrionalis* exposed to the herbicide atrazine no longer respond to the odour cues of predators (Hrsam et al. 2016). Losing the ability to adaptively respond to predators can have important consequences for individual survival and, in turn, the population as a whole (Wong and Candolin 2015, Saaristo et al. 2018). Anthropogenic pollutants can also affect intraspecific interactions. Some heavy metals and pesticides, for instance, can bind to pheromones and disrupt pheromonal systems. An example of this is seen in the palmate newt *Lissotriton helveticus*, where the scents of male exposed to nitrates were less attractive to females than the scents produced by unexposed males (Secondi et al. 2009). Similarly, males of the newt *Notophthalmus viridescens* exposed to the insecticide endosulfan took longer to react to female olfactory signals, resulting in reduced mating success (Park and Propper 2002). Such examples underscore how various pollutants can disrupt, not only the signal itself, but also the sensorial capacity of the receiver to detect such signals. This can have important population-level consequences if it leads to individuals making suboptimal mating decisions that could, in turn, affect the quality and quantity of offspring (Candolin and Wong 2019, Aulsebrook et al. 2020). In this regard, if anthropogenic disturbances have a disproportionate effect on particular channels of communication, it is possible that the relative importance of other sensory modalities might have to increase (e.g., Heuschele et al. 2009, van der Sluijs et al. 2011).

Another obvious effect of human activities is the destruction and transformation of natural habitats, which can also have negative effects on amphibian chemical communication. Examples include the transformation of natural forests to exotic plantations, and the consequent changes in pond chemistry related to the input of leaf litter and the release of leachates into aquatic systems. For instance, males of the palmate newt *Lissotriton helveticus* exposed to exotic eucalypt plantations and leachates had difficulties discriminating between artificial ponds containing chemical cues from predators and potential mates (Iglesias-Carrasco et al. 2017a). Exposure to eucalypts also disrupted olfactory-based mate preferences, with females no longer associating preferentially with high quality males in contaminated water (Iglesias-Carrasco et al. 2017b).

It is important to realise that, in a rapidly changing world, there are also opportunities to harness our knowledge of chemical communication in amphibians to effect positive management and conservation outcomes. An excellent case in point comes from efforts to control the invasive cane toad *Rhinella marina* in Australia. Specifically, researchers are currently exploring two potential ways of using the chemicals produced by the cane toads themselves to manage the invasive pest. The first exploits the cannibalistic tendencies of tadpoles to locate and prey on the eggs of conspecifics by following chemical toxins that are released by the developing eggs (Crossland et al. 2012, Crossland and Shine 2012). Studies have shown that both the eggs of the cane toad and the exudate extracted from the parotoid glands of adults (McCann et al. 2019) can be used as an attractant to lure tadpoles into artificial traps. The second way in which chemicals can potentially be exploited for biological control relies on suppression cues, which are waterborne cues used by conspecifics to supress the development and viability of recently laid eggs to limit competition for food or space (Crossland and Shine 2011). Studies have found that exposure to suppression cues of older conspecifics reduces survival and growth of hatchlings, which can potentially help control field tadpole populations (Clarke et al. 2015, 2016). Such possibilities underscore the fact that innovative approaches might be useful as a control method for some of the most invasive organisms on the planet.

Conclusions and Future Directions

In this chapter, we have highlighted the importance of chemical cues and signals in the daily lives of amphibians. We have discussed how chemical communication can mediate crucial interactions among conspecifics in a wide range of contexts, from territoriality and competition, to mate attraction and parental care. We have shown how chemical cues can influence the way individuals interact with their surroundings, with consequences for foraging success, predator avoidance and deterrence, and habitat selection. Lastly, we have shown how chemical communication interacts with other sensory modalities to maximise individual fitness, and how chemical signals and cues can be disrupted in an increasingly human-dominated world.

Yet, despite what we currently know about the role of chemical communication in shaping the behaviour and life history of amphibians, there is still much that remains unknown. This is especially true of certain amphibian groups, such as caecilians. Due to their nocturnal and fossorial habits, caecilians have proven to be especially challenging to study. Indeed, the ecology, natural history and behaviour of caecilians are among the least well known of the vertebrates (Jared et al. 2019). That said, the specialised tentacular organ, burrowing habits and secretive lives of caecilians suggest that chemical communication is likely to play a vital role in this group and, as a result, warrants greater research effort.

In addition, we need to further explore the chemical composition of the cues and signals used by amphibians, as well as the structure of the different glands and receptors involved in chemical communication. To date, very few of the compounds used in chemical communication have been identified and described, and most of the research has focussed on only a handful of species. However, given the diversity of amphibians and the environments in which they inhabit, a better understanding of the molecules and receptors involved in chemical communication could offer important insights into the evolution of this ancient form of communication.

Lastly, as human activities alter the chemical environment, we face the challenges of understanding and anticipating how wild populations might respond to anthropogenic disturbances and whether we have the capacity to come up with effective conservation and management strategies. In particular, given its importance to survival and reproduction, we need to better understand which pollutants have the most detrimental effects on chemical communication. Research is also urgently needed to understand whether disruption of chemical communication can lead, in the short term, to plastic and adaptive shifts in the use of different sensory channels or whether disturbances to chemical communication results in negative population level effects. Such insights will be critical for understanding the fate of amphibians in a world of rapid and unprecedented change.

Acknowledgements

We thank David Duchene for providing comments to earlier versions of the chapter. This work was supported by funding from the Andalusia Government (postdoctoral fellowship to MIC) and the Australian Research Council (DP190100642, DP220100245 and FT190100014 to BBMW).

References

Aragón, P. 2009. Conspecific male chemical cues influence courtship behaviour in the male newt *Lissotriton boscai*. Behaviour 146: 1137–1151.
Aulsebrook, L.C., M.G. Bertram, J.M. Martin, A.E. Aulsebrook, T. Brodin, J.P. Evans et al. 2020. Reproduction in a polluted world: Implications for wildlife. Reproduction 160: 13–23.
Baum, M.J. and J.A. Cherry. 2015. Processing by the main olfactory system of chemosignals that facilitate mammalian reproduction. Horm. Behav. 68: 53–64.
Belanger, R.M. and L.D. Corkum. 2009. Review of aquatic sex pheromones and chemical communication in anurans. J. Herpetol. 43: 184–191.
Binckley, C.A. and W.J. Resetarits. 2003. Functional equivalence of non-lethal effects: Generalized fish avoidance determines distribution of gray treefrog, *Hyla chrysoscelis*, larvae. Oikos 102: 623–629.
Blaustein, A.R. and B. Waldman. 1992. Kin recognition in anuran amphibians. Anim. Behav. 44: 207–221.

Bossuyt, F., L.M. Schulte, M. Maex, S. Janssenswillen, P.Y. Novikova, S.D. Biju et al. 2019. Multiple independent recruitment of sodefrin precursor-like factors in anuran sexually dimorphic glands. Mol. Biol. Evol. 36: 1921–1930.

Bro-Jørgensen, J. 2010. Dynamics of multiple signalling systems: animal communication in a world in flux. Trends Ecol. Evol. 25: 292–300.

Brodie, Jr E. and Jr. D. Formanowicz. 1981. Larvae of the predaceous diving beetle *Dytiscus verticalis* acquire an avoidance response to skin secretions of the newt *Notophthalmus viridescens*. Herpetologica 37: 172–177.

Brown, J.L., V. Morales and K. Summers. 2008. Divergence in parental care, habitat selection and larval life history between two species of Peruvian poison frogs: An experimental analysis. J. Evol. Biol. 21: 1534–1543.

Brown, J.L., V. Morales and K. Summers. 2009. Tactical reproductive parasitism via larval cannibalism in Peruvian poison frogs. Biol. Lett. 5: 148–151.

Brunetti, A.E., M.L. Lyra, W.G.P. Melo, L.E. Andrade, P. Palacios-Rodríguez, B.M. Prado et al. 2019. Symbiotic skin bacteria as a source for sex-specific scents in frogs. Proc. Natl. Acad. Sci. 116: 2124–2129.

Buxton, V.L. and J.H. Sperry. 2017. Reproductive decisions in anurans: A review of how predation and competition affects the deposition of eggs and tadpoles. Bioscience. 67: 26–38.

Byrne, P.G. and J.S. Keogh. 2007. Terrestrial toadlets use chemosignals to recognize conspecifics, locate mates and strategically adjust calling behaviour. Anim. Behav. 74: 1155–1162.

Candolin, U. 2003. The use of multiple cues in mate choice. Biol. Rev. 78: 575–595.

Candolin, U. and B.B.M. Wong. 2019. Mate choice in a polluted world: Consequences for individuals, populations and communities. Philos. Trans. R. Soc. B 374: 20180055.

Chivers, D.P. and J.F. Smith. 1998. Chemical alarm signalling in aquatic predator-prey systems: a review and prospectus. Écoscience 5: 338–352.

Chivers, D.P., J.M. Kiesecker, A. Marco, J. Devito, M.T. Anderson and A.R. Blaustein. 2001. Predator-induced life history changes in amphibians: Egg predation induces hatching. Oikos 92: 135–142.

Chivers, D.P., R.S. Mirza and J.G. Johnston. 2002. Learned recongnition of herterospecific alarm cues enhances survival duraing encounters with predators. Behaviour 139: 929–938.

Chouinard, A.J. 2012. Rapid onset of mate quality assessment via chemical signals in a woodland salamander (*Plethodon cinereus*). Behav. Ecol. Sociobiol. 66: 765–775.

Clarke, G.S., M.R. Crossland, C. Shilton and R. Shine. 2015. Chemical suppression of embryonic cane toads *Rhinella marina* by larval conspecifics. J. Appl. Ecol. 52: 1547–1557.

Clarke, G.S., M.R. Crossland and R. Shine. 2016. Can we control the invasive cane toad using chemicals that have evolved under intraspecific competition? Ecol. Appl. 26: 463–474.

Crossland, M.R. and R. Shine. 2011. Cues for cannibalism: Cane toad tadpoles use chemical signals to locate and consume conspecific eggs. Oikos 120: 327–332.

Crossland, M.R., T. Haramura, A.A. Salim, R.J. Capon and R. Shine. 2012. Exploiting intraspecific competitive mechanisms to control invasive cane toads (*Rhinella marina*). Proc. R. Soc. B 279: 3436–3442.

Crossland, M.R. and R. Shine. 2012. Embryonic exposure to conspecific chemicals suppresses cane toad growth and survival. Biol Lett. 8: 226–229.

Dantzer, B.J. and R.G. Jaeger. 2007. Detection of the sexual identity of conspecifics through volatile chemical signals in a territorial salamander. Ethology 113: 214–222.

Elias, D.O., A.C. Mason, W.P. Maddison and R.R. Hoy. 2003. Seismic signals in a courting male jumping spider (Araneae: Salticidae). J. Exp. Biol. 206: 4029–4039.

Eluvathingal, L.M., B.A. Shanbhag and S.K. Saidapur. 2009. Association preference and mechanism of kin recognition in tadpoles of the toad *Bufo melanostictus*. J. Biosci. 34: 435–444.

Ferrari, M.C.O., F. Messier and D.P. Chivers. 2007. Degradation of chemical alarm cues under natural conditions: Risk assessment by larval woodfrogs. Chemoecology 17: 263–266.

Ferrari, M.C.O., B. Wisenden and D.P. Chivers. 2010. Chemical ecology of predator–prey interactions in aquatic ecosystems: a review and prospectus. Can. J. Zool. 88: 698–724.

Ferreira, R.B., R. Lourenço-de-Moraes, C. Zocca, C. Duca, K.H. Beard and E.D. Brodie. 2019. Antipredator mechanisms of post-metamorphic anurans: a global database and classification system. Behav. Ecol. Sociobiol. 73: 2–21.

Forester, D. 1986. The recognition and use of chemical signals by a nesting salamander. pp. 205–220. *In*: Duvall, D., D. Muller-Schwarze and R. Silverstein [eds.]. Chemical Signals in Vertebrates, 4. Ecology, Evolution and Comparative Biology. Plenum Press, New Yor, USA.

Forester, D. and A. Wisnieski. 1991. The significance of airborne olfactory cues to the recognition of home area by the dart-poison frog *Dendrobates pumilio*. J. Herpetol. 25: 502–504.

Gamble, L.R., K. McGarigal and B.W. Compton. 2007. Fidelity and dispersal in the pond-breeding amphibian, *Ambystoma opacum*: Implications for spatio-temporal population dynamics and conservation. Biol. Conserv. 139: 247–257.

Gautier, P., K. Olgun, N. Uzum and C. Miaud C. 2006. Gregarious behaviour in a salamander: Attraction to conspecific chemical cues in burrow choice. Behav. Ecol. Sociobiol. 59: 836–841.

Gonzalo, A., C. Cabido, P. Galán, P. López and J. Martín. 2006. Predator, but not conspecific, chemical cues influence pond selection by recently metamorphosed Iberian green frogs, *Rana perezi*. Can. J. Zool. 84: 1295–1299.

Gosling, L.M. and S.C. Roberts. 2001. Scent-marking by male mammals: Cheat-proof signals to competitors and mates. Adv. Study Behav. 30: 169–217.

Halfwerk, W., R.A. Page, R.C. Taylor, P.S. Wilson and M.J. Ryan. 2014. Crossmodal comparisons of signal components allow for relative-distance assessment. Curr. Biol. 24: 1751–1755.

Halfwerk, W. and H. Slabbekoorn. 2015. Pollution going multimodal: The complex impact of the human-altered sensory environment on animal perception and performance. Biol. Lett. 11: 20141051.

Halliday, T.R. 1975. On the biological significance of certain morphological characters in males of the smooth newt *Triturus vulgaris* and of the palmate newt *Triturus helveticus* (Urodela: Salamandridae). Zool. J. Linn. Soc. 56: 291–300.

Hamer, R., F.L. Lemckert and P.B Banks. 2011. Adult frogs are sensitive to the predation risks of olfactory communication. Biol. Lett. 7: 361–363.

Hayes, R.A., M.R. Crossland, M. Hagman, R.J. Capon and R. Shine. 2009. Ontogenetic variation in the chemical defenses of cane toads (*Bufo marinus*): Toxin profiles and effects on predators. J. Chem. Ecol. 35: 391–399.

Heuschele, J. and U. Candolin. 2007. An increase in pH boosts olfactory communication in sticklebacks. Biol. Lett. 3: 411–413.

Heuschele, J., M. Mannerla, P. Gienapp and U. Candolin. 2009. Environment-dependent use of mate choice cues in sticklebacks. Behav. Ecol. 20: 1223–1227.

Higham, J.P. and E.A. Hebets. 2013. An introduction to multimodal communication. Behav. Ecol. Sociobiol. 67: 1381–1388.

Himstedt. W. and D. Simon. 1995. Sensory basis of foraging behavior in caecilians (Amphibia, Gymnophiona). Herpetol. J. 5: 266–270.

Houck, L.D. 2009. Pheromone communication in amphibians and reptiles. Annu. Rev. Physiol. 71: 161–76.

Hrsam, M.A.E., S.A.A.K. Nutie and J.A.R.R. Ohr. 2016. The herbicide atrazine induces hyperactivity and compromises tadpole detection of predator chemical cues. Environ. Toxicol. Chem. 35: 2239–2244.

Iglesias-Carrasco, M., M.L. Head, M.D. Jennions and C. Cabido. 2016. Condition-dependent trade-offs between sexual traits, body condition and immunity: the effect of novel habitats. BMC Evol. Biol. 16: 1–10.

Iglesias-Carrasco, M., M.L. Head, M.D. Jennions, J. Martín and C. Cabido. 2017a. Leaf extracts from an exotic tree affect responses to chemical cues in the palmate newt, *Lissotriton helveticus*. Anim. Behav. 127: 243–251.

Iglesias-Carrasco, M., M.L. Head, M.D. Jennions and C. Cabido. 2017b. Secondary compounds from exotic tree plantations change female mating preferences in the palmate newt (*Lissotriton helveticus*). J. Evol. Biol. 30: 1788–1795.

Ishii, S., K. Kubokawa, M. Kikuchi and H. Nishio. 1995. Orientation of the toad, *Bufo japonicus*, toward the breeding pond. Zool. Sci. 12: 475–484.

Jaeger, R.G. 1981. Dear enemy recognition and the costs of aggression between salamanders. Am. Nat. 117: 962–974.

Jaeger, R.G. and S. Wise. 1991. A reexamination of the male salamander "sexy faeces hypothesis". J. Herpetol. 25: 370–373.

Jared, C., P.L. Mailho-Fontana, S.G.S. Jared, A. Kupfer, J.H.C. Delabie, M. Wilkinson et al. 2019. Life history and reproduction of the neotropical caecilian *Siphonops annulatus* (Amphibia, Gymnophiona, Siphonopidae), with special emphasis on parental care. Acta Zool. 100: 292–302.

Johansson, B.G. and T.M. Jones. 2007. The role of chemical communication in mate choice. Biol. Rev. 82: 265–289.

Kam, Y.C., Z.S. Chuang and C.F. Yen. 1996. Reproduction, oviposition-site selection, and tadpole oophagy of an arboreal nester, *Chirixalus eiffingeri* (Rhacophoridae). J. Herpetol. 30: 52–59.

Kam, Y.C. and H.W. Yang. 2002. Female-offspring communication in a Taiwanese tree frog, *Chirixalus eiffingeri* (Anura: Rhacophoridae). Anim. Behav. 64: 881–886.

Kikuyama, S., F. Toyoda, Y. Ohmiya, K. Matsuda, S. Tanaka and H. Hayashi. 1995. Sodefrin: a female attracting peptide pheromone in newt cloacal glands. Science 267: 1643–1645.

King, J.D., L.A. Rollins-Smith, P.F. Nielsen, A. John and J.M. Conlon. 2005. Characterization of a peptide from skin secretions of male specimens of the frog, *Leptodactylus fallax* that stimulates aggression in male frogs. Peptides 26: 597–601.

Kloskowski, J. 2020. Better desiccated than eaten by fish: Distribution of anurans among habitats with different risks to offspring. Freshw. Biol. 65: 2124–2134.

Largen, W. and S.K. Woodley. 2008. Cutaneous tail glands, noxious skin secretions, and scent marking in a terrestrial salamander (*Plethodon shermani*). Herpetologica 64: 270–280.

Lee, J.S.F. and B. Waldman. 2002. Communication by fecal chemosignals in an archaic frog, *Leiopelma hamiltoni*. Copeia 2002: 679–686.

Lüddecke, T., S. Schulz, S. Steinfartz and M. Vences. 2018. A salamander's toxic arsenal: review of skin poison diversity and function in true salamanders, genus *Salamandra*. Sci. Nat. 105: s00114-018-1579-4.

Lürling, M. and M. Scheffer. 2007. Info-disruption: pollution and the transfer of chemical information between organisms. Trends Ecol. Evol. 22: 374–379.

Maex, M., D. Treer, H. De Greve, P. Proost, I. Van Bocxlaer and F. Bossuyt. 2018. Exaptation as a mechanism for functional reinforcement of an animal pheromone system. Curr. Biol. 28: 2955–2960.e5.

Mailho-Fontana, P.L., C. Jared, M.M. Antoniazzi, J.M. Sciani, D.C. Pimenta, A.N. Stokes et al. 2019. Variations in tetrodotoxin levels in populations of *Taricha granulosa* are expressed in the morphology of their cutaneous glands. Sci. Rep. 9: 1–8.

Marco, A., D.P. Chivers, J.M. Kiesecker and A.R. Blaustein. 1998. Mate choice by chemical cues in western redback (*Plethodon vehiculum*) and Dunn's (*P. dunni*) Salamanders. Ethology 104: 781–788.

McCann, S., M. Crossland, M. Greenlees and R. Shine. 2019. Invader control: factors influencing the attraction of cane toad (*Rhinella marina*) larvae to adult parotoid exudate. Biol. Invasions 21: 1895–1904.

Michaels, C.J., S. Das, Y.M. Chang and B. Tapley. 2018. Modulation of foraging strategy in response to distinct prey items and their scents in the aquatic frog *Xenopus longipes* (Anura: Pipidae). Herpetol. Bull. 143: 1–6.

Moreno, N., R. Morona, J.M. López, L. Dominguez, M. Muñoz and A. González. 2008. Anuran olfactory bulb organization: Embryology, neurochemistry and hodology. Brain Res. Bull. 75: 241–245.

Nunes, A.L., A. Richter-Boix, A. Laurila and R. Rebelo. 2013. Do anuran larvae respond behaviourally to chemical cues from an invasive crayfish predator? A community-wide study. Oecologia 171: 115–127.

Ogurtsov, S. and V. Bastakov. 2001. Imprinting on native pond odour in the pool frog, *Rana lessonae*. pp. 433–438. *In:* Marchlewska-Koj, A., J. Lepri and D. Müller-Schwarze [eds.]. Chemical Signals in Vertebrates 9. Springer, Berlin, Germany.

Park, D. and C.R. Propper. 2001. Repellent function of male pheromones in the red-spotted newt. J. Exp. Zool. 289: 404–408.

Park, D. and C.R. Propper. 2002. Endosulfan affects pheromonal detection and glands in the male red-spotted newt, *Notophthalmus viridescens*. Bull. Environ. Contam. Toxicol. 69: 609–616.

Partan, S.R. and P. Marler. 2005. Issues in the classification of multimodal communication signals. Am. Nat. 166: 231–245.

Perry, D. 2014. Proteins of parotoid gland secretions from toads of the genus *Bufo*. Contemp. Herpetol. 2000: 1–5.

Peterson, J. and A. Blaustein. 1991. Unpalatability in anuran larvae as a defense against natural salamander predators. Ethol. Ecol. Evol. 3: 63–71.

Pfennig, D., H. Reeve and P. Sherman. 1993. Kin recognition and cannibalism in spadefoot toad tadpoles. Anim. Behav. 46: 87–94.

Poelman, E.H. and M. Dicke. 2007. Offering offspring as food to cannibals: Oviposition strategies of Amazonian poison frogs (*Dendrobates ventrimaculatus*). Evol. Ecol. 21: 215–227.

Polo-Cavia, N. and I. Gomez-Mestre. 2014. Learned recognition of introduced predators determines survival of tadpole prey. Funct. Ecol. 28: 432–439.

Relyea, R.A. and J.R. Auld. 2005. Predator- and competitor-induced plasticity: How changes in foraging morphology affect phenotypic trade-offs. Ecology 86: 1723–1729.

Rieger, J.F., C.A. Binckley and W.J. Resetarits. 2004. Larval performance and oviposition site preference along a predation gradient. Ecology 85: 2094–2099.

Saaristo, M., T. Brodin, S. Balshine, M.G. Bertram, B.W. Brooks, S.M. Ehlman et al. 2018. Direct and indirect effects of chemical contaminants on the behaviour, ecology and evolution of wildlife. Proc. R. Soc. B 285: 20181297.

Schmidt, A. and M.H. Wake. 1990. Olfactory and vomeronasal systems of caecilians (Amphibia: Gymnophiona). J. Morphol. 205: 255–268.

Schoeppner, N.M. and R.A. Relyea. 2005. Damage, digestion, and defence: the roles of alarm cues and kairomones for inducing prey defences. Ecol. Lett. 8: 505–512.

Schoeppner, N.M. and R.A. Relyea. 2009. Interpreting the smells of predation: How alarm cues and kairomones induce different prey defences. Funct. Ecol. 23: 1114–1121.

Schulte, L.M., J. Yeager, R. Schulte, M. Veith, P. Werner, L.A. Beck et al. 2011. The smell of success: Choice of larval rearing sites by means of chemical cues in a Peruvian poison frog. Anim. Behav. 81: 1147–1154.

Secondi, J., E. Hinot, Z. Djalout, S. Sourice and A. Jadas-Hécart. 2009. Realistic nitrate concentration alters the expression of sexual traits and olfactory male attractiveness in newts. Funct. Ecol. 23: 800–808.

Sever, D.M. 2003. Courtship and mating glands. *In*: Sever, D. [ed]. Reproductive biology and phylogeny of urodela (Amphibia). Science Publishers, Enfield, NH, USA.

Sever, D.M. 1988. The ventral gland in female salamander *Eurycea bislineata* (Amphibia: Plethodontidae). Copeia 1988: 572–579.

Shinn, A. and J. Dole. 2018. Evidence for a role for olfactory cues in the feeding response of leopard frogs, *Rana pipiens*. Herpetologica 34: 167–172.

Sinsch, U. 2006. Orientation and navigation in Amphibia. Mar. Freshw. Behav. Physiol. 39: 65–71.

Smith, M.A. and D.M. Green. 2005. Dispersal and the metapopulation paradigm in amphibian ecology and conservation: are all amphibian populations metapopulations? Ecography 28: 110–128.

Sousa-Filho, L.M., C.D.T. Freitas, M.D.P. Lobo, A.C.O. Monteiro-Moreira, R.O. Silva, L.A.B Santana et al. 2016. Biochemical profile, biological activities, and toxic effects of proteins in the *Rhinella schneideri* parotoid gland secretion. J. Exp. Zool. Part A Ecol. Genet. Physiol. 325: 511–523.

South, J., T.L. Botha, N.J. Wolmarans, V. Wepener and O.L.F. Weyl. 2020. Playing with food: detection of prey injury cues stimulates increased functional foraging traits in *Xenopus laevis*. Afr. Zool. 55: 25–33.

Starnberger, I., D. Poth, P.S. Peram, S. Schulz, M. Vences, J. Knudsen et al. 2013. Take time to smell the frogs: Vocal sac glands of reed frogs (Anura: Hyperoliidae) contain species-specific chemical cocktails. Biol. J. Linn. Soc. 110: 828–838.

Stynoski, J.L. 2009. Discrimination of offspring by indirect recognition in an egg-feeding dendrobatid frog, *Oophaga pumilio*. Anim. Behav. 78: 1351–1356.

Summers, K. and C.S. McKeon. 2004. The evolutionary ecology of phytotelmata use in Neotropical poison frogs. Misc. Publ. Museum Zool. Univ. Michigan 193: 55–73.

Toledo, R.C. and C. Jared. 1995. Cutaneous granular glands and amphibian venoms. Comp. Biochem. Physiol. - Part A Physiol. 111: 1–29.

Touchon, J.C., I. Gomez-Mestre and K.M. Warkentin. 2006. Hatching plasticity in two temperate anurans: Responses to a pathogen and predation cues. Can. J. Zool. 84: 556–563.

Toyoda, F., S. Tanaka, K. Matsuda and S. Kikuyama. 1994. Hormonal control of response to and secretion of sex attractants in Japanese newts. Physiol. Behav. 55: 569–576.

VanBocxlaer, I., D. Treer, M. Maex, W. Vandebergh, S. Janssenswillen, G. Stegen et al. 2015. Side-by-side secretion of Late Palaeozoic diverged courtship pheromones in an aquatic salamander. Proc. R. Soc. B. 282: 20142960.

van der Sluijs, I., S.M. Gray, M.C.P. Amorim, I. Barber, U. Candolin, A.P. Hendry et al. 2011. Communication in troubled waters: responses of fish communication systems to changing environments. Evol. Ecol. 25: 623–640.

Veeranagoudar, D.K., B.A. Shanbhag and S.K. Saidapur. 2004. Mechanism of food detection in the tadpoles of the bronze frog *Rana temporalis*. Acta Ethol. 7: 37–41.

Vences, M., G. Wahl-Boos, S. Hoegg, F. Glaw, E. Spinelli Oliveira, A. Meyer et al. 2007. Molecular systematics of mantelline frogs from Madagascar and the evolution of their femoral glands. Biol. J. Linn. Soc. 92: 529–539.

Villinger, J. and B. Waldman. 2008. Self-referent MHC type matching in frog tadpoles. Proc. R. Soc. B 275: 1225–1230.

Waldman, B. 1985. Olfactory basis of kin recognition in toad tadpoles. J. Comp. Physiol. A. 156: 565–577.

Warbeck, A., I. Breiter and J. Parzefall. 1996. Evidence for chemical communication in the aquatic caecilian *Typhlonectes natans*. Memories de Biospeologie 23: 37–41.

Warbeck, A. and J. Parzefall. 2001. Mate recognition via waterbone chemical cues in the viviparous caecilian *Typhlonectes natans* (Amphibia: Gymnophiona). pp. 263–268. *In*: Marchlewska-Koj, A., J. Lepri and D. Müller-Schwarze [eds.]. Chemical Signals in Vertebrates 9. Springer, Berlin, Germany.

Watson, R.T., A. Mathis and R. Thompson. 2004. Influence of physical stress, distress cues, and predator kairomones on the foraging behavior of Ozark zigzag salamanders, *Plethodon angusticlavius*. Behav. Process. 65: 201–209.

Weiss, L., I. Manzini and T. Hassenklöver. 2021. Olfaction across the water–air interface in anuran amphibians. Cell Tissue Res. 383: 301–325. doi.org/10.1007/s00441-020-03377-5.

Wells, K. 2007. The Ecology and Behavior of Amphibians. University of Chicago Press, Chicago, USA.

Wong, B.B.M. and U. Candolin. 2015. Behavioral responses to changing environments. Behav. Ecol. 26: 665–673.

Woodley, S.K. 1994. Plasma androgen levels, spermatogenesis, and secondary sexual characteristics in two species of plethodontid salamanders with dissociated reproductive patterns. Gen. Comp. Endocrinol. 96: 206–214.

Woodley, S.K. 2010. Pheromonal communication in amphibians. J. Comp. Physiol. A 196: 713–727.

Woodley, S.K. 2014. Chemical signaling in amphibians. *In*: Mucignat-Caretta, C. [ed.]. Neurobiology of Chemical Communication. CRC Press, Boca Raton, FL, USA.

Woodley, S.K. and N.L. Staub. 2021. Pheromonal communication in urodelan amphibians. Cell Tissue Res. 383: 327-345. 10.1007/s00441-020-03408-1.

Wyatt, T.D. 2017. Pheromones. Curr. Biol. 27: 739–743.

CHAPTER 11

Acoustic Communication in Anurans

Sandra Goutte

INTRODUCTION

During the rainy season, a flooded clearing in the forest becomes the stage of a loud concert as the sun sets. A sudden movement in the water or in the air, and the forest falls silent. The silence lasts only for a moment, and the symphony of croaks, trills and squeaks resumes, louder than ever. Similar concerts are heard from tropical forests to high-elevation grasslands, from savannas to buzzing city centers, and across all continents (except Antarctica). The ubiquitous and sometimes deafening sounds produced by frogs and toads have intrigued behavioral ecologists and evolutionary biologists for centuries. What are the functions of these sounds? Do species produce more than one type of call? How complex can acoustic communication be in anurans? What are the evolutionary forces shaping the diversity of anuran calls?

A tremendous amount of research has been conducted on anuran communication, from natural history observations to single-cell neurological experiments. This chapter aim is not to review all these studies, but to give an overview of the current knowledge on acoustic communication in anurans, and identify areas of research that need more attention. The first section describes the anuran acoustic repertoire and provides examples of the social contexts in which acoustic signals are produced. The vast majority of anuran acoustic communication being reproductive, the second section focuses on the behavioral ecology of acoustic advertisement. In order to fully evaluate the impact of environmental conditions on acoustic communication, it is important to understand its underlying physiology. The third section briefly describes the morphological and physiological mechanisms of call production and hearing in anurans. Call diversity across the Anura and the causes and consequences of inter- and intraspecific variation in vocalizations are then presented (fourth and fifth sections). Finally, the present and future of anuran acoustic communication in an environment ever-more altered by human activities is discussed (sixth section).

Experimental Research Building (C1), New York University Abu Dhabi, Saadiyat island campus, Abu Dhabi, United Arab Emirates
Email: froggologist@gmail.com

Functions of Anurans Acoustic Signals

Most anurans are highly vocal and acoustic signals are used in diversity of social contexts. These calls assume a variety of functions, from signaling their presence to potential mates, defend a territory or scare off a predator. Toledo et al. (2015) reviewed the calls reported in the herpetological literature and categorized them into three major functional groups: reproductive, aggressive and defensive. Evidence for the function of each type of anuran vocalizations emerged from observational data and, in some cases, behavioral experiments. These different categories are succinctly reviewed below, with some remarks on the lesser-known calls produced by females, tadpoles and juveniles.

Reproductive Calls

Reproductive calls encompass several subcategories: advertisement, courtship, release, amplectant and rain calls (Toledo et al. 2015). Except for rain calls, all reproductive calls are produced during the mating season. Advertisement calls serve to attract potential mates typically over long distances and are produced by males, except in a few species (see section on female calls). This call type is the most commonly heard acoustic signal produced by anurans and is subject to sexual and natural selective pressures (Ryan 1985, Gerhardt 1994). Additionally, these vocalizations are species-specific and highly variable across the order, and thus carry a taxonomic value (Fig. 1; Köhler et al. 2017). Consequently, advertisement calls have been at the center of acoustic research in anurans. Advertisement is multimodal in several distantly related species breeding near noisy torrent habitats and involves acoustic and visual signals (Lindquist and Hetherington 1996, Haddad and Giaretta 1999, Wogel et al. 2004, Grafe 2007, Preininger et al. 2009, 2013, Brunner and Guayasamin 2020). In these species, vocalizations can be viewed as "alerting calls" allowing orientation towards the general direction of the caller, followed by visual displays such as arm waving or foot-flagging, which permit a more precise localization (Grafe 2007). These alerting calls are typically low-complexity high-pitched calls, in the form of trills (*Atelopus zeteki, Hylodes* spp.), single (*Sachatamia orejuela*) or series (*Staurois* spp.) of short notes.

Courtship calls are close-range communication signals between a male and a female, often heard just before amplexus (e.g., Marquez and Verrell 1991, Emerson 1992, Nali et al. 2021). In some species with biparental care such as *Ranitomeya imitator* or *Colestethus beebei*, males regularly produce courtship calls near a tadpole's individual pool to initiate the production of an unfertilized egg by the female, which will feed the tadpole (Bourne et al. 2001, Brown et al. 2010). A single species may have multiple types of courtship calls and those are generally of lower amplitude than advertisement calls.

Release calls are emitted by males or unreceptive females when clasped by a male in order to be released. These are typically short notes repeated irregularly and with highly variable structure (Leary 1999, Castellano et al. 2002, Grenat and Martino 2013, Stănescu et al. 2019). Although not as commonly recorded as advertisement calls, release calls can be used for species identification as species-specificity was shown between congeners (Di Tada et al. 2001, Castellano et al. 2002, Grenat and

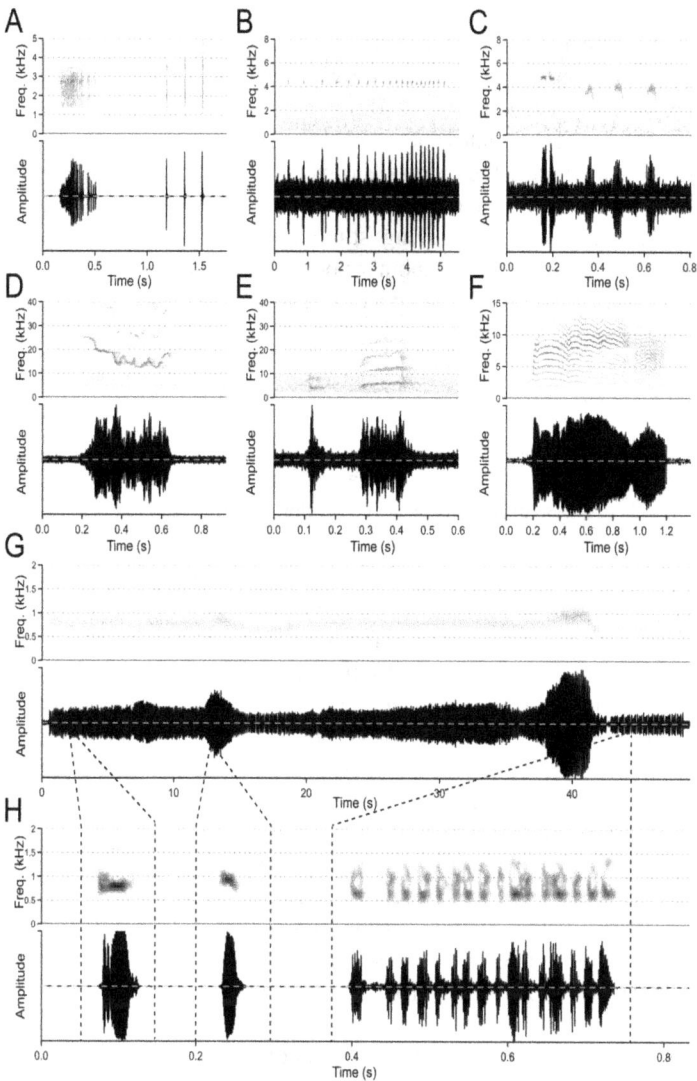

Figure 1. Examples of anuran calls. **A.** Advertisement call of the Harlequin tree frog *Rhacophorus pardalis* (Brunei), composed of a pulsed note producing a "squeak", followed by three clicks with broad frequency bandwidth. **B.** Advertisement call of the green-spotted foot-flagging frog *Staurois tuberilinguis* (Malaysia), composed of short tonal beeps with decreasing inter-notes intervals. The call's dominant frequency (~ 4.3 kHz) is above most of the torrent's noise spectral energy. **C.** Aggressive call of *S. tuberilinguis* (Malaysia), containing two short tonal notes and three lower-frequency pulsed notes. **D.** Advertisement call of the hole-in-the-head frog *Huia cavitympanum* (Malaysia), consisting in a high-frequency whistle including ultrasonic harmonics and rapid frequency modulations. **E.** Advertisement call of the grass odorous frog *Odorrana graminea* (China) composed of a first note produced mouth shut, and a second mouth opened. The second note is longer and contains more energy in high-frequency harmonics than the note produced mouth shut. **F.** Distress call of Izecksohn's treefrog *Bokermannohyla izecksohni* (Brazil) produced mouth opened and containing many harmonics, including at high frequencies. **G.** Advertisement call of the Wallace's flying frog *Rhacophorus nigropalmatus* (Brunei), consisting in a long train of notes produced at a high rate. **H.** Three note types composing the advertisement call of *R. nigropalmatus*. This is a graded call, where note types morph into one-another during the call.

Martino 2013, Köhler et al. 2017). Release calls could thus be an example of inter-specific communication limiting hybridization between sympatric species, and the occurrence of character displacement in release calls has been suggested (Stănescu et al. 2019, but see Castellano et al. 2002).

During the amplexus, males may produce calls, before (amplectant calls) or after oviposition, before ending the amplexus (post-oviposition male release call; Toledo et al. 2015). Little research has been conducted on these calls and their functions remain unclear, although egg laying stimulation was proposed as function for amplectant calls. The rain call is another poorly-known vocalization. It is produced by males outside the typically reproductive season or hours in case of rain (Toledo et al. 2015).

Aggressive Calls

The aggressive call category encompasses territorial, encounter and fighting calls. After the advertisement call, the territorial call is the most commonly heard anuran vocalization (Fig. 1). Social context only can determine that a call is territorial, as its structure varies importantly from one species to the next. Although the production of aggressive calls is constrained by the same morphological and physiological parameters as advertisement calls, the divergence between the two call types is sometimes striking. For example, the advertisement call of the spring peeper *Pseudacris crucifer* is a single tonal note, while its territorial call is a long trill (Owen 2003); in *Colostethus panamensis* the situation is reversed, with short trill as advertisement and a tonal territorial call (Wells 1980). In other species, the territorial call is a variation of the advertisement call, with elements repeated a greater number of times, at a higher rate, and/or with a lower dominant frequencies (Wells and Schwartz 2006). Aggressive calls are graded in many species, intensifying with concurrent males approaching. The hourglass frog *Dendropsophus ebraccatus* is a well-studied example, where males modulate the aggressive and advertisement elements of their calls depending on the immediate social context (Wells 1989). Females may also emit aggressive calls when males approach them (Bard and Wells 1987). In the black-spotted rock frog *Staurois guttatus*, females emit calls mouth open and display foot-flagging behavior when males approach closer than 30 cm from them, which causes them to retreat (Preininger et al. 2016).

A male entering the territory of another may result in physical combat. Prior and during the fight, males can utter specific vocalizations, denoted encounter and fighting calls, respectively (Toledo et al. 2015). As observations of males combats are rare, very little is known about these calls. Toledo et al. (2015) note that fighting calls are generally shorter and with a lower amplitude than other call types, possibly because they are close-range signals directed to the opponent only.

Defensive Calls

This call category corresponds to sounds emitted in response to a predator's approach or attack. Three subcategories are distinguished: distress, warning, and alarm calls. Distress calls, sometimes qualified as screams, are loud sounds emitted when an individual is grasped by a predator (Fig. 1; Hödl and Gollmann 1986, Toledo et al.

2009, Stănescu et al. 2019). The obvious function of the distress call is to startle the predator, which might let go of the frog. Distress calls are similar in structure across species, usually emitted mouth opened (as opposed to most other call types) and can be produced by males, females and juveniles alike.

Warning calls are produced by frogs under immediate threat of a predator, but before the actual attack. Such calls aim to discourage predators from attacking, and are known in species presenting defenses such as toxicity or fangs (Toledo et al. 2015). Alarm calls can resemble distress or warning calls, but are directed towards conspecifics or other animals in the immediate surroundings rather than the predator (e.g., Capranica 1968, Leary and Razafindratsita 1998). The intent of the frog being preyed upon to alarm other animals is difficult to demonstrate, and the reactions reported from neighboring frogs could be viewed as a type of eavesdropping on public information as there is no obvious benefit of this vocalization to the caller. Capranica (1968) also suggests that the sound emitted just before jumping to evade a predator may simply result from a rapid expulsion of the air from the lungs, which provides a diving or landing advantage to the fleeing frog.

Female Reproductive Calls

Female reproductive (advertisement or courtship) calls are known to occur in a few species and are poorly known compared to male vocalizations. In most of these species, females respond to males' calls when engaged in a courtship by a short, low-amplitude call (Toledo et al. 2015). Two functions have been proposed for these female vocalizations: (*i*) Aid mate localization, as for example in the Iberian midwife toad *Alytes cisternasii*, where males call hidden from underground burrows (Marquez and Verrell 1991). These males show phonotactic behavior in response to simulated female calls outside their burrows (Marquez and Verrell 1991). The ultrasonic calls of female concave-eared torrent frogs *Odorrana tormota* also elicit highly accurate phonotactic behavior in males. (*ii*) Identify the caller as females to territorial males, which may be aggressive towards other males. This is possibly the case in the giant river frog *Limnonectes leporinus*, where females approach a male's territory vocalizing softly, which responds by a soft courtship call (Emerson 1992). Another, non-exclusive, hypothetical function for the female vocalization would be the indication of the female reproductive state. Indeed, hormonal levels have been shown to control advertisement call behavior in males as well as in females, with androgen treatment inducing calling behavior in both sexes (Penna et al. 1992, Kime et al. 2021). Interestingly, circulating androgen levels increase importantly during egg maturation and are higher in females *L. leporinus* than breeding conspecific males (Emerson 1992).

Female anurans producing spontaneous advertisement calls are even rarer in the literature and this behavior has been described in only a handful of species. Given the distribution of female vocalizing species in the phylogeny of Anura, female vocalizations seem to be a derived character which has evolved multiple times (Emerson and Boyd 1999). Such calls seem to procure a fitness advantage to females when the probability of mate encounter is low or if resources provided by males are the limiting factor. In the smooth guardian frog *Limnonectes palavanensis*, males are sole providers of parental care and females form lek-like aggregations to advertise for a potential mate (Goyes Vallejos et al. 2017). In this species, males also vocalize but at a much lower rate than females. The Darwin's frog *Rhinoderma darwinii* is another example of sex role reversal in anurans. In this species, males brood tadpoles in their vocal pouch and both males (pregnant and non-pregnant) and females produced similar advertisement calls (Serrano et al. 2020). The acoustic communication system is still poorly understood in this species, but Serrano et al. (2020) suggest that call monomorphism may be linked to sex differences in phenology and operating sex ratio through the reproductive season. In the Mexican chirping frog *Eleutherodactylus cystignathoides*, females incorporate calling aggregation and produce calls similar to males' advertisement calls, possibly to gauge their agonistic response (Serrano and Penna 2018). Female reproductive vocalizations and systems where both males and females vocalize may be more common than previously thought, and more research is needed to uncover the full diversity of reproductive behaviors in anurans.

Acoustic Communication in Tadpoles and Juveniles

The tadpoles of three species are currently known to produce acoustic communication signals. The tadpoles of two species of the genus *Ceratophrys* emit a short train of metallic sounding notes when touched or bitten by a conspecific larvae or touched by an object held by the experimenter (Natale et al. 2011, Costa et al. 2014). Interestingly, these sounds are not produced when coming into contact with other frog species larvae, which the *Ceratophrys* tadpoles prey upon. The emission of these sounds may prevent or minimize the incidence of cannibalism between *Ceratophrys* larvae (Natale et al. 2011), although more research on this topic is needed. It is worth noting that experiments on tadpoles of another member of the family Ceratophryidae, *Lepidobatrachus llanensis*, demonstrated the existence of a mechanism of cannibalism avoidance but the absence of production of sound when interacting with conspecifics (Salgado Costa et al. 2016). The carnivorous tadpoles of *Gephyromantis corvus* produce single clicks or click series in a different context: while these calls are produced when the larvae are kept alone, they are more frequent in the presence of a prey, and even more so in the presence of conspecific tadpole (Reeve et al. 2011). The larvae of *G. corvus* emit sounds particularly when attacking a prey, a conspecific or an intruder while displaying aggressive behavior (Reeve et al. 2011).

Newly-metamorphosed juveniles spadefoot toad *Pelobates fuscus*, lesser swimming frog *Pseudis minuta* and Burmeister's frog *Phyllomedusa burmeisteri* have been reported to emit vocalizations other than release calls (Toledo et al. 2015, Goldberg et al. 2016, ten Hagen et al. 2016). The function(s) of these calls remain uncertain but could be linked to feeding behavior as in tadpoles (ten Hagen et al. 2016). In the case of the lesser swimming frog *Pseudis minuta*, gonadal differentiation is accelerated compared to other anurans, and secondary sexual characters such as the vocal sac develop early after metamorphosis. Juvenile males are thus found to vocalize alongside mature males in calling aggregation (Goldberg et al. 2016). Acoustic communication in anuran larvae and juveniles is largely unknown and further research is needed to understand their evolution and roles in community dynamics.

Behavioral Ecology of Acoustic Advertisement

Anurans exhibit a range of reproductive behaviors (Chapter 9), from males calling in isolation to attract females to the site of oviposition, to lek-like systems (e.g., Grafe 1997, Friedl and Klump 2005), male-male countersinging or complex multimodal courtships (Grafe 2007, Nali et al. 2021). The choice of a calling post providing a good broadcasting platform and resources such as an adequate oviposition site impacts the reproductive success of individual males, and as such is a resource on its own. Most species have an aquatic larval stage and thus are bound to reproduce at or near a water source. Male stream frogs tend to call relatively isolated from one-another or in small aggregations. Even in such low-density conditions, certain species exhibit territorial and complex reproductive behaviors. For example, males of the giant river frog *Limnonectes leporinus* mentioned above build a crater-like nest in the gravel river bed where they produce soft calls to attract females (Emerson 1992). The fangs and the unusual sexual size dimorphism (males are larger than females) in this species indicate that territorial behavior and combats occur between males.

Puddles, ponds or flooded grasslands can be crowded with calling anurans, often creating multispecies choruses. Choruses provide a reaching advantage to the signaling males, as the sound amplitude is multiplied and the advertisement reaches further potential mates. Larger choruses are louder and should thus attract more females than smaller ones (Ryan et al. 1981, Wagner and Sullivan 1992). This relationship is however not always verified (Gerhardt et al. 1987, Telford and van Sickle 1989, Henzi et al. 1995), as other factors may explain the number of females at a breeding site. Choruses also impose several challenges to signaling males. Within a chorus, the proximity of multiple males allows females to quickly assess males' quality, which promotes male-male competition (Arak 1983). In these dense assemblages, conspecific and heterospecific acoustic competition is high and the probability of interference increases with the number of signaling individuals. In several species, experiments demonstrated that males are capable of fine-scale temporal adjustments, adapting their calling rate to vocalize within the short silent intervals in background noise or between synthetic calls (Zelick and Narins 1985, Schwartz 1993, Grafe 2005, Capshaw et al. 2020). Additionally, in many species, males space themselves within the aggregation, and use aggressive calls to maintain the distance with their neighbors, and will thus only directly compete with a limited number of neighbors (Whitney and Krebs 1975, Wells 2007, Berec 2017).

Calling within an aggregation also provides shelter from eavesdropping predators and parasites through several processes (reviewed in Trillo et al. 2019). Indeed, although advertising in group increases signal conspicuousness and thus encounter rate, predation or parasitism risk may actually be reduced for each individual calling male compared to an isolated calling individual through dilution or confusion effects, or an increased detectability of the danger (Trillo et al. 2019). Additionally, in mixed species choruses, frogs can use sympatric heterospecific calls as public information regarding the presence or absence of predators (Phelps et al. 2007).

In mature acoustic communities, spectral niche partitioning, which limits masking between species calling simultaneously, may occur (Krause 1993, Villanueva-Rivera 2014, Arriaga-Jaramillo et al. 2021). However, empirical data does not always support the acoustic niche hypothesis and further studies are needed to establish which community characteristics (age of the assemblage, duration of simultaneous calling through the night/season, phylogenetic proximity among species, etc...) may drive spectral niche partitioning (Garey et al. 2018, Ulloa et al. 2019). Additionally, certain species are capable of adjusting their calling frequencies to avoid spectral overlap with heterospecific calls in novel assemblages. For example, males of the white-banded treefrog *Boana albomarginata* switch to higher frequencies when an acoustic invasion is simulated by broadcasting the call of the American bullfrog *Lithobates catesbeianus* (Both and Grant 2012). When faced with a new acoustic competitor at the breeding site, plastic responses vary across species and do not always avoid spectral overlap (Bleach et al. 2015, Medeiros et al. 2017). Avoidance of acoustic interference between sympatric species can be achieved through the occupancy of different micro-habitats, the use of different calling posts or call timing (Drewry and Rand 1983, Garcia-Rutledge and Narins 2001, Bignotte-Giró et al. 2019). For example, the five *Eleutherodactylus* species of within a Puerto Rican assemblage occupy different acoustic niches differing in at least one of the acoustic dimensions (frequency, microhabitat and timing; Bignotte-Giró et al. 2019).

Call Production and Hearing in Anurans

Call Production

Anurans can produce vocalizations using three possible mechanisms: expiratory, inspiratory or laryngeal calling. The most common mechanism produces calls during expiration, that is when air is pushed from the lungs through the larynx into the buccal cavity. As the air passes through the larynx with pressure, and elicits vibration of the vocal cords, thereby producing the sound. Vocal cord mass and morphology determines the fundamental frequency and pulse shape of the calls. As the mass of the vocal cords is generally proportional to the body mass of the individual, call fundamental frequency is generally inversely correlated with body size (e.g., Gingras et al. 2013). Flank muscles contractions may be continuous or repetitive, in which case pulsed sounds are emitted. The air pushed into the buccal cavity may inflate vocal sacs, if those are present. Only males possess vocal sacs, and their morphology and size is highly variable across species (Fig. 2). Vocal sacs serve as sound radiator for many vocalizing anurans, and increase the efficiency of sound production (Wells and Schwartz 2006). By placing a calling male in helium, Rand and Dudley (1993) showed that the vocal sac is not a sound resonator, as it was previously suspected. The very large tympanum of the American bullfrog *Lithobates catesbeianus* has shown to serve as resonator as well (Purgue 1997). After the emission of the sound, air is normally pushed back from the vocal sac or the buccal cavity and recycled back to the lungs, going through the open glottis silently. Distress calls and, in some species, advertisement and courtship calls, are produced with the mouth open, typically with a single or a few important expirations. This causes the sound energy to be distributed over a greater frequency range than when the call is filtered through the buccal cavity and the vocal sac (Gridi-Papp 2008). In giant frogs *Conraua* spp., the air passing through the narrow space between two fang-like projections in the middle of the lower jaw produces a whistle (Amiet and Goutte 2017). In the grass

Figure 2. Diversity of vocal sac morphology in anurans. **A.** Yalden's treefrog *Leptopelis yaldeni* (Ethiopia), **B.** Amhara grass frog *Ptychadena amharensis* (Ethiopia), **C.** Hourglass treefrog *Dendropsophus ebraccatus* (Panama), **D.** Túngara frog *Engystomops pustulosus* (Panama), **E.** Dhofar toad *Duttaphrynus dhufarensis* (United Arab Emirates), **F.** Luidan stream toad *Ansonia platysoma* (Malaysia), **G.** Long-nosed reed frog *Hyperolius nasutus* (Ethiopia), **H.** Sarawak slender litter frog *Leptobrachella gracilis* (Malaysia).

odorous frog *Odorrana graminea*, males combine notes produced with the vocal sac and the mouth open (unpublished data). The notes produced mouth open are richer in harmonics including in high and ultrasonic frequency ranges (Fig. 1). The aggressive call produced by female black-spotted rock frogs *Staurois guttatus* mouth open also contains a greater number of harmonics reaching high-frequencies compared to the calls produced by males, mouth shut (Preininger et al. 2016).

Fire belly toads *Bombina* spp. produce inspiratory calls: the sound is produced when air is pushed with force from the buccal cavity to the lungs. These frogs do not have vocal sacs, but a buccal floor layered with three different muscles, which provide the power to push the air towards the lungs with sufficient pressure. The painted frogs *Discoglossus* spp. produce both expiratory and inspiratory calls. Finally, the fully aquatic clawed frogs of the family Pipidae, which includes *Xenopus* spp., are deprived of vocal cords and do not use air movements to produce sounds. Instead, two cartilaginous discs in the larynx are rapidly moved to produce clicks, although the exact mode of sound production is still under exploration (Yager 1992, Kwong-Brown et al. 2019).

Hearing

Most anurans rely heavily on acoustic communication for reproduction and their hearing structures are well adapted to detect conspecific calls on land. Their middle ear, which in its most complete form includes a tympanic membrane, air-filled middle ear cavity, extra-stapes and stapes, mechanically amplifies and transmits airborne sound to the inner ear (Van Dijk et al. 2011). This transmission pathway is particularly important at sound frequencies above one kilohertz (kHz), where the energy loss during transmission of vibrations between air and body tissues is high. The sensitivity range of the middle ear typically matches the dominant frequency of the species vocalizations, allowing them to discriminate conspecific calls from background noise and heterospecific calls occurring at different frequencies (Gerhardt and Schwartz 2001). Within the inner ear, two sensory organs are responsible for airborne sound hearing: the amphibian papilla (AP), sensitive to low-frequency sounds ranging from ~ 200 Hz to ~ 1 kHz, and the basilar papilla (BP), which sensitivity varies amongst species but typically encompasses frequencies above 1.2 kHz (Schoffelen et al. 2008). The dominant frequencies of anuran calls may fall into either of these two organs' sensitivity ranges, but typically matches one of the two sensitivity peaks (Yang et al. 2019).

Tympanum size and hearing sensitivity is sexually dimorphic in certain species, with males having proportionally larger tympanums and a greater auditory sensitivity (e.g., Shen et al. 2011b). Individuals' hearing sensitivity also varies according to their hormonal levels and body temperature (Penna et al. 1992, Meenderink and van Dijk 2006, Sun et al. 2020). While the basilar papilla (BP) seems insensitive to temperature changes, the function of the amphibian papilla (AP) is affected by temperature in three ways: when temperature increases, the organ's sensitivity and best frequency increase, while the response latency decreases (Stiebler and Narins 1990, van Dijk et al. 1990, Meenderink and van Dijk 2006, Sun et al. 2020). Thus, anurans' low-frequency hearing ability overall improves but their sensitivity peak also shifts towards higher frequencies.

In some anurans, the tympanic middle ear is absent or incompletely developed, but inner ears are retained (Pereyra et al. 2016). These "earless" frogs may channel sound to their inner ears through extra-tympanic pathways involving the lungs, mouth cavity or cranial bones (Hetherington and Lindquist 1999, Boistel et al. 2013), and most "earless" species are able to hear their own vocalizations (Lindquist et al. 1998). However, the pumpkin toadlet *Brachycephalus ephippium* and its close relative *B. pitanga* are insensitive to sounds in the frequency range of their own vocalizations (Goutte et al. 2017). In this group of brightly colored frogs, the production of low amplitude calls appears to be vestigial, and visual communication may have replaced acoustic communication. In the earless New Zealand frogs *Leiopelma* spp., no acoustic communication is known and these frogs rely mostly on chemical communication cues (Waldman and Bishop 2004).

Inter-specific Variations in Advertisement Calls

Diversity of Anuran Calls

A striking feature of advertisement calls is their diversity and species-specificity. Indeed, most species may be recognized by their advertisement calls, and taxonomists have long been using the acoustic traits as diagnostic characters in their species descriptions (Köhler et al. 2017). Calls traits are particularly useful taxonomic characters in species complexes, where morphological characters alone do not permit species identification. This is the case, for example, for the sympatric species of African grass frogs *Ptychadena doro* and *Ptychadena delphina*, which are morphologically indistinguishable but produce very distinct advertisement calls, one reminiscent of a hen call (*P. doro*; *doro* means *chicken* in Amharic), and the other one of a dolphin's clicking (*P. delphina*; Goutte et al. 2021).

Anuran advertisement calls are incredibly diverse, from bird-like tonal tweets to long insect-like trills, simple repeats of metallic clicks, barking noises or complex frequency-modulated chirps (Fig. 1). However, compared to mammals and insects, anuran calls' spectral range is relatively limited, with most species vocalizing between 100 Hz and 10 kHz. Certain ranids have evolved ultrasonic calls, that is, of frequencies above 20 kHz, the human hearing threshold (Feng and Narins 2008, Arch et al. 2009, Shen et al. 2011a). Although in the same family, ultrasounds-producing frogs do not form a monophyletic clade, and this extreme call characteristic is thought to have converged in species living in noisy torrent habitats, allowing them to evade masking (see section on the Acoustic Adaptation Hypothesis below).

Call traits can be categorized as either static or dynamic, although most are intermediate on the static-dynamic continuum (Gerhardt 1991, Köhler et al. 2017). Call traits considered the most static present the least intra-individual and intra-specific variability but have an important inter-specific variation, and are favored for taxonomic use. Köhler et al. (2017) reviewed the call characters the most commonly used and provided recommendation on the use of advertisement calls in anuran taxonomy. They find that the duration of calls or notes, pulse rate within notes, number of pulses and dominant frequency are the most static call traits, and thus useful for taxonomic use. Establishing which call traits are most static permits the development of automated or semi-automated sound analysis methods allows the use of acoustic specific signatures for non-invasive assessment of anuran diversity and the long-term monitoring of populations (e.g., Llusia et al. 2013).

Although most anurans produce stereotyped calls with a single or a few note types, certain species show an extended vocal repertoire (e.g., *Boophis madagascariensis* (Narins et al. 2000, Christensen-Dalsgaard et al. 2002); *Gracixalus quangi* (Rowley et al. 2011); *Odorrana tormota* (Feng et al. 2002); *Polypedates leucomystax* (Christensen-Dalsgaard et al. 2002)), with up to 28 note types for the Madagascar bright-eyed frog *Boophis madagascariensis*. The function of these large call repertoires is unclear, but Narins et al. (2000) propose several non-exclusive hypotheses borrowed from similar behaviors in birds: (*i*) larger repertoires, similarly to more complex calls, are more attractive to females and thus selected for, (*ii*) in territorial species, a male producing multiple note types could create the illusion of a crowded site, thereby repelling new males from settling in, (*iii*) repertoires allow matching notes in countersinging duets between neighboring males. Increasing the number of note type would permit countersinging with more neighbors, which could be important in territorial species. These three hypotheses imply that males would benefit from an increased vocal repertoire if they are in close proximity form each-other, either because females compare males' repertoires, or for acoustic territorial defense. All the species mentioned above call in small aggregations or are known to be territorial, and individual males could thus benefit from having a greater repertoire than their neighbors.

Environmental Selection: The Acoustic Adaptation Hypothesis

Advertisement calls transmitted over a greater range and with better content integrity (less distortion) should confer a fitness advantage to the caller as they recruit more potential mates than distorted or short-reaching calls. Sounds are attenuated and distorted by elements of the environment they are transmitted through, such as wind, vegetation and temperature gradients (Everest and Pohlmann 2009). Each habitat thus has specific acoustic properties impacting sounds differently. Additionally, the distortion imposed by the environment depends on the properties of the sound itself, mainly its spectral properties. For example, low-frequency sounds are transmitted further in environments containing multiple small-scale obstacles (such as trees), which cause diffraction and affect high-frequency sounds more. The Acoustic Adaptation Hypothesis (AAH) proposes that acoustic signals should be shaped through natural selection for maximal reach and minimal distortion in the home habitat of the caller (Morton 1975, Rothstein and Fleischer 1987, Wilkins et al. 2013). Based on the bioacoustics physics, a set of predictions can be drawn for favorable acoustic characteristics for each habitat (see Ey and Fischer 2009). The AAH was first developed and tested in birds (Morton 1975, Martens and Geduldig 1990), but it has received an increasing amount of attention in anuran bioacoustics in recent years (Bosch and De la Riva 2004, Erdtmann and Lima 2013, Goutte et al. 2018, Velásquez et al. 2018, Röhr et al. 2020).

As in other vertebrates, tests of the AAH in anurans has yield mixed results, with most studies failing to support, or supporting only partially the AAH (Ey and Fischer 2009, Erdtmann and Lima 2013). Frogs and toads living in torrents constitute an example of corroboration of the AAH in anurans (Goutte et al. 2016, 2018, Röhr et al. 2016). Males trying to attract mates near waterfalls or fast flowing streams are faced with the challenge of being heard over the loud noise of the flowing water, which can mask their vocalizations. Torrent noise is low-frequency, typically below 2 kHz, so vocalizations with higher frequencies should escape masking and stand out better in this environment. This hypothesis was verified in several groups of torrent frogs using phylogenetically-informed comparative analyses across (Goutte et al. 2016, 2018) and within species (Vargas-Salinas and Amézquita 2013; but see Vargas-Salinas and Amézquita 2014).

The absence of systematic support of the AAH can be explained by several factors: (*i*) habitats are often classified in a few broad categories that may not capture the diverse acoustic conditions callers encounter, (*ii*) confounding effects such as body size, species relatedness or temperature are not always accounted for and may mask any signal of the AAH, (*iii*) other evolutionary forces such as sexual selection may work in opposite direction and outweigh the effects of natural selection for greater propagation of the acoustic signals. It is therefore not surprising that the AAH is supported in extreme cases only, such as the noisy environments of torrents, when a major acoustic constraint is imposed by the environment on signal transmission and creates a strong selective pressure.

Hybrid Calls

Advertisement calls are often considered as a major prezygotic barrier for hybridization between closely related species, with female choice based on call features being the main determinant of mate pairing. Phonotaxis experiments have indeed shown that females anurans can discriminate conspecific from heterospecific calls, and preferentially direct themselves towards their own species under natural circumstances (Wells 1977, Gerhardt 1995). However, hybrids and multispecies pairings, sometimes between species very different in size and morphology, are not rare in nature, especially between species breeding in high density assemblages. Even complex courtship behaviors, territoriality and female choosiness does not prevent hybridization between closely related species (e.g., Nali et al. 2022). The low proportion of hybrids in sympatric populations is thus likely mostly due to lower fitness of the hybrids (e.g., slower development rate, lower viability, or lower reproductive success) rather than the call acting as a prezygotic barrier (Höbel and Gerhardt 2003, Lemmon and Lemmon 2010). Divergence between advertisement calls of sympatric species may thus be regarded as the result of reinforcement certainly minimizing hybridization, but not fully preventing it.

Because vocalizations are not learnt in anurans, studying hybrid calls can help understand the mechanisms of call traits inheritance. Additionally, as hybridization does occur in nature, understanding the role of hybrids in the reproductive ecology of populations may be important to understand populations' demographic dynamics. Under certain circumstances, females may benefit from mating with a closely related species. This is the case of the female plains' spadefoot toad *Spea bombifrons* which produce offspring developing faster when mating with males of the closely related Mexican spadefoot toad *Spea multiplicata*, which is advantageous when the pond in which tadpoles develop is destined to dry out quickly (Pfennig and Simovich 2002).

Hybrid call traits can be similar to one of the parent species, intermediate between the two parent species, or unique to the hybrids. For example, hybrids resulting from crosses between a gray treefrog *Hyla chrysoscelis* and a pinewoods treefrog *Hyla femoralis*, or between a *H. chrysoscelis* and a bird-voiced treefrogs *Hyla avioca*, produce calls with pulse rates intermediate between their two parent species, but with other traits closer to those of pure gray treefrog (Gerhardt 1974, Gerhardt et al. 1994). Fire-belly toad hybrids (*Bombina bombina* x *Bombina variegata*) produce calls with intermediate call rate, call duration, inter-call intervals and fundamental frequency (Schneider and Eichelberg 1974). Interestingly, within a species pair, hybrid call traits do not seem to vary according to which is the maternal or the paternal species are, suggesting that genes related to call are not on sex chromosomes or linked to sex genes (Doherty and Gerhardt 1983). Additionally, *H. femoralis x H. chrysoscelis* hybrid females prefer hybrid calls compared to calls of either parent species (Doherty and Gerhardt 1983, 1984). This suggests a potential "genetic coupling" involving the same neural pathways for the production and perception of temporal call traits. The genetic mechanisms underlying anuran vocalizations remains largely unknown to date. However, methodological advances and increased accessibility to both genomic sequencing and audio recording should allow exciting new discoveries in the coming years.

Causes and Consequences of Intraspecific Variation

There is no evidence of cultural evolution of calls in frogs, as it is known in some populations of birds (Baker and Cunningham 1985, Williams and Lachlan 2022). Intraspecific variation in calls thus reflects differences in genetic makeup, individual condition (such as body weight or hormonal level), or environment. Population-specific differences in calls may be significant, especially in species with a broad distribution range and occurring across multiple habitats (e.g., Ryan and Wilczynski 1991). In such species, inter-population differences may reflect genetic drift, local adaptation for better sound transmission, environmental conditions, character displacement, or a combination of several of these processes.

Direct Impact of the Environment

The dominant frequency of the call is the best-known indicator of body size in vertebrates, among and within species (Fitch and Hauser 2003, Gingras et al. 2013, Garcia et al. 2017) and this relationship has been shown in many groups of frogs, although some exceptions exist (Goutte et al. 2016, Tonini et al. 2020). However, frogs calling from the water are found to call at lower frequencies than their relatives calling from the ground or vegetation, a trend found consistent across the three frog families examined (Ranidae, Leptodactylidae and Hylidae; Muñoz et al. 2020). Additionally, a recent experiment showed that males túngara frogs *Engystomops pustulosus* produce lower-frequency vocalizations when calling from deeper puddles, where they can inflate their lungs and vocal sacs to the fullest (Goutte et al. 2020). Fully-inflated males also produce more complex calls, which are preferred by females (Halfwerk et al. 2019). As smaller males can inflate more fully in shallower puddles than larger males, these less attractive males produce more attractive calls when only shallow water bodies are available, which in turn will have repercussions on the gene pool of the next generation.

Temperature also plays an important role in anuran acoustic communication, affecting call phenology, production and perception. Ambient temperature affects calling activity of most anuran species in a predictable manner, and each species/ population has a peak activity temperature below or above which calling activity decreases (Saenz et al. 2006, Steelman and Dorcas 2010, Llusia et al. 2013, Ospina et al. 2013). This is particularly true for species breeding in temperate climate (e.g., Gage and Farina 2017), while tropical species' calling activity may be better predicted by rainfall (Plenderleith et al. 2018). A change of temperature may thus differentially affect species, by shifting, prolonging or shortening their breeding period.

As temperature increases, so does calling males' metabolic rate, leading to more rapid muscle contractions and resulting in increased pulse and call rates (e.g., Gayou 1984, Navas and Bevier 2001, Wong et al. 2004, Ziegler et al. 2016). Increased call rate is typically accompanied by shorter call duration and shorter inter-call intervals (e.g., Lörcher 1969, Heinzmann 1970, Gayou 1984, Navas and Bevier 2001). Interestingly, female gray treefrogs *Hyla versicolor* prefer calls with pulse rates produced by males at their own current temperature, suggesting a "temperature coupling" of the production and the processing of calls (Gerhardt 1978). As pulse rate is considered a static and useful acoustic character for species recognition (Köhler et al. 2017), such temperature coupling of temporal call traits may have an important role in reproductive success of populations living in environments with fluctuating temperatures. Although call dominant frequency is considered to be relatively insensitive to temperature changes, a slight positive correlation between pitch and ambient temperature has been reported in several species (Blair 1958, Lörcher 1969, Heinzmann 1970, Schneider and Eichelberg 1974, Gayou 1984). Biomechanical links between call rate and pitch might explain these variations in call frequency, which could then be considered as a side-effect of the increase in call rate. Temperature thus mostly affects temporal parameters of advertisement calls, and seldom, and to a much lower extent, spectral parameters.

Air temperature also affects signal transmission, e.g., through attenuation or distortion (Everest and Pohlmann 2009). Warmer air temperatures cause a faster transmission of sound waves, but also an increase of air turbulences, which in turn increase attenuation. Finally, temperature affects receivers' hearing physiology. While the basilar papilla (BP) seems relatively insensitive to temperature changes, the function of the amphibian papilla (AP) is affected by temperature in three ways: when temperature increases, the organ's sensitivity and best frequency increase, while the response latency decreases (Stiebler and Narins 1990, van Dijk et al. 1990, Meenderink and van Dijk 2006, Sun et al. 2020). Thus, anurans' low-frequency hearing ability overall improves but their sensitivity peak shifts towards higher frequencies. Depending on whether call elements conveying species identity and male fitness fall within the AP (low-frequency) or BP (high-frequency) hearing range, temperature-dependent peak sensitivity shifts may have important repercussions on mate choice. In the little torrent frog *Amolops torrentis*, female preference is not affected by temperature change, as species call falls in the BP range (4.3 kHz; Sun et al. 2020). Green treefrogs *Dryophytes cinereus*, on the other hand, produce a bimodal call including a 900 Hz low-frequency element, which falls in the AP range and is important for species recognition (Gerhardt and Mudry 1980). Females of this species predictably change their preference to lower pitched calls in colder temperatures, which becomes closer to a sympatric species and may be a cause of hybridization (Gerhardt and Mudry 1980).

Reproductive Character Displacement

Reproductive character displacement (RCD) is the process by which sexual traits, such as advertisement calls, evolve through selection to minimize hybridization with sympatric species (Pfennig and Pfennig 2009). The advertisement calls of sympatric and allopatric populations of a given species can thus diverge significantly. This phenomenon has been documented in several systems in frogs (e.g., *Scinax madeira* (Jansen et al. 2016), *Ranoidea genimaculata* (Hoskin et al. 2005)). In the green treefrog *Dryophytes cinereus*, RCD also involves reproductive behaviors such as the choice of calling site and the strength of female preference for conspecific calls (Höbel and Gerhardt 2003). In chorus frogs *Pseudacris feriarum* and *P. nigrita,* different call traits are displaced in each contact zone, and the rarer species in a given site experiences the strongest RCD (Lemmon 2009). Lab-based experiments demonstrated that reinforcement occurs in this system, as hybrid males are less fertile and females systematically prefer conspecific calls over hybrid calls (Lemmon and Lemmon 2010).

Eavesdropping Enemies

Conspicuous acoustic signals may attract unwanted attention to the caller from sound-oriented predators or parasites (Zuk and Kolluru 1998). The frog-eating bat *Trachops cirrhosis* preys on multiple frog species by eavesdropping on their advertisement calls and can distinguish between the calls of edible and non-palatable local species (Tuttle and Ryan 1981, Jones et al. 2020). Mosquitoes and frog-biting midges have also been shown to locate their victims using their calls (Bernal et al. 2006, Legett et al. 2021) and advertising males are sometimes found covered with them while silent frogs lack mosquitoes altogether (personal observation). Beside the loss of blood, midges can transmit trypanosome parasites to the bitten frog (Bernal and Pinto 2016). Eavesdropping predators and parasites thus impose a selective pressure on advertisement calls, which typically leads to less conspicuous signals (Trillo et al. 2019). An increased selective pressure from eavesdroppers may thus lead to changes (through plasticity or evolution) in certain call traits or calling behaviors (reviewed in Trillo et al. 2019). Habitat features and social contexts, such as chorusing behavior modulate the pressure of eavesdroppers (Trillo et al. 2016, Halfwerk et al. 2019) and its impact on call evolution.

Sexual Selection

Phonotaxis experiments have demonstrated that females, all other things being equal, often prefer males producing calls with a lower pitch, reflecting a larger body size, and more complex or with a higher pulse rate (e.g., Ryan 1980, Gerhardt 1991, Ryan and Keddy-Hector 1992). The direction of this sexual selection favors more conspicuous signals, which increases the risk of predation or parasitism. Differences in local predation/parasitism pressures and sexual selection strength can thus lead to a different optimal phenotypic equilibrium in different populations. This is the case in urban and forest populations of túngara frogs *Engystomops pustulosus*. In urban habitats, the level of predation by frog-eating bats and attacks by blood-sucking midges is greatly reduced (Halfwerk et al. 2019). Released from this natural selection pressure, urban frogs produce more complex calls than forest males, thereby increasing their attractiveness to females (Halfwerk et al. 2019).

Anuran Acoustic Communication in the Anthropocene

Many of human-induced environmental changes, either global or local, impact anuran acoustic communication signals and hearing directly or indirectly (Fig. 3). A glass building doesn't reflect the sound the same way as a tree or a patch of reeds, and a paved street doesn't attenuate sounds the way mud or the forest floor would. Additionally, the noise produced by a buzzing city is typically much louder than non-urbanized areas and is concentrated in low frequencies. Species able to survive in anthropogenically modified habitats may undergo rapid and radical changes to adapt, through plasticity and/or evolution. For example, ground vibrations caused by road traffic and wind turbines cause midwife toads *Alytes obstetricans* to decrease their call rate (Caorsi et al. 2019).

Anthropogenic Noise

Anthropogenic noise is a major source of acoustic communication disruption in urban and peri-urban populations. A growing interest in the question in the past decade has revealed that each species responds differently to anthropogenic noise and few predictions can be made. Species for which calls spectrally overlap with the low-frequency noise caused by human activities are more likely to undergo a significant drop in signal-to-noise ratio and reach distance, and thus benefit from modifying certain call traits. Changes in calls due to anthropogenic noise are mostly plastic, short-term adjustments, with call traits reverting to baseline values after a noisy event such as a car or airplane passing (e.g., Kruger and Du Preez 2016, Higham et al. 2021), although in some field-based studies, it is impossible to tease apart whether those adjustments are plastic or adaptive traits (e.g., Grenat et al. 2019). An increase in call pitch is the most common response to anthropogenic noise. By increasing their call pitch in noisy conditions, males avoid masking and increase their signal-to-noise ratio (Cunnington and Fahrig 2010, Kruger and Du Preez 2016, Grenat et al. 2019). For example, brown treefrogs *Litoria ewingii* significantly increase their pitch immediately after the noise of a passing vehicle, even several hundred meters from the road (Higham et al. 2021). Interestingly, Higham et al. (2021) also found that the pitch was overall higher in choruses near the road than in choruses further away, indicating a long-term effect in addition to plasticity.

Call rate is also affected by anthropogenic noise, even in species for which there is minimal spectral overlap with ambient noise. This is the case of the gray treefrog *Hyla versicolor* and the European treefrog *Hyla arborea* for example, which decrease their call rate in the presence of noise (Lengagne 2008, Cunnington and Fahrig 2010).

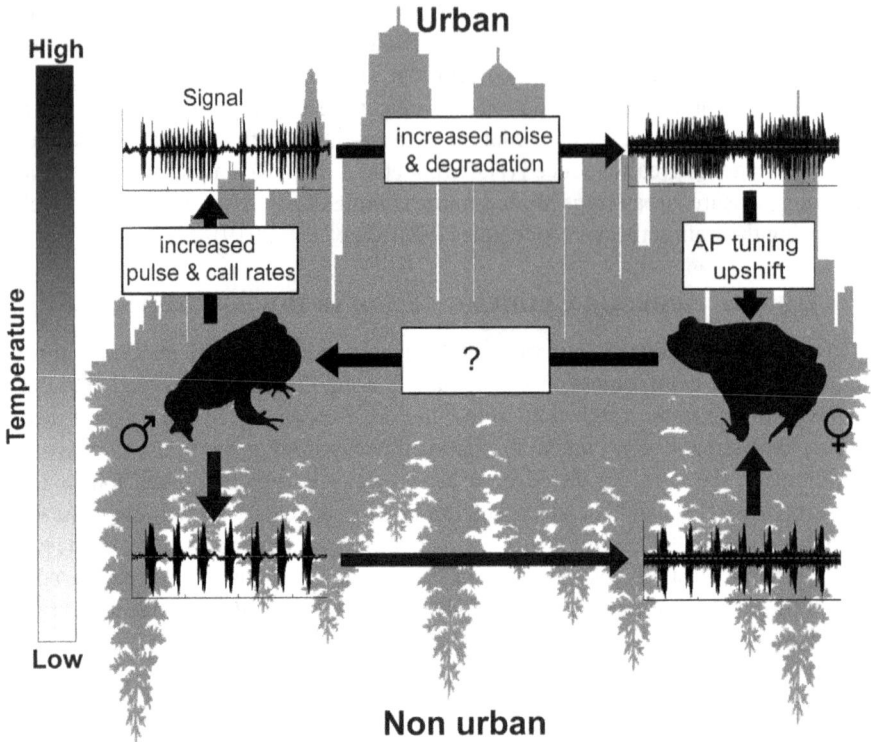

Figure 3. Anuran acoustic communication in the Anthropocene. In natural habitats, male's advertisement call is degraded and partially masked by biotic and abiotic noise (e.g., insects stridulation, wind, flowing water) through its transmission. Gravid females within acoustic reach perceive and process the signal, display phonotactic behavior and assess the suitability of the male as a mate. In urban habitats, the Urban Heat Island effect causes an increase in male's body temperature, impacting temporal traits of his calls such as pulse and call rates. Anthropogenic noise and increased sound diffraction caused by built structures cause a greater attenuation and degradation of the signal, which consequently has a shorter effective reach. The warmer ambient temperature also causes an upshift of the best frequency of females' amphibian papilla (AP). If the species' advertisement calls falls into the AP range, this auditory tuning shift may impact females' ability to identify conspecific calls or change her preferences. Females' preferred call rate may also be affected by temperature changes.

In other species, call rate increases during and/or shortly after anthropogenic noise events (Halfwerk et al. 2016, Kruger and Du Preez 2016, Grenat et al. 2019). In the túngara frog *Engystomops pustulosus*, the impact of anthropogenic noise on male calling behavior seems to be similar to that of the sound of a conspecific chorus, and males increase their call rate, complexity and amplitude when exposed to low-frequency noise (Halfwerk et al. 2016). Noise can affect female choice as well, by increasing the choice latency and impeding orientation towards the calling male (Bee and Swanson 2007), although very little data is available to date.

Global and Local Climate Warming

Global land surface temperature was estimated to be between 1.34°C and 1.83°C higher in 2011–2020 than 1850–1900 and will continue to rise in the next decades with, among other things, an increase in the number of heat waves and high-heat-stress nights (Masson-Delmotte et al. 2021). Climate warming is even more dramatic in cities, where solar energy is absorbed by built and paved structures during the day and radiated at night. As a result, ambient temperature and soil surface temperature are higher in urban areas, especially at night when most anuran are acoustically active (Sakakibara and Owa 2005, Kardinal Jusuf et al. 2007). This "Urban Heat Island" effect (UHI; Taha 1997, Chen et al. 2006) often leads to a 3–5°C difference between city centers and surrounding rural areas (Sakakibara and Owa 2005, Rajagopalan et al. 2014) and up to 10°C in some extreme cases such as Athens, Greece (Santamouris et al. 2001).

Global and local climate warming will affect acoustic signal production, transmission and perception, and likely impact mating behavior (see Fig. 3). In turn, these temperature-related changes may affect signal evolution, leading to divergences between populations and ultimately result in speciation. While the effects of temperature changes on call production and phenology have been relatively well documented, more research is needed to expand our knowledge on hearing sensitivity shifts in warming conditions and their consequences on mate choice, populations' reproductive health and the impact of globally and locally rising temperatures on anuran populations. Monitoring the impact of climate change on calling behavior in anurans will also require long-term standardized data, which is increasingly accessible as methods, software and hardware are developing (e.g., Gibb et al. 2019, Matsumoto et al. 2019).

References

Amiet, J.-L. and S. Goutte. 2017. Chants d'amphibiens du Cameroun. J.-L. Amiet and Éditions Petit Génie, Saint-Nazaire, France.

Arak, A. 1983. Sexual selection by male–male competition in natterjack toad choruses. Nature 306: 261–262.

Arch, V.S., T.U. Grafe, M. Gridi-Papp and P.M. Narins. 2009. Pure ultrasonic communication in an endemic Bornean frog. PLoS One 4: e5413.

Arriaga-Jaramillo, F.G., O.M. Cuellar-Valencia, I. García-Gómez, I. Ceballos-Castro, W. Bolívar-García, D.A. Velásquez-Trujillo et al. 2021. Acoustic segregation of five sympatric and syntopic species of genus *Pristimantis* (Anura: Strabomantidae) from Western Colombia. Stud. Neotrop. Fauna Environ. DOI: 10.1080/01650521.2021.1944758.

Baker, M.C. and M.A. Cunningham. 1985. The biology of bird-song dialects. Behav. Brain Sci. 8: 85–100.

Bard, K.M. and K.D. Wells. 1987. Vocal communication in a neotropical treefrog, *Hyla Ebraccata*: responses of females to advertisement and aggressive calls. Behaviour 101: 200–210.

Bee, M.A. and E.M. Swanson. 2007. Auditory masking of anuran advertisement calls by road traffic noise. Anim. Behav. 74: 1765–1776.

Berec, M. 2017. Where is my place? Quick chorus structure assembly in the European tree frog. Acta Herpetol. 12: 109–112.

Bernal, X.E., A.S. Rand and M.J. Ryan. 2006. Acoustic preferences and localization performance of blood-sucking flies (*Corethrella* Coquillett) to túngara frog calls. Behav. Ecol. 17: 709–715.

Bernal, X.E. and C.M. Pinto. 2016. Sexual differences in prevalence of a new species of trypanosome infecting túngara frogs. Int. J. Parasit. 5: 40–47.

Bignotte-Giró, I., G.A. Fong and G.M. López-Iborra. 2019. Acoustic niche partitioning in five Cuban frogs of the genus *Eleutherodactylus*. Amphibia-Reptilia 40: 1–11.

Blair, W.F. 1958. Mating call in the speciation of anuran amphibians. Am. Nat. 92: 27–51.

Bleach, I.T., C. Beckmann, C. Both, G.P. Brown and R. Shine. 2015. Noisy neighbours at the frog pond: effects of invasive cane toads on the calling behaviour of native Australian frogs. Behav. Ecol. Sociobiol. 69: 675–683.

Boistel, R., T. Aubin, P. Cloetens, F. Peyrin, T. Scotti, P. Herzog et al. 2013. How minute sooglossid frogs hear without a middle ear. Proc. Natl. Acad. Sci. U.S.A. 110: 15360–15364.

Bosch, J. and I. De la Riva. 2004. Are frog calls modulated by the environment? An analysis with anuran species from Bolivia. Can. J. Zool. 82: 880–888.

Both, C. and T. Grant. 2012. Biological invasions and the acoustic niche: the effect of bullfrog calls on the acoustic signals of white-banded tree frogs. Biol. Lett. 8: 714–716.

Bourne, G.R., A.C. Collins, A.M. Holder and C.L. McCarthy. 2001. Vocal communication and reproductive behavior of the frog *Colostethus beebei* in Guyana. J. Herpetol. 35: 272–281.

Brown, J.L., V. Morales and K. Summers. 2010. A key ecological trait drove the evolution of biparental care and monogamy in an amphibian. Am. Nat. 175: 436-446.

Brunner, R.M. and J.M. Guayasamin. 2020. Nocturnal visual displays and call description of the cascade specialist glassfrog *Sachatamia orejuela*. Behaviour 157: 1257–1268.

Caorsi, V., V. Guerra, R. Furtado, D. Llusia, L.R. Miron, M. Borges-Martins et al. 2019. Anthropogenic substrate-borne vibrations impact anuran calling. Sci. Rep. 9: 19456.

Capranica, R.R. 1968. The vocal repertoire of the bullfrog (*Rana catesbeiana*). Behaviour 31: 302–325.

Capshaw, G., A.P. Foss-Grant, K. Hartmann, J.F. Sehuanes and C.F. Moss. 2020. Timing of the advertisement call of the common tink frog (*Diasporus diastema*) shifts with the acoustic behaviour of local conspecifics. Bioacoustics 29: 79–96.

Castellano, S., L. Tontini, C. Ggiacoma, A. Lattes and E. Balletto. 2002. The evolution of release and advertisement calls in green toads (*Bufo viridis* complex). Biol. J. Linn. Soc. 77: 379–391.

Chen, X.-L., H.-M. Zhao, P.-X. Li and Z.-Y. Yin. 2006. Remote sensing image-based analysis of the relationship between urban heat island and land use/cover changes. Remote Sens. Environ. 104: 133–146.

Christensen-Dalsgaard, J., T.A. Ludwig and P.M. Narins. 2002. Complex vocal communication in the southeast asian frog *Polypedates leucomystax*. Bioacoustics 13: 80.

Costa, C.S., M.C. Pereyra, L. Alcalde, R. Herrera, V.L. Trudeau and G.S. Natale. 2014. Underwater sound emission as part of an antipredator mechanism in *Ceratophrys cranwelli* tadpoles. Acta Zool. 95: 367–374.

Cunnington, G.M. and L. Fahrig. 2010. Plasticity in the vocalizations of anurans in response to traffic noise. Acta Oecol. 36: 463–470.

Di Tada, I.E., A. Martino and U. Sinsch. 2001. Release vocalizations in neotropical toads (*Bufo*): ecological constraints and phylogenetic implications. J. Zool. Syst. Evol. Res. 39: 13–23.

Doherty, J.A. and H.C. Gerhardt. 1983. Hybrid tree frogs: vocalizations of males and selective phonotaxis of females. Science 220: 1078–1080.

Doherty, J.A. and H.C. Gerhardt. 1984. Acoustic communication in hybrid treefrogs: sound production by males and selective phonotaxis by females. J. Comp. Physiol. A 154: 319–330.

Drewry, G.E. and A.S. Rand. 1983. Characteristics of an acoustic community: Puerto Rican frogs of the genus *Eleutherodactylus*. Copeia 1983: 941–953.

Emerson, S.B. 1992. Courtship and nest-building behavior of a Bornean frog, *Rana blythi*. Copeia 1992: 1123–1127.

Emerson, S.B. and S.K. Boyd. 1999. Mating vocalizations of female frogs: Control and evolutionary mechanisms. Brain Behav. Evol. 53: 187–197.

Erdtmann, L.K. and A.P. Lima. 2013. Environmental effects on anuran call design: what we know and what we need to know. Ethol. Ecol. Evol. 25: 1–11.

Everest, F.A. and K.C. Pohlmann. 2009. Master Handbook of Acoustics. Tab Electronics, New York, USA.

Ey, E. and J. Fischer. 2009. The 'acoustic adaptation hypothesis'—a review of the evidence from birds, anurans and mammals. Bioacoustics 19: 21–48.

Feng, A.S., P.M. Narins and C.-H. Xu. 2002. Vocal acrobatics in a Chinese frog, *Amolops tormotus*. Naturwissenschaften 89: 352–356.

Feng, A.S. and P.M. Narins. 2008. Ultrasonic communication in concave-eared torrent frogs (*Amolops tormotus*). J. Comp. Physiol. A 194: 159–167.

Fitch, W.T. and M.D. Hauser. 2003. Unpacking "honesty": vertebrate vocal production and the evolution of acoustic signals. pp. 65–137. *In:* Simmon, A.M., R.R. Fay and A.N. Popper [eds.]. Acoustic Communication. Springer, New York, USA.

Friedl, T.W.P. and G.M. Klump. 2005. Sexual selection in the lek-breeding European treefrog: body size, chorus attendance, random mating and good genes. Anim. Behav. 70: 1141–1154.

Gage, S. and A. Farina. 2017. The role of sound in terrestrial ecosystems: three case examples from Michigan, USA. pp. 31–60. *In:* Farina, A. and S.H. Gage [eds.]. Ecoacoustics: The Ecological Role of Sounds. Wiley, Hoboken, NJ, USA.

Garcia, M., C.T. Herbst, D.L. Bowling, J.C. Dunn and W.T. Fitch 2017. Acoustic allometry revisited: morphological determinants of fundamental frequency in primate vocal production. Sci. Rep. 7: 10450.

Garcia-Rutledge, E.J. and P.M. Narins. 2001. Shared acoustic resources in an old world frog community. Herpetologica 57: 104–116.

Garey, M.V., D.B. Provete, T. Gonçalves-Souza, L.S. Ouchi-Melo, C.F.B. Haddad and D.C. Rossa-Feres. 2018. Phylogenetic and adaptive components of the anuran advertisement call correlate with temporal species co-occurrence. Biol. J. Linn. Soc. 125: 292–301.

Gayou, D.C. 1984. Effects of temperature on the mating call of *Hyla versicolor*. Copeia 1984: 733–738.

Gerhardt, H.C. 1974. The vocalizations of some hybrid treefrogs: acoustic and behavioral analyses. Behaviour 49: 130–151.

Gerhardt, H.C. 1978. Temperature coupling in the vocal communication system of the gray tree frog, *Hyla versicolor*. Science 199: 992–994.

Gerhardt, H.C. and K.M. Mudry. 1980. Temperature effects on frequency preferences and mating call frequencies in the green treefrog, *Hyla cinerea* (Anura: Hylidae). J. Comp. Physiol. 137: 1–6.

Gerhardt, H.C., R.E. Daniel, S.A. Perrill and S. Schramm. 1987. Mating behaviour and male mating success in the green treefrog. Anim. Behav. 35: 1490–1503.

Gerhardt, H.C. 1991. Female mate choice in treefrogs: static and dynamic acoustic criteria. Anim. Behav. 42: 615–635.

Gerhardt, H.C. 1994. The evolution of vocalization in frogs and toads. Annu. Rev. Ecol. Syst. 25: 293–324.

Gerhardt, H.C., M.B. Ptacek, L. Barnett and K.G. Torke. 1994. Hybridization in the diploid-tetraploid Treefrogs *Hyla chrysoscelis* and *Hyla versicolor*. Copeia 1994: 51–59.

Gerhardt, H.C. 1995. Phonotaxis in female frogs and toads: execution and design of experiments. pp. 209–220. *In:* Klump, G.M., R.J. Dooling, R.R. Fay and W.C. Stebbins [eds.]. Methods in Comparative Psychoacoustics. Birkhäuser, Basel, Switzerland.

Gerhardt, H.C. and J.J. Schwartz. 2001. Auditory tuning and frequency preferences in anurans. pp. 73–85. *In:* Ryan, M.J. [ed.]. Anuran Communication. Smithsonian Institution Press, Washington DC, USA.

Gibb, R., E. Browning, P. Glover-Kapfer and K.E. Jones. 2019. Emerging opportunities and challenges for passive acoustics in ecological assessment and monitoring. Methods Ecol. Evol. 10: 169–185.

Gingras, B., M. Boeckle, C.T. Herbst and W.T. Fitch. 2013. Call acoustics reflect body size across four clades of anurans. J. Zool. 289: 143–150.

Goldberg, J., D.A. Barrasso, M.G. Agostini and S. Quinzio. 2016. Vocal sac development and accelerated sexual maturity in the lesser swimming frog, *Pseudis minuta* (Anura, Hylidae). Zoology 119: 489–499.

Goutte, S., A. Dubois, S.D. Howard, R. Marquez, J.J.L. Rowley, J.M. Dehling et al. 2016. Environmental constraints and call evolution in torrent-dwelling frogs. Evolution 70: 811–826.

Goutte, S., M.J. Mason, J. Christensen-Dalsgaard, F. Montealegre-Z, B.D. Chivers, F.A. Sarria-S et al. 2017. Evidence of auditory insensitivity to vocalization frequencies in two frogs. Sci. Rep. 7: 12121.

Goutte, S., A. Dubois, S.D. Howard, R. Marquez, J.J.L. Rowley, J.M. Dehling et al. 2018. How the environment shapes animal signals: a test of the Acoustic Adaptation Hypothesis in frogs. J. Evol. Biol. 148–158.

Goutte, S., M.I. Muñoz, M.J. Ryan and W. Halfwerk. 2020. Floating frogs sound larger: environmental constraints on signal production drives call frequency changes. Sci. Nat. 107: 41.

Goutte, S., J. Reyes-Velasco, X. Freilich, A. Kassie and S. Boissinot. 2021. Taxonomic revision of grass frogs (Ptychadenidae, *Ptychadena*) endemic to the Ethiopian highlands. ZooKeys 1016: 77–141.

Goyes Vallejos, J., T. Ulmar Grafe, H.H. Ahmad Sah and K.D. Wells. 2017. Calling behavior of males and females of a Bornean frog with male parental care and possible sex-role reversal. Behav. Ecol. Sociobiol. 71: 95.

Grafe, T.U. 1997. Costs and benefits of mate choice in the lek-breeding reed frog, *Hyperolius marmoratus*. Anim. Behav. 53: 1103–1117.

Grafe, T.U. 2005. Anuran choruses as communication networks. pp. 277–299. *In:* McGregor, P.K. [ed.]. Animal Communication Networks. Cambridge University Press, Cambridge, UK.

Grafe, T.U. 2007. Multimodal signaling in male and female foot-flagging frogs *Staurois guttatus* (Ranidae): an alerting function of calling. Ethology 113: 772–781.

Grenat, P.R. and A.L. Martino. 2013. The release call as a diagnostic character between cryptic related species *Odontophrynus cordobae* and *O. americanus* (Anura: Cycloramphidae). Zootaxa 3635: 583–586.

Grenat, P.R., F.E. Pollo, M.A. Ferrero and A.L. Martino. 2019. Differential and additive effects of natural biotic and anthropogenic noise on call properties of *Odontophrynus americanus* (Anura, Odontophryinidae): Implications for the conservation of anurans inhabiting noisy environments. Ecol. Indic. 99: 67–73.

Gridi-Papp, M. 2008. The structure of vocal sounds produced with the mouth closed or with the mouth open in treefrogs. J. Acoust. Soc. Am. 123: 2895–2902.

Haddad, C.F.B. and A.A. Giaretta. 1999. Visual and acoustic communication in the Brazilian torrent frog, *Hylodes asper* (Anura: Leptodactylidae). Herpetologica 55: 324–333.

Halfwerk, W., A.M. Lea, M.A. Guerra, R.A. Page and M.J. Ryan. 2016. Vocal responses to noise reveal the presence of the Lombard effect in a frog. Behav. Ecol. 27: 669–676.

Halfwerk, W., M. Blaas, L. Kramer, N. Hijner, P.A. Trillo, X.E. Bernal et al. 2019. Adaptive changes in sexual signalling in response to urbanization. Nat. Ecol. Ecol. 3: 374.

Heinzmann, U. 1970. Untersuchungen zur bio-akustik und ökologie der geburtshelferkröte, *Alytes o. obstetricans* (Laur.). Oecologia 5: 19–55.

Henzi, S.P., M.L. Dyson, S.E. Piper, N.E. Passmore and P. Bishop. 1995. Chorus attendance by male and female painted reed frogs (*Hyperolius marmoratus*): environmental factors and selection pressures. Funct. Ecol. 9: 485–491.

Hetherington, T.E. and E.D. Lindquist. 1999. Lung-based hearing in an "earless" anuran amphibian. J. Comp. Physiol. A 184: 395–401.

Higham, V., N.D.S. Deal, Y.K. Chan, C. Chanin, E. Davine, G. Gibbings et al. 2021. Traffic noise drives an immediate increase in call pitch in an urban frog. J. Zool. 313: 307–315.

Höbel, G. and H.C. Gerhardt. 2003. Reproductive character displacement in the acoustic communication system of green tree frogs (*Hyla cinerea*). Evolution 57: 894–904.

Hödl, W. and G. Gollmann. 1986. Distress calls in neotropical frogs. Amphibia-Reptilia 7: 11–21.

Hoskin, C.J., M. Higgie, K.R. McDonald and C. Moritz. 2005. Reinforcement drives rapid allopatric speciation. Nature 437: 1353–1356.

Jansen, M., M. Plath, F. Brusquetti and M.J. Ryan. 2016. Asymmetric frequency shift in advertisement calls of sympatric frogs. Amphibia-Reptilia 37: 137–152.

Jones, P.L., T.J. Divoll, M.M. Dixon, D. Aparicio, G. Cohen, U.G. Mueller et al. 2020. Sensory ecology of the frog-eating bat, *Trachops cirrhosus*, from DNA metabarcoding and behavior. Behav. Ecol. 31: 1420–1428.

Kardinal Jusuf, S., N.H. Wong, E. Hagen, R. Anggoro and Y. Hong. 2007. The influence of land use on the urban heat island in Singapore. Habitat Int. 31: 232–242.

Kime, N.M., S. Goutte and M.J. Ryan. 2021. Arginine vasotocin affects vocal behavior but not selective responses to conspecific calls in male túngara frogs. Horm. Behav. 128: 104891.

Köhler, J., M. Jansen, A. Rodríguez, P.J.R. Kok, L.F. Toledo, M. Emmrich et al. 2017. The use of bioacoustics in anuran taxonomy: theory, terminology, methods and recommendations for best practice. Zootaxa 4251: 1–124.

Krause, B. 1993. The niche hypothesis: A virtual symphony of animal sounds, the origins of musical expression and the health of habitats. The Soundscape Newsletter 6: 6–10.

Kruger, D.J.D. and L.H. Du Preez. 2016. The effect of airplane noise on frogs: a case study on the Critically Endangered Pickersgill's reed frog (*Hyperolius pickersgilli*). Ecol. Res. 31: 393–405.

Kwong-Brown, U., M.L. Tobias, D.O. Elias, I.C. Hall, C.P. Elemans and D.B. Kelley. 2019. The return to water in ancestral *Xenopus* was accompanied by a novel mechanism for producing and shaping vocal signals. eLife 8: e39946.

Leary, C.J. and V.R. Razafindratsita. 1998. Attempted predation on a hylid frog, *Phrynohyas venulosa*, by an indigo snake, *Drymarchon corais*, and the response of conspecific frogs to distress calls. Amphibia-Reptilia 19: 442–446.

Leary, C.J. 1999. Comparison between release vocalizations emitted during artificial and conspecific amplexus in *Bufo americanus*. Copeia 1999: 506–508.

Legett, H.D., I. Aihara and X.E. Bernal. 2021. Within host acoustic signal preference of frog-biting mosquitoes (Diptera: Culicidae) and midges (Diptera: Corethrellidae) on Iriomote Island, Japan. Entomol. Sci. 24: 116–122.

Lemmon, E.M. 2009. Diversification of conspecific signals in sympatry: geographic overlap drives multidimensional reproductive character displacement in frogs. Evolution 63: 1155–1170.

Lemmon, E.M. and A.R. Lemmon. 2010. Reinforcement in chorus frogs: lifetime fitness estimates including intrinsic natural selection and sexual selection against hybrids. Evolution 64: 1748–1761.

Lengagne, T. 2008. Traffic noise affects communication behaviour in a breeding anuran, *Hyla arborea*. Biol. Conserv. 141: 2023–2031.

Lindquist, E.D. and T.E. Hetherington. 1996. Field studies on visual and acoustic signaling in the 'earless' Panamanian golden frog, *Atelopus zeteki*. J. Herpetol. 30: 347–354.

Lindquist, E.D., T.E. Hetherington and S.F. Volman. 1998. Biomechanical and neurophysiological studies on audition in eared and earless harlequin frogs (*Atelopus*). J. Comp. Physiol. A 183: 265–271.

Llusia, D., R. Márquez, J.F. Beltrán, M. Benítez and J.P. do Amaral. 2013. Calling behaviour under climate change: geographical and seasonal variation of calling temperatures in ectotherms. Global Change Biol. 19: 2655–2674.

Lörcher, K. 1969. Vergleichende bio-akustische Untersuchungen an der Rot- und Gelbbauchunke, *Bombina bombina* (L.) und *Bombina v. variegata* (L.). Oecologia 3: 84–124.

Marquez, R. and P. Verrell. 1991. The courtship and mating of the Iberian midwife toad *Alytes cisternasii* (Amphibia: Anura: Discoglossidae). J. Zool. 225: 125–139.

Martens, J. and G. Geduldig. 1990. Acoustic adaptations of birds living close to Himalayan torrents. pp. 123–131. *In:* Current Topics in Avian Biology. Proceedings of the 100th International Meeting of the Deutsche Ornithologische Gesellschaft. Bonn.

Masson-Delmotte, V., P. Zhai, A. Pirani, S.L. Connors, C. Péan, S. Berger et al. 2021. IPCC, 2021: Climate Change 2021: The Physical Science Basis. Contribution of Working Group I to the Sixth Assessment Report of the Intergovernmental Panel on Climate Change. Cambridge University Press, Cambridge, UK.

Matsumoto, H., J. Haxel, B. Kahn, L. Roche, R.P. Dziak, A. Turpin et al. 2019. Field testing and performance evaluation of the Long-term Acoustic Real-Time Sensor for Polar Areas (LARA). OCEANS 2019 MTS/IEEE SEATTLE. DOI:10.23919/OCEANS40490.2019.8962602.

Medeiros, C.I., C. Both, T. Grant and S.M. Hartz. 2017. Invasion of the acoustic niche: variable responses by native species to invasive American bullfrog calls. Biol. Invasions 19: 675–690.

Meenderink, S.W.F. and P. van Dijk. 2006. Temperature dependence of anuran distortion product otoacoustic emissions. J. Assoc. Res. Otolaryngol. 7: 246–252.

Morton, E. 1975. Ecological sources of selection on avian sounds. Am. Nat. 109: 17–34.

Muñoz, M.I., S. Goutte, J. Ellers and W. Halfwerk. 2020. Environmental and morphological constraints interact to drive the evolution of communication signals in frogs. J. Evol. Biol. 33: 1749–1757.

Nali, R., R. Turin and C. Prado. 2021. The courtship call of *Bokermannohyla ibitiguara* (Anura: Hylidae) and details of its mating behavior. Caldasia DOI: https://doi.org/10.15446/caldasia.v44n2.90725.

Nali, R.C., K.R. Zamudio and C.P.A. Prado. 2022. Hybridization despite elaborate courtship behavior and female choice in Neotropical tree frogs. Integr. Zool. DOI: 10.1111/1749-4877.12628.

Narins, P.M., E.R. Lewis and B.E. McClelland. 2000. Hyperextended call note repertoire of the endemic Madagascar treefrog *Boophis madagascariensis* (Rhacophoridae). J. Zool. 250: 283–298.

Natale, G.S., L. Alcalde, R. Herrera, R. Cajade, E.F. Schaefer, F. Marangoni et al. 2011. Underwater acoustic communication in the macrophagic carnivorous larvae of *Ceratophrys ornata* (Anura: Ceratophryidae). Acta Zool. 92: 46–53.

Navas, C.A. and C.R. Bevier. 2001. Thermal dependency of calling performance in the eurythermic frog *Colostethus subpunctatus*. Herpetologica 57: 384–395.

Ospina, O.E., L.J. Villanueva-Rivera, C.J. Corrada-Bravo and T.M. Aide. 2013. Variable response of anuran calling activity to daily precipitation and temperature: implications for climate change. Ecosphere 4: art47.

Owen, P.C. 2003. The Structure, Function, and Evolution of Aggressive Signals in Anuran Amphibians. Ph.D. Thesis, University of Connecticut, Connecticut, USA.

Penna, M., R. Capranica and J. Somers. 1992. Hormone-induced vocal behavior and midbrain auditory sensitivity in the green treefrog, *Hyla cinerea*. J. Comp. Physiol. A 170: 73–82.

Pereyra, M.O., M.C. Womack, J.S. Barrionuevo, B.L. Blotto, D. Baldo, M. Targino et al. 2016. The complex evolutionary history of the tympanic middle ear in frogs and toads (Anura). Sci. Rep. 6: 34130.

Pfennig, K.S. and M.A. Simovich. 2002. Differential selection to avoid hybridization in two toad species. Evolution 56: 1840–1848.

Pfennig, K.S. and D.W. Pfennig. 2009. Character displacement: ecological and reproductive responses to a common evolutionary problem. Q. Rev. Biol. 84: 253–276.

Phelps, S.M., A.S. Rand and M.J. Ryan. 2007. The mixed-species chorus as public information: túngara frogs eavesdrop on a heterospecific. Behav. Ecol. 18: 108–114.

Plenderleith, T.L., D. Stratford, G.W. Lollback, D.G. Chapple, R.D. Reina and J.-M. Hero. 2018. Calling phenology of a diverse amphibian assemblage in response to meteorological conditions. Int. J. Biometeorol. 62: 873–882.

Preininger, D., M. Boeckle, A. Freudmann, I. Starnberger, M. Sztatecsny and W. Hödl. 2013. Multimodal signaling in the Small Torrent Frog (*Micrixalus saxicola*) in a complex acoustic environment. Behav. Ecol. Sociobiol. 67: 1449–1456.

Preininger, D., M. Boeckle and W. Hödl. 2009. Communication in noisy environments ii: visual signaling behavior of male foot-flagging frogs *Staurois latopalmatus*. Herpetologica 65: 166–173.

Preininger, D., S. Handschuh and W. Hödl. 2016. Comparison of female and male vocalisation and larynx morphology in the size dimorphic foot-flagging frog species *Staurois guttatus*. Herpetol. J. 26: 187–197.

Purgue, A.P. 1997. Tympanic sound radiation in the bullfrog *Rana catesbeiana*. J. Comp. Physiol. A 181: 438–445.

Rajagopalan, P., K.C. Lim and E. Jamei. 2014. Urban heat island and wind flow characteristics of a tropical city. Solar Energy 107: 159–170.

Rand, A.S. and R. Dudley. 1993. Frogs in helium: the anuran vocal sac is not a cavity resonator. Physiol. Zool. 66: 793–806.

Reeve, E., S.H. Ndriantsoa, A. Strauss, R.-D. Randrianiaina, T.R. Hiobiarilanto, F. Glaw et al. 2011. Acoustic underwater signals with a probable function during competitive feeding in a tadpole. Naturwissenschaften 98: 135–143.

Röhr, D.L., G.B. Paterno, F. Camurugi F.A. Juncá and A.A. Garda. 2016. Background noise as a selective pressure: stream-breeding anurans call at higher frequencies. Org. Divers. Evol. 16: 269–273.

Röhr, D.L., F. Camurugi, P.A. Martinez, R.S. Sousa-Lima, F.A. Juncá and A.A. Garda. 2020. Habitat-dependent advertisement call variation in the monkey frog *Phyllomedusa nordestina*. Ethology 126: 651–659.

Rothstein, S.I. and R.C. Fleischer. 1987. Vocal dialects and their possible relation to honest status signalling in the brown-headed cowbird. Condor 89: 1–23.

Rowley, J., Đ. Quang Vinh, T. Nguyen, T. Cao and S. Nguyen. 2011. A new species of Gracixalus (Anura: Rhacophoridae) with a hyperextended vocal repertoire from Vietnam. Zootaxa 22–38.

Ryan, M. and W. Wilczynski. 1991. Evolution of intraspecific variation in the advertisement call of a cricket frog (acris-Crepitans, Hylidae). Biol. J. Linn. Soc. 44: 249–271.

Ryan, M.J. 1980. Female mate choice in a neotropical frog. Science 209: 523–525.

Ryan, M.J., M.D. Tuttle and L.K. Taft. 1981. The costs and benefits of frog chorusing behavior. Behav. Ecol. Sociobiol. 8: 273–278.

Ryan, M.J. 1985. The Tungara Frog. A Study in Sexual Selection and Communication. University of Chicago Press, Chicago, USA and London, UK.

Ryan, M.J. and A. Keddy-Hector. 1992. Directional patterns of female mate choice and the role of sensory biases. Am. Nat. 139: S4–S35.

Saenz, D., L.A. Fitzgerald, K.A. Baum and R.N. Conner. 2006. Abiotic correlates of anuran calling phenology: the importance of rain, temperature, and season. Herpetol. Monogr. 20: 64–82.

Sakakibara, Y. and K. Owa. 2005. Urban–rural temperature differences in coastal cities: influence of rural sites. Int. J. Climatol. 25: 811–820.

Salgado Costa, C., V.L. Trudeau, A.E. Ronco and G.S. Natale. 2016. Exploring antipredator mechanisms: new findings in ceratophryid tadpoles. J. Herpetol. 50: 233–238.

Santamouris, M., N. Papanikolaou, I. Livada, I. Koronakis, C. Georgakis, A. Argiriou et al. 2001. On the impact of urban climate on the energy consumption of buildings. Solar Energy 70: 201–216.

Schneider, H. and H. Eichelberg. 1974. The mating call of hybrids of the fire-bellied toad and yellow-bellied toad (*Bombina bombina* (L.), *Bombina v. variegata* (L.), Discoglossidae, Anura). Oecologia 16: 61–71.

Schoffelen, R.L.M., J.M. Segenhout and P. van Dijk. 2008. Mechanics of the exceptional anuran ear. J. Comp. Physiol. A 194: 417–428.

Schwartz, J.J. 1993. Male calling behavior, female discrimination and acoustic interference in the neotropical treefrog *Hyla microcephala* under realistic acoustic conditions. Behav. Ecol. Sociobiol. 32: 401–414.

Serrano, J.M. and M. Penna. 2018. Sexual monomorphism in the advertisement calls of a Neotropical frog. Biol. J. Linn. Soc. 123: 388–401.

Serrano, J.M., M. Penna, A. Valenzuela-Sánchez, M.A. Mendez and C. Azat. 2020. Monomorphic call structure and dimorphic vocal phenology in a sex-role reversed frog. Behav. Ecol. Sociobiol. 74: 127.

Shen, J.-X., Z.-M. Xu, A.S. Feng and P.M. Narins. 2011a. Large odorous frogs (*Odorrana graminea*) produce ultrasonic calls. J. Comp. Physiol. A 197: 1027–1030.

Shen, J.-X., Z.-M. Xu, Z.-L. Yu, S. Wang, D.-Z. Zheng and S.-C. Fan. 2011b. Ultrasonic frogs show extraordinary sex differences in auditory frequency sensitivity. Nat. Comm. 2: 342.

Stănescu, F., L.R. Forti, D. Cogălniceanu and R. Márquez. 2019. Release and distress calls in European spadefoot toads, genus *Pelobates*. Bioacoustics 28: 224–238.

Steelman, C.K. and M.E. Dorcas. 2010. Anuran calling survey optimization: developing and testing predictive models of anuran calling activity. J. Herpetol. 44: 61–68.

Stiebler, I.B. and P.M. Narins. 1990. Temperature-dependence of auditory nerve response properties in the frog. Hear. Res. 46: 63–81.

Sun, X., L. Zhao, Q. Chen, J. Wang and J. Cui. 2020. Auditory sensitivity changes with diurnal temperature variation in little torrent frogs (*Amolops torrentis*). Bioacoustics 29: 684–696.

Taha, H. 1997. Urban climates and heat islands: albedo, evapotranspiration, and anthropogenic heat. Energy Build. 25: 99–103.

Telford S.R. and J. van Sickle. 1989. Sexual selection in an African toad (*Bufo gutteralis*): the roles of morphology, amplexus displacement and chorus participation. Behaviour 110: 62–75.

ten Hagen, L., A. Rodríguez, N. Menke, C. Göcking, M. Bisping, K.-H. Frommolt et al. 2016. Vocalizations in juvenile anurans: common spadefoot toads (*Pelobates fuscus*) regularly emit calls before sexual maturity. Sci. Nat. 103: 75.

Toledo, L.F., I.A. Martins, D.P. Bruschi, M.A. Passos, C. Alexandre and C.F.B. Haddad. 2015. The anuran calling repertoire in the light of social context. Acta Ethol. 18: 87–99.

Toledo, L.F.D., C. Fernando and B. Haddad. 2009. Defensive vocalizations of neotropical anurans. South Am. J. Herpetol. 4: 25–42.

Tonini, J.F.R., D.B. Provete, N.M. Maciel, A.R. Morais, S. Goutte, L.F. Toledo et al. 2020. Allometric escape from acoustic constraints is rare for frog calls. Ecol. Evol. 10: 3686–3695.

Trillo, P.A., X.E. Bernal, M.S. Caldwell, W.H. Halfwerk, M.O. Wessel and R.A. Page. 2016. Collateral damage or a shadow of safety? The effects of signalling heterospecific neighbours on the risks of parasitism and predation. Proc. R. Soc. B 283: 20160343.

Trillo, P.A., C.S. Benson, M.S. Caldwell, T.L. Lam, O.H. Pickering and D.M. Logue. 2019. The influence of signaling conspecific and heterospecific neighbors on eavesdropper pressure. Front. Ecol. Evol. 7: 292.

Tuttle, M.D. and M.J. Ryan. 1981. Bat predation and the evolution of frog vocalizations in the Neotropics. Science 214: 677–678.

Ulloa, J.S., T. Aubin, D. Llusia, É.A. Courtois, A. Fouquet, P. Gaucher et al. 2019. Explosive breeding in tropical anurans: environmental triggers, community composition and acoustic structure. BMC Ecol. 19: 28.

van Dijk, P., E.R. Lewis and H.P. Wit. 1990. Temperature effects on auditory nerve fiber response in the American bullfrog. Hear. Res. 44: 231–240.

van Dijk, P., M.J. Mason, R.L.M. Schoffelen, P.M. Narins and S.W.F. Meenderink. 2011. Mechanics of the frog ear. Hear. Res. 273: 46–58.

Vargas-Salinas, F. and A. Amézquita. 2013. Stream noise, hybridization, and uncoupled evolution of call traits in two lineages of poison frogs: *Oophaga histrionica* and *Oophaga lehmanni*. PLoS ONE 8: e77545.

Vargas-Salinas, F. and A. Amézquita. 2014. Abiotic noise, call frequency and stream-breeding anuran assemblages. Evol. Ecol. 28: 341–359.

Velásquez, N.A., F.N. Moreno-Gómez, E. Brunetti and M. Penna. 2018. The acoustic adaptation hypothesis in a widely distributed South American frog: Southernmost signals propagate better. Sci. Rep. 8: 6990.

Villanueva-Rivera, L.J. 2014. *Eleutherodactylus* frogs show frequency but no temporal partitioning: implications for the acoustic niche hypothesis. PeerJ 2: e496.

Wagner, W.E. and B.K. Sullivan. 1992. Chorus organization in the gulf coast toad (*Bufo valliceps*): male and female behavior and the opportunity for sexual selection. Copeia 1992: 647–658.

Waldman, B. and P.J. Bishop. 2004. Chemical communication in an archaic anuran amphibian. Behav. Ecol. 15: 88–93.

Wells, K.D. 1977. The social behaviour of anuran amphibians. Anim. Behav. 25: 666–693.

Wells, K.D. 1980. Behavioral ecology and social organization of a dendrobatid frog (*Colostethus inguinalis*). Behav. Ecol. Sociobiol. 6: 199–209.

Wells, K.D. 1989. Vocal communication in a neotropical treefrog, *Hyla ebraccata*: responses of males to graded aggressive calls. Copeia 1989: 461–466.

Wells, K.D. and J.J. Schwartz. 2006. The behavioral ecology of anuran communication. pp. 44–86. *In:* Narins, P.M., A.S. Feng, R.R. Fay and A.N. Popper [eds.]. Hearing and Sound Communication in Amphibians. Springer, New York, USA.

Wells, K.D. 2007. The Ecology and Behavior of Amphibians. The University of Chicago Press, Chicago, USA.

Whitney, C.L. and J.R. Krebs. 1975. Spacing and calling in Pacific tree frogs, *Hyla regilla*. Can. J. Zool. 53: 1519–1527.

Wilkins, M.R., N. Seddon and R.J. Safran. 2013. Evolutionary divergence in acoustic signals: causes and consequences. Trends Ecol. Evol. 28: 156–166.

Williams, H. and R.F. Lachlan. 2022. Evidence for cumulative cultural evolution in bird song. Phil. Trans. R. Soc. B 377: 20200322.

Wogel, H., P.A. Abrunhosa and L.N. Weber. 2004. The tadpole, vocalizations and visual displays of *Hylodes nasus*. Amphibia-Reptilia 25: 219–227.

Wong, B.B.M., A.N.N. Cowling, R.B. Cunningham, C.F. Donnelly and P.D. Cooper. 2004. Do temperature and social environment interact to affect call rate in frogs (*Crinia signifera*)? Austral Ecol. 29: 209–214.

Yager, D.D. 1992. A unique sound production mechanism in the pipid anuran *Xenopus borealis*. Zool. J. Linn. Soc. 104: 351–375.

Yang, Y., B. Zhu, J. Wang, S.E. Brauth, Y. Tang and J. Cui. 2019. A test of the matched filter hypothesis in two sympatric frogs, *Chiromantis doriae* and *Feihyla vittata*. Bioacoustics 28: 488–502.

Zelick, R. and P.M. Narins. 1985. Characterization of the advertisement call oscillator in the frog *Eleutherodactylus coqui*. J. Comp. Physiol. A 156: 223–229.

Ziegler, L., M. Arim and F. Bozinovic. 2016. Intraspecific scaling in frog calls: the interplay of temperature, body size and metabolic condition. Oecologia 181: 673–681.

Zuk, M. and G.R. Kolluru. 1998. Exploitation of sexual signals by predators and parasitoids. Q. Rev. Biol. 73: 415–438.

CHAPTER 12

Amphibian Coloration: Proximate Mechanisms, Function, and Evolution

Bibiana Rojas,[1,]* *J.P. Lawrence*[2] and *Roberto Márquez*[3]

INTRODUCTION

Amphibians have been commonly studied in relation to their physiology and life history strategies, due to the fascinating changes they undergo throughout their complex life cycle, and to their communication, predominantly acoustic for anurans and chemical for urodeles. Furthermore, some species, such as the axolotl (*Ambystoma mexicanum*) and frogs of the genus *Xenopus*, have become important models for biomedical research. Since the late 19th and early 20th century, there has also been considerable interest in the ontogenetic and cellular/molecular aspects of amphibian pigmentation (e.g., DuShane 1935, Cabello Ruz 1943, Bagnara et al. 1968), as well as the frequent intraspecific variation in color patterns (e.g., Hargitt 1912, Hairston and Pope 1948, Dunlap 1955). More recently, and after a relatively slow start, the number of studies on the adaptive value of amphibian coloration has also been rapidly increasing, particularly in the context of anti-predator strategies (reviewed in Rudh and Qvarnström 2013, Rojas 2017).

Some of the first studies on amphibian coloration were related mostly to the incredible polymorphism observed in some cryptic frogs (reviewed in Hoffman and Blouin 2000). The great diversity in vibrant and conspicuous coloration of some

[1] Konrad Lorenz Institute of Ethology, Department of Interdisciplinary Life Sciences, University of Veterinary Medicine Vienna, Savoyenstraße 1, 1160, Vienna, Austria; University of Jyvaskyla, Department of Biological and Environmental Science, PO Box 35, 40014, Jyväskylä, Finland.
[2] Michigan State University, Lyman Briggs College, Holmes Hall, East Lansing, MI 48825, USA.
[3] University of Michigan, Department of Ecology and Evolutionary Biology, and Michigan Society of Fellows, 1105 North University Ave. Ann Arbor, MI. 48109, USA; University of Chicago, Department of Ecology and Evolution, 1101 E. 57th St. Chicago, IL. 60637, USA.
* Corresponding author: bibiana.rojas@vetmeduni.ac.at

Neotropical poison frogs also started to be documented recurrently in taxonomic descriptions, travel diaries and natural history notes as early as 1874 (e.g., Belt 1874), but it was only until the late 1990's when the function and evolution of such diversity started to be increasingly addressed. Studies on the function of salamander warning coloration, on the other hand, seem to have bloomed a little earlier (Huheey 1960, Huheey and Brandon 1974, Brandon and Huheey 1975, Brandon et al. 1979a, b, Brodie and Brodie 1980).

Over the last two decades, research on the (potential) role of coloration in amphibian sexual selection has also been proliferating, albeit with the overrepresentation of some taxa (e.g., *Oophaga pumilio*, the strawberry poison frog, Maan and Cummings 2008, 2009, Crothers and Cummings 2013). There have been numerous accounts of the existence of sexual dichromatism (e.g., Bell and Zamudio 2012, Bell et al. 2017) and, in some cases, the influence of colour patterns on mate choice or intra-sexual competition (e.g., Maan and Cummings 2008, Crothers et al. 2011, Acord et al. 2013), both in anurans and urodeles. Coloration in gymnophionans has been studied much less but, given their fossorial lifestyle, instances of conspicuous coloration in species of this group (Wollenberg and Measey 2009) continue to puzzle researchers.

Here, we provide an overview of the research conducted on amphibian coloration to understand its functional basis, antipredator function, and role in sexual selection. We review the available knowledge and provide suggestions for avenues of research yet to be explored when seeking to understand color in amphibians.

The Functional Basis of Amphibian Coloration

A long history of physiological, behavioral, ecological, and evolutionary studies of coloration phenotypes in animals, together with the ease with which they are scored, has resulted in a profound understanding of the mechanisms governing pigmentation development and evolution in vertebrate model systems (Bennett and Lamoreux 2003, Sturm 2009, Irion et al. 2016, Patterson and Parichy 2019) and, increasingly, in non-model systems (Hofreiter and Schöneberg 2010, Hubbard et al. 2010, Kronforst et al. 2012, Thibaudeau and Altig 2012, Kuriyama et al. 2020). Since at least the late 1800s, studies using amphibians have contributed to our understanding of several mechanistic aspects of vertebrate pigmentation, such as the ontogenetic origin (Ehrmann 1892, Eycleshymer 1906, DuShane 1935), structure and function of pigment cells (Bagnara et al. 1968, Bagnara and Matsumoto 2006), the role of hormones in pigmentation (Allen 1916, Smith 1916, Houssay 1925, Parker 1948), the mechanisms underlying rapid color change (Parker 1948, Tuma and Gelfand 1999, Sköld et al. 2013), and the role of non-cellular chromophores (i.e., color-producing molecules) in green (Cabello Ruz 1943, Barrio 1965) and fluorescent (Taboada et al. 2017a, b) coloration. In this section, we review the current knowledge and open questions on the molecular, cellular, and physiological mechanisms underlying color patterns and their evolution in amphibians.

Chromatophores: the Building Blocks of Vertebrate Coloration

In vertebrates, skin coloration is produced by chromatophores, a group of cells that originate in the neural crest (DuShane 1935, Fujii 1993), and can be classified in two broad groups based on the physical mechanisms through which they produce color (Bagnara et al. 1968, Ligon and McCartney 2016). First are pigmentary chromatophores, which produce color via the absorption of particular wavelengths of light by pigments within them. Three kinds of pigmentary chromatophores are predominant in vertebrates: melanophores, which contain melanic pigments and are black or brown, and xanthophores and erythrophores, which are yellow or red, respectively, and contain carotenoid and pteridine pigments. The second type are structural chromatophores, which contain cellular nanostructures that produce color by scattering and reflecting certain wavelengths of light. Iridophores, which contain purine crystalline platelets, are the most widespread kind of structural chromatophore in vertebrates (Bagnara 1966, Fujii 1993, Bagnara and Matsumoto 2006). In most vertebrates, chromatophores are arranged three-dimensionally within the dermis, forming structures known as dermal chromatophore units, which consist of stacked xanthophores, iridophores and melanophores. Xanthophores are at the outermost end (i.e., closest to the epidermis) followed by iridophores and finally melanophores, which are found near the base of the dermis. Melanophores, in addition, have dendritic processes that extend around and over iridophores (Fig. 1A; Bagnara et al. 1968). A notable exception to this pattern are mammals and birds, in which pigment cells (mostly melanocytes, homologous to melanophores) are in the dermis, and transport pigments into neighboring keratinocyte cells, which may retain them or deposit them into hairs or feathers (Weiner et al. 2007, Mills and Patterson 2009).

Chromatophores dictate an animal's coloration in two ways: through the actual colors produced by their pigments or nanostructures (i.e., color), and through the spatial organization of these colors in the skin (i.e., pattern). In pigmentary chromatophores, color is determined by the metabolic pathways involved in either the production of pigments, or their uptake from the diet and posterior modification (Goodwin 1984, d'Ischia et al. 2015). In structural chromatophores, in contrast, color is determined by the morphology and organization of the reflective platelets within cells (Fujii 1993, Ligon and McCartney 2016). Below we detail how color is produced in the main chromatophore types.

Melanophores

Melanic pigments are produced within melanophores, by organelles called melanosomes, through a pathway that is relatively conserved across vertebrates (d'Ischia et al. 2015). Melanosomes can produce different types of related pigments, whose relative quantities produce different colors. The best-studied among them are eumelanin and pheomelanin, which reflect different colors and are responsible for multiple instances of coat and plumage color variation in mammals and birds (Bennett and Lamoreux 2003, Hoekstra 2006, Hubbard et al. 2010). Melanocytes that produce higher levels of the darker eumelanin relative to pheomelanin result in darker coat or feather colors, and vice versa. Amphibians were traditionally thought to produce only eumelanin, leading to the view that variation in melanic pigmentation is solely due to the amount, color and distribution of eumelanin in melanophores (reviewed by D'Alba and Shawkey 2019). This, however, contrasts with the notion that pheomelanin is a more straightforward product of the melanogenesis pathway, and requires less regulatory/catalytic proteins than eumelanogenesis (Schallreuter

2007, Ito and Wakamatsu 2008, McNamara et al. 2021). In fact, pheomelanin was recently uncovered in the skin of the pipid frog *Hymenochirus boettgeri* (Wolnicka-Glubisz et al. 2012), suggesting that this pigment may play an unappreciated role in amphibian coloration.

In some anuran species the pigment pterorhodin, which produces a wine-red coloration, is produced in a specialized kind of melanophore (Bagnara et al. 1973). This pigment is thought to be unique to frogs, and has only been found in species of the Neotropical subfamily Phyllomedusinae and the Australo-Papuan family Pelodryadidae, which appear to have convergently evolved pterorhodin-containing melanophores (Bagnara et al. 1973, Bagnara 2003). Although these melanosomes display similar cellular structure and early development to melanin-containing ones, very little is known about pigment synthesis in these organelles (Bagnara 2003), or their prevalence among other groups of amphibians or vertebrates in general. Future pigment surveys of amphibian skins will most likely reveal the extent to which pheomelanin, pterorhodin, and perhaps other uncharacterized melanosome-derived pigments contribute to coloration in amphibians.

Xanthophores

Pteridines are also synthesized metabolically in xanthophores (Obika and Bagnara 1964, Stackhouse 1966), and variation in the pteridine synthesis pathways has been shown to produce noticeable changes in amphibian pigmentation (Bagnara et al. 1978). For example, experimentally inhibiting the enzyme *xdh* (xanthine dehydrogenase) in tiger salamanders (*Ambystoma tigrinum*) results in reduced pterin content in xanthophores, which translates to a much darker appearance, devoid of the characteristic yellow pigmentation of this species (Frost and Bagnara 1979). Recent work has shown differential expression of genes involved in pteridin synthesis between different color morphs of some frog species (Stuckert et al. 2019, Rodríguez et al. 2020, Stuckert et al. 2021).

Carotenoids, on the other hand, cannot be synthesized by most animals, except for a few arthropod species that have acquired carotenoid synthesis genes via horizontal transfer from fungi (Moran and Jarvik 2010, Altincicek et al. 2012, Cobbs et al. 2013). They are instead obtained from dietary sources, and sometimes modified before deposition in their target tissue (Goodwin 1984). In vertebrates, variation in carotenoid-based coloration is usually achieved either through modulation of pigment uptake and/or deposition rates, or through chemical modification of carotenoids to alter their light absorption properties (Goodwin 1984, Toews et al. 2017). For instance, within a population of *Oophaga pumilio*, the skin concentration of some carotenoids (but not others) is strongly correlated with skin brightness (Crothers et al. 2016). In the polytypic frog *Ranitomeya sirensis*, red and yellow populations differ markedly in skin expression levels of the cytochrome P450 enzyme CYP3A80, which is part of a protein family known to convert yellow dietary carotenoids into red ketocarotenoids. The higher expression of this enzyme in red frogs is thought to result in much higher accumulation of red than yellow carotenoids in xanthophores (Twomey et al. 2020a), providing a clear example of how punctual changes in carotenoid metabolism pathways can produce coloration differences. Similar mechanisms have been shown or suggested to underlie red/yellow/orange variation across vertebrate taxa, including birds (Lopes et al. 2016, Mundy et al. 2016, Twyman et al. 2018, Hooper et al. 2019), squamates (Andrade et al. 2019), and fish (O'Quin et al. 2013), which suggests that metabolic carotenoid modification may be a common route in vertebrate color evolution.

Iridophores

While pigmentary chromatophores produce color by absorbing light at specific wavelengths, iridophores do so by reflecting or refracting white or specific spectra of light. These cells contain stacks of reflective platelets, usually made of guanine, which are effectively transparent but have very high refractive indices. Their stacked arrangement in the cytosol, which has a lower refractive index, results in particular wavelengths being reflected. Multiple aspects of the guanine platelets, such as their width, separation, and orientation, or the height of platelet stacks determine the wavelengths reflected by iridophores (Bagnara 1966, Land 1972). For example, in dendrobatid poison frogs, iridophores have been shown to be a major contributor to hue variation, with thicker guanine platelets being associated with longer wavelength (i.e., redder) hues, and thinner platelets associated with shorter wavelength (i.e., bluer) hues (Twomey et al. 2020b).

In the majority of documented cases, the color of a patch of skin results from the combined action of the three chromatophore types. For instance, green can be obtained by combining the absorption of short wavelengths by yellow xanthophores and long wavelengths by melanophores with the amplification of medium wavelengths by iridophores. Blue is often achieved when melanophores absorb long wavelengths and iridophores selectively reflect short wavelengths (Bagnara et al. 2007). Therefore, to understand the mechanisms underlying color variation, it is necessary to consider both the nature of individual chromatophores and their interactions.

Physiological Chlorosis and Fluorescence

Although the majority of variation in amphibian coloration is thought to be mediated by the arrangement of chromatophores in the epidermis, there is mounting evidence that other mechanisms play an important role in the coloration of multiple species. Chief among these is physiological chlorosis: the accumulation of the biliverdin in the blood, lymph, bones, and soft tissues, whose green/blue-green color is visible through translucent skin (Fig. 1B), which at least partially lacks chromatophores (Cabello Ruz 1943, Barrio 1965). Biliverdin is a green pigment that results from the catabolism of heme from senescent red blood cells. In humans, this pigment accounts for the greenish color of aging bruises. In most vertebrates, biliverdin is either directly excreted (fish, amphibians, birds, non-avian reptiles; Colleran and O'Carra 1977, Cornelius 1991) or reduced to bilirubin and further broken down for excretion (mammals; Colleran and O'Carra 1977). In chlorotic frogs, on the other hand, biliverdin is accumulated, with some species exhibiting biliverding plasma concentrations orders of magnitude higher than those found in non-cholortic vertebrates (Barrio 1965). This is achieved through the action of a recently identified group of serpin proteins that bind biliverdin, presumably preventing its excretion. Furthermore, these biliverdin-binding serpins modify the spectral properties of biliverdin, which at least in one species (*Boana punctata*) results in a fine-tuning of the frogs' color to very closely match the leaves on which they perch, most likely to enhance camouflage (Taboada et al. 2020).

Chlorosis mediated by biliverdin and its associated serpins has been experimentally demonstrated in 11 species of the families Hylidae and Centrolenidae, but at least 430 species across 11 families are known to accumulate green pigments in their blood, lymph, bones, or soft tissues. The phylogenetic distribution of these species points to more than 40 convergent origins of physiological chlorosis in anurans (Taboada et al. 2017b, Taboada et al. 2020). Many of the biliverdin-binding serpins that have been identified in Hylid frogs appear to be paralogous, suggesting that different, closely related proteins have repeatedly evolved the ability to bind biliverdin in different lineages (Taboada et al. 2020). This scenario points to mechanisms other than chromatophore-based coloration as underappreciated avenues for the evolution of coloration in anurans, and perhaps other amphibians.

In addition to green coloration, chemicals in the lymph and other tissues have recently been shown to produce fluorescence in two species of hylid frogs, *Boana punctata* and *B. atlantica* (Taboada et al. 2017a,b). A combination of newly discovered fluorophore molecules called hyloins, found in the frogs' lymph and serous cutaneous glands, produce green fluorescence when excited by UV-A/blue light. Under dawn and moonlight conditions, fluorescence was estimated to account for 18-30% of a frog's reflectance (Taboada et al. 2017a). Based on other traits that appear to be correlated with fluorescence, Taboada et al. (2017b) speculated that this phenomenon may be present in over 200 species across seven anuran families. Since then, several additional examples of fluorescence have been reported across the amphibian phylogeny (Goutte et al. 2019, Lamb and Davis 2020), but the mechanisms underlying fluorescence in these cases remain largely unstudied. The fact that entirely new groups of fluorescent molecules and biliverdin-binding proteins associated with amphibian coloration have been discovered in the last few years underscores the importance of functional research to acquire a more complete understanding of evolution of coloration (and other phenotypes).

The Molecular Basis of Macroscopic Color Patterns

Color patterning, the spatial organization of an organism's color patches, is driven by differential expression of genes involved in pigmentation (or other coloration) pathways, and by chromatophore interactions in different regions of the skin. Although less well understood than the mechanisms governing color itself, the molecular basis of some common patterns have been characterized (Mills and Patterson 2009, Irion et al. 2016). For example, in the frogs *Xenopus laevis* and *Rana forreri*, ventrally-expressed inhibitors suppress ventral melanization, resulting in much lighter ventral than dorsal pigmentation (Fukuzawa and Ide 1988, Fukuzawa et al. 1995). This same mechanism underlies the lighter ventral patterns of multiple species of mice and fish (Millar et al. 1995, Zuasti 2002). In both mice and fish the gene *agouti* (or its orthologs) has been identified as the melanization suppressor (Millar et al. 1995, Zuasti 2002). This also seems to be the case in at least some species of frogs (Fukuzawa et al. 1995, Goutte et al. 2021), but has not been fully confirmed.

The molecular mechanisms underlying striped color patterns in amphibians and fish have received considerable attention. In several species of salamanders, newts, and zebrafish of the genus *Danio*, dark stripes are produced by dense aggregations of melanophores, which are absent or nearly absent in the lighter inter-stripes (Mills and Patterson 2009, Irion et al. 2016). For some salamander species, the melanophore-free region is formed as melanophores migrate away from the lateral line primordium during larval development (Parichy 1996a,b). In zebrafish, stripe position is also achieved through chromatophore migration during development. In addition to interactions between chromatophores and the tissue environment, interactions between different chromatophore types, both over short (Watanabe et al. 2006, Eom et al. 2012) and long (Eom et al. 2015) distances, play an important role in defining the spatial localization of color patches. Iridophores initially establish the location of the inter-stripe, and serve as a reference for posterior xanthophore and melanophore migration and differentiation to produce the final color pattern (Maderspacher and Nüsslein-Volhard 2003, Frohnhöfer et al. 2013, Patterson and Parichy 2019). Although the cellular and molecular processes responsible for defining color patterns have not been studied in detail in most species of amphibians, we anticipate that the renewed interest in the functional underpinnings of coloration among herpetologists, facilitated by recent technological advances for functional biology, should yield important advances on this subject in the near future. This can be evidenced by recently increasing influx of studies on the genetic (e.g., Goutte et al. 2021) and transcriptomic (e.g., Burgon et al. 2020, Rodríguez et al. 2020, Stuckert et al. 2021) bases of amphibian color pattern variation.

Coloration Change through Ontogeny

The life cycle of most amphibian species is marked by drastic phenotypic changes over time given its biphasic nature, in which larvae transition from an aquatic to a terrestrial lifestyle and body plan. Accordingly, coloration changes widely across an individual's lifetime, with larval and adult forms displaying considerably different colorations. Embryos acquire visible coloration patterns early in development, as chromatophore cells originating in the neural crest, become differentiated, and migrate to different parts of the body to produce a given color pattern (Bagnara et al. 1978). As tadpoles hatch and continue to grow, they eventually reach their "characteristic" larval coloration, with darker, more complex, or more contrasting coloration tending to appear at later stages of development (Thibaudeau and Altig 2012). As with other phenotypes, tadpoles of some species exhibit considerable phenotypic plasticity in coloration. For example, tadpoles of the hylid frog *Dryophytes chrysoscelis* develop considerably different tail pigmentation when reared in the presence or absence of predators or their cues (McCollum and Van Buskirk 1996, McCollum and Leimberger 1997), and larvae of the Tiger Salamander *Ambystoma mavoritum nebulosum* develop darker or paler colorations depending on the substrate and water turbidity where they are reared (Fernandez and Collins 1988).

The metamorphic climax involves an organism-wide remodeling of tissues and functions as tadpoles transition into frogs. In the stages surrounding metamorphosis, as the skin is remodeled, pigmentation changes drastically (Kemp 1961, Heatwole and Barthalmus 1995, (Parichy et al. 2006). During this period, both existing and newly differentiated pigment cells migrate extensively, organizing themselves in dermal chromatophore units (Fig. 1A), and usually culminating in the individual's adult coloration (e.g., DuShane 1943, Stearner 1946, Bagnara et al. 1968, 1978, Yasutomi 1987, Lechaire and Denefle 1991). Although the precise mechanisms

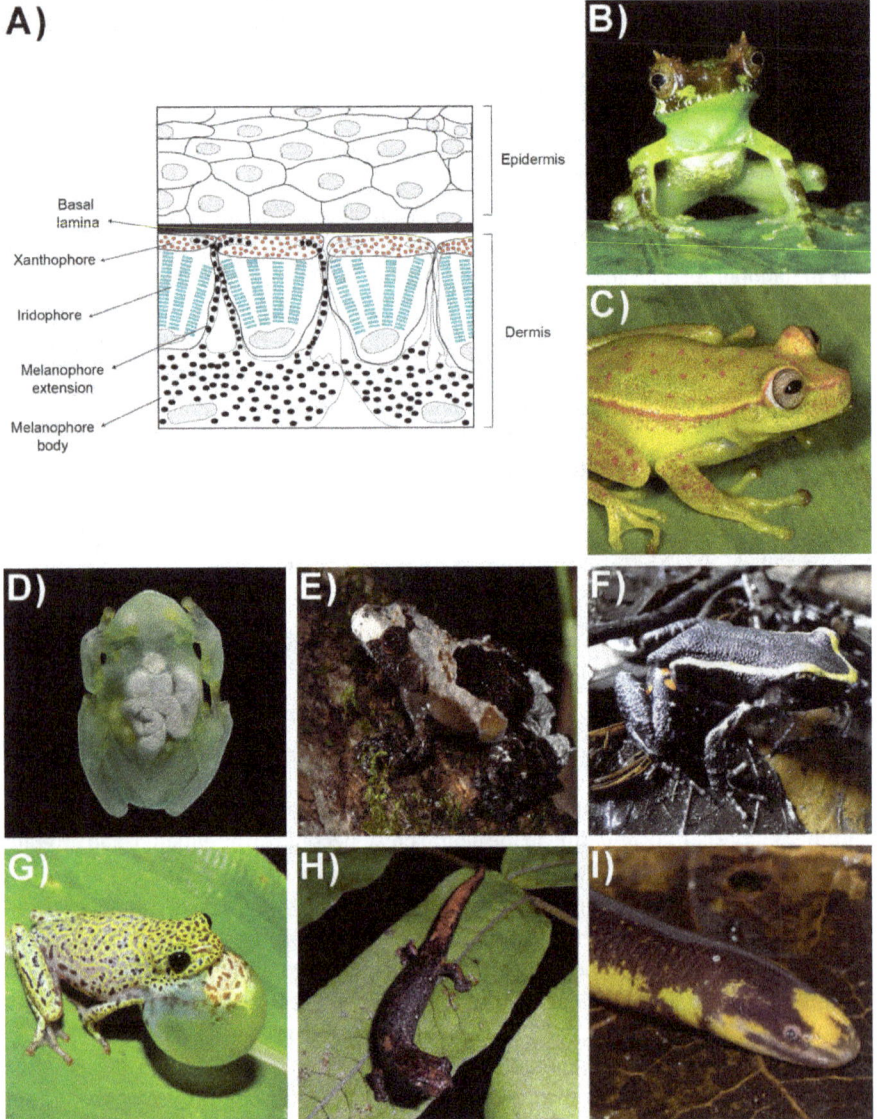

Figure 1. (A) Structure of dermal chromatophore units on a sagittal skin section. Melanosomes on the left melanophore are scattered, while those on the right are retracted. (B) Physiological chlorosis in *Gracixalus supercornutus*; (C) Biofluorescence in *Boana punctata*; (D) Transparent skin in *Hyalinobatrachium fleishmanni;* (E) *Theloderma asperum* displaying bird-dropping-like coloration; (F) Putative mimetic coloration of *Leptodactylus lineatus*; (G) Conspicuous sac coloration in *Hyperolius* sp. (H) Conspicuous coloration in *Bolitoglossa tamaense* I) Conspicuous coloration in *Rhinatrema bivittatum*. Jodi Rowley B, E; J.P. Lawrence C, D; Bibiana Rojas F; Daniela Rößler G; Aldemar A. Acevedo H; Antoine Fouquet I.

underlying these transformations have not been characterized, they are thought to be mediated by endocrine hormones, many of which mediate metamorphic changes in the skin (Heatwole and Barthalmus 1995, Parichy et al. 2006). For instance, thyroxine, one of the thyroid hormones known to trigger several metamorphic changes, induces melanophore migration from the dermis to the epidermis of frog integument tissue cultures (Yasutomi 1987). Work on zebrafish showing that thyroid hormones regulate pigment cell maturation (Saunders et al. 2019) is in line with this notion.

Although most amphibian species do not experience considerable changes in coloration past metamorphosis, there are several documented instances of post-metamorphic color change in frogs (e.g., Richards and Nace 1983, Duellman and Ruiz-Carranza 1986, Vaissi et al. 2018, Bueno-Villafane et al. 2020, and reviewed by Hoffman and Blouin 2000) and salamanders (e.g., Fernandez and Collins 1988). In most cases, changes in coloration consist of a general darkening or lightening of an individual's colors, or the loss/gain of pattern elements such as spots or stripes (Hoffman and Blouin 2000), with some species experiencing quite drastic changes. The majority of ontogenetic coloration change reports are in the family Hyperoliidae, especially the genus *Hyperolius*. Juveniles of multiple species in this genus are pale green and transition to dark-green or brown coloration in adulthood, often gaining or losing pattern elements. In other species, juveniles are brown with an hourglass pattern which fades or is otherwise modified over time (reviewed in Hoffman and Blouin 2000). It is frequent for only one sex to experience ontogenetic color change in hyperoliids (Schiøtz 1967, Richards 1976). Studies on the hormonal underpinnings of sexual dimorphisim in these species have shown that the adult pattern can be induced at metamorphosis by exposing tadpoles to estrogen and/or testosterone, suggesting that ontogenetic color change is intrinsically linked to sexual maturation (Richards 1982, Hayes and Menendez 1999).

Aside from *Hyperolius*, there are a few noteworthy examples of drastic ontogenetic coloration change in other taxa. Perhaps the most commonly cited case is that of the newt *Notophthalmus viridescens*, where post-metamorphic juveniles (i.e., efts) are bright-red or orange, while adults display dark olive-green coloration (Kelly 1878, Gage 1891). Color change is accompanied by morphological change in several other traits, including tail and limb morphology. In fact, efts and adults differ so much in appearance that taxonomists considered them separate species for over half a century (Rafinesque 1820, Kelly 1878). Given the toxins present in both eft and adult skins, eft coloration is often considered aposematic (Hurlbert 1970, see Aposematism section below for further information). Similarly, in the microhylid frog *Oreophryne ezra* juveniles display yellow spots contrasting against a glossy black background, and gradually transition into plain apricot/brown adults (Kraus and Allison 2009). Adults are nocturnal and call from concealed perches while juveniles are diurnal and perch on conspicuously exposed locations. Although they do not seem to secrete toxic/distasteful chemicals, their coloration closely resembles that of a putatively toxic beetle of the genus *Pantorhytes*, which has led to the idea that *O. ezra* mimic *Pantorhytes* sp. as juveniles, and transition to a cryptic lifestyle as adults (Bulbert et al. 2018). A third case occurs in poison frogs of the genus *Phyllobates*, where metamorphs are black with a thin yellow, green or orange dorsolateral stripe.

In two species, *P. bicolor* and *P. terribilis*, the black background gradually gives way to a yellow suffusion and eventually a solid bright-yellow coloration, while in all other species coloration remains unchanged after metamorphosis (Myers et al. 1978, Silverstone 1976). A similar situation is present in the centrolenid genus *Nymphargus*, where *N. anomalus* and *N. rosada* display the chlorotic green skin and bone coloration characteristic of glass frogs, but over time become tan brown with white bones (Rada et al. 2017). In both *Phyllobates* (Márquez et al. 2020) and *Nymphargus* (Castroviejo-Fisher et al. 2014, Guayasamin et al. 2020), very similar instances of ontogenetic color change appear to have evolved independently in closely related (but not sister) species, suggesting that relatively small modifications of the developmental program of these species may trigger appreciable ontogenetic coloration changes.

Although the molecular and cellular basis of chromatophore differentiation and migration during embryonic and larval development have been studied to a considerable extent (see citations in previous paragraphs), the mechanisms underlying coloration ontogeny around and (especially) after metamorphosis still remain an open field. Evidence from *Ambystoma* salamanders suggests that pre- and post-metamorphic color patterns may be functionally decoupled (Parichy 1998), suggesting that larval and adult coloration may be regulated by at least partially distinct genetic networks. Recent work on *Salamandra* salamanders however showed that environmentally-induced changes in larval coloration influence adult coloration (Sanchez et al. 2019), suggesting at least some degree of coupling. Further studies on the functional basis of coloration ontogeny across amphibian life cycles have the potential to yield interesting insights on multiple aspects of evolutionary biology, ranging from the evolution of life stage-specific ecological strategies to the role of ontogenetic modularity in phenotypic evolution.

Short-term Plastic Color Change

In addition to ontogenetic change, vertebrate color patterns can also change over shorter time scales, and in a reversible manner. Famous examples of color change include the many polar birds and mammals that seasonally molt their coats or feathers to match their highly seasonal environment, or reptiles (e.g., chameleons) that can drastically change their coloration within seconds. In birds and mammals, most color change occurs through the shedding and posterior re-growth of hairs or feathers, which allows for the formation of a new color pattern at each molt (Zimova et al. 2018). In all other vertebrate groups, rapid color change is generally achieved through the transport and rearrangement of chromophores (i.e., pigments, platelets), usually within chromatophores (Sköld et al. 2012, Ligon and McCartney 2016), or through changes in chromatophore cell shape.

The best studied example of this phenomenon is melanosome aggregation and scattering, in which melanosomes are transported between the melanophore body and its periphery through a network of microtubules (Nascimento et al. 2003, Aspengren et al. 2009). When melanosomes are scattered throughout the cytosol, the cells convey a darker appearance than when melanosomes are aggregated at the center. This effect is accentuated if melanosomes travel up the cell's dendritic processes and occlude the iridophores and xanthophores above them (Bagnara et al. 1968, Tuma and Gelfand 1999, Aspengren et al. 2009, Sköld et al. 2012). This process has been replicated *in vitro*, and is often used as a model system to study intracellular transport and molecular motors (reviewed by Tuma and Gelfand 1999). In amphibians, melanophore movement has been shown to underlie rapid color change in multiple contexts, such as the lightening and darkening of frog skin during background matching (e.g., Bagnara et al. 1968, Nielsen 1978), and possibly thermoregulation (e.g., Fernandez and Bagnara 1991), as well as changes from dull to bright colors during breeding (e.g., Kindermann and Hero 2016).

Xanthophores can also produce short-term color changes through microtubule-guided scattering and aggregation dynamics similar to those observed in melanophores, which are also mediated by a similar set of hormones (reviewed by Ligon and McCartney 2016). However, the movement patterns of pigment-containing vesicles seem to be different to those of melanophores, possibly due to differences in microtubule arrangement (Byers and Porter 1977, Beckerle and Porter 1983). The majority of research on this subject has, however, occurred on non-amphibian taxa, mostly teleost fish. Iridophores, on the other hand, can effect rapid color change through the sliding and re-orientation of platelets, as well as their vertical expansion or compression. For example, changes in iridophore shape and platelet orientation accompany plastic changes from light to darker green colors in *Hyla cinerea* and *H. arborea* (Nielsen 1978). Again, beyond a few examples (e.g., *H. cinerea* and *H. arborea*; Nielsen 1978, Nielsen and Dyck 1978), most studies on iridophore-mediated color change have focused on fish (Sköld et al. 2012, Ligon and McCartney 2016) or reptiles (e.g., Teyssier et al. 2015), while amphibian systems remain relatively unstudied.

The Genetic Basis of Variation in Coloration

In addition to the functional mechanisms that produce phenotypic traits, from an evolutionary perspective, it is important to understand the genetic basis of variation in such traits, for example between species or within a population. In amphibians, most studies of the genetic basis of variation in coloration have focused on species with discrete morphs that vary within populations (i.e., polymorphic species), and have been aimed at elucidating the mode of inheritance of these morphs (e.g., Moriwaki 1953, Goin 1950, Fogleman et al. 1980, Richards and Nace 1983, Summers et al. 2004, O'Neill and Beard 2010, Richards-Zawacki et al. 2012, reviewed by Hoffman and Blouin 2000). A generality that can be drawn from these studies is that the absence of certain colors (e.g., in albinistic morphs) or pattern elements (e.g., spots or stripes) tends to be recessive, which points to an association with loss-of-function alleles. The majority of studied polymorphisms seem to be controlled by one or a few loci, which is expected considering that the maintenance of discrete morphs is greatly facilitated by simple (i.e., Mendelian) modes of inheritance. The genetic basis of more complex (i.e., quantitative) variation has received much less attention, even though it putatively accounts for a large percentage of variation both within and between populations. Notably, Vestergaard and co-authors (2015) studied color pattern variation in a hybrid zone between two morphs of *Ranitomeya imitator* to estimate the number of genes that underlie color and pattern differences in this system. Their results indicate that one or two loci of major effect may be responsible for most of the pattern differences between morphs. Whether this is a general pattern among amphibians remains unknown.

Beyond characterizations of their mode of inheritance, the genetic basis of differences in coloration has been sparsely studied in amphibians. A salient avenue of research in this regard is identifying the genetic elements that underlie coloration variation. The identity and history of the loci underlying phenotypic variation provide key insights on multiple aspects of phenotypic evolution, such as the origin of adaptive variants (e.g., mutation, introgression, selection on standing variants) or the timing and nature of selection they've experienced, and can inform functional studies by pointing to candidate genes or pathways. To the best of our knowledge, to date, only two studies have identified candidate genes for color pattern differences within species in the wild. Posso-Terranova and Andrés (2017) sequenced a region of the *mc1r* gene, a well-known member of the vertebrate melanogenesis pathway that underlies melanin-based coloration variants in multiple species, in the poison frogs *Oophaga histrionica* and *O. lehmani.* They found an association between background color (black vs. brown) and *mc1r* alleles, with black frogs in both species carrying alleles exhibiting premature stop codons, which resulted in considerably shorter proteins that lacked multiple functional domains present in most other vertebrates. Truncated *mc1r* proteins have been mostly associated to albinistic or otherwise pale-colored forms in other species (e.g., Everts et al. 2000, Gross et al. 2009, Xiao et al. 2019) and are usually considered to be non-functional. Further work to understand how the truncated *Oophaga mc1r* alleles function in melanin pathways could provide interesting insights into the functional basis of melanic pigmentation, and trans-membrane receptor proteins in general. Goutte et al. (2021), on the other hand, conducted a genome-wide association study aimed at identifying the genetic basis of vertebral stripe presence/absence in the grass frog *Ptychadena neumanni,* and found a strong signature of association between vertebral stripes and genetic variants surrounding an ortholog of *agouti,* which is also part of the melanogenesis pathway, and has been associated to melanic pigmentation variants across vertebrate taxa. Although vertebral stripes are polymorphic in several species of *Ptychadena,* the history of *agouti* variants in the genus suggests that the alleles behind each morph are not the same across species, pointing to a dynamic evolution of this polymorphism over the genus's history.

The size and repetitive content of amphibian genomes (e.g., Rogers et al. 2018, Sun et al. 2020) make the genome-wide association approaches often used to map phenotypes to genomic regions costly and methodologically challenging. A commonly used approach to partially circumvent this limitation is comparing gene expression profiles between relevant tissues (different regions of the skin, for example) or developmental stages (e.g., Burgon et al. 2020, Stuckert et al. 2021). Although these studies are limited in the extent to which they can pinpoint the genetic variants directly responsible for (heritable) phenotypic variation, they can certainly identify candidate genetic pathways and even genes for further exploration. Furthermore, they can also provide functional validation for genetic association studies. As sequencing and functional genetic (e.g., genome editing) technologies continue to advance, we expect to see an increase in functional evolutionary genetic studies of coloration in amphibians, which should lead to more solidly bridging our understandings of the molecular,developmental, ecological, and evolutionary drivers of coloration evolution.

Interactions Between Predators and Amphibian Prey

Considerable research on amphibian coloration has focused on its ecological function, with particular emphasis on antipredator defenses. Coloration can broadly be viewed on a gradient from cryptic to conspicuous, and the opposite ends of the spectrum could be considered camouflage or aposematism, although recent research has highlighted the multi-functionality of coloration that may belie this spectrum as overly simplistic (Reynolds and Fitzpatrick 2007, Bell and Zamudio 2012, Barnett et al. 2018). Because of their ubiquity and wide range of functional coloration, amphibians are model systems for understanding the evolution and function of color. For example, a large part of what is currently known on the evolution of aposematism derives from research on dendrobatid poison frogs (e.g., see Summers and Clough 2001, Santos et al. 2003, Saporito et al. 2007, Wang and Shaffer 2008, Maan and Cummings 2012, Rojas et al. 2014a,b, Santos et al. 2014, Lawrence et al. 2019, Carvajal-Castro et al. 2021). Other aspects of coloration, such as deimatic displays, have been considerably less studied despite appearing to be somewhat common in amphibians (Fig. 4), particularly in some taxa (e.g., tree frogs, myobatrachid frogs). In this section, we will discuss some of the particularly notable examples of antipredator coloration, the research surrounding them, and what is yet unknown.

Camouflage

Camouflage, where species display coloration meant to avoid detection or recognition by predators (Nokelainen and Stevens 2016), can largely be classified in three main categories: background matching, disruptive coloration, and masquerade (but see Ruxton et al. 2018 for further description of camouflage). In short, animals that employ background matching aim to blend in with their surroundings (e.g., Fig. 2A), while those that employ disruptive coloration aim to prevent their shape from being discerned by predators (e.g., Fig. 2E). Those that employ masquerade, in contrast, aim to look like an inedible object in the environment (i.e., stick, rock, leaf; Skelhorn et al. 2010). While camouflage is the most common anti-predator strategy among amphibians, what type of camouflage they exhibit is largely variable among different groups of amphibians.

Frogs that employ camouflage are extremely variable in body plan, which allows them to capitalize on different strategies. When at rest, for example, frogs in the genus *Cruziohyla* look like leaves, a resemblance further accentuated by fringed extensions on their legs that make them appear to be the insect-damaged edges of leaves and light blotches reminiscent of lichens on their dorsum (Toledo and Haddad 2009). These particular frogs however, when active, may display startle coloration as well (see below) as their flanks have bright orange coloration (Gally et al. 2014), as in other phyllomedusine frogs (but see Robertson and Greene 2017 for possible social function of these colors). Frogs such as cricket frogs (*Acris*) exhibit impressive

Figure 2. Examples of (putative) camouflage in amphibians. (A) *Pseudophryne guentheri*, (B) *Megophrys nasuta* (C) *Acris crepitans*, (D) *Hyla cinerea*, (E) *Pristimantis zeuctotylus*, (F) *Rhinella aff. margaritifera*, (G) *Crinia signifera*, (H) *Ceratophrys cornuta*, (I) *Cryptobranchus alleganiensis*. Image Credits: J.P. Lawrence A, B, C, D, G, H. Todd Pierson I; Bibiana Rojas E, F.

background matching that is accentuated by jump behavior. These frogs will jump a long distance, and upon landing, will change their body orientation. This feature, coupled with their cryptic coloration makes them exceptionally difficult for predators to detect (Cox et al. 2020).

   While frogs seem to show greater variability in body plan which can help them camouflage, salamanders have relatively little variation in body form (Bonett and Blair 2017). This may be because most salamanders are fossorial or aquatic, thus extra extensions of skin might make those lifestyles difficult. As such, camouflage in salamanders could perhaps be best characterized as masquerade or background matching. With their elongated and slender shape, many resemble sticks or roots, and likely can avoid detection from would-be predators. Perhaps the most impressive example of putative masquerade in salamanders occurs in the cryptobranchids, which include the Chinese and Japanese Giant Salamanders, and the North American Hellbender (*Cryptobranchus alleganiensis*, Fig. 2I). These aquatic salamanders have fringes of skin along their flanks meant to increase surface area for oxygen exchange in cool waters (Guimond and Hutchison 1973). However, coupled with their mottled brown and black coloration, these salamanders appear to be little more than waterlogged pieces of decaying wood. Thus far, however, the function of this coloration is largely speculative and warrants further research.

As mentioned above, a particular case of camouflage is masquerade, where an organism looks like an inanimate object in its environment (Skelhorn et al. 2010). Masquerade is common in insects (i.e., leaf-mimicking katydids and walking sticks), and, while it appears to also be rather common in a number of species of amphibians (i.e., *Phyllomedusa*, *Cruziohyla*, and *Theloderma*), formal examination has largely been overlooked. An exception is *Rhinella margaritifera* (Barnett et al. 2021), an Amazonian toad whose coloration and body shape make it resemble dead leaves in the leaf litter (Fig. 4F). Other species appear to mimic bird droppings (i.e., *Theloderma asperum* or *Dendropsophus marmoratus*, Fig. 1E), but no research to our knowledge has tested whether this is indeed an example of masquerade.

As camouflage relies on the idea that predators will have difficulty in detecting or recognising prey, within-population variability in prey color (i.e., polymorphism) is evolutionarily favorable (Bond and Kamil 2002). Variability in phenotype among individuals results in predators having difficulty in developing a search image for prey species, thus making detection more difficult (Bond and Kamil 2002, Bond and Kamil 2006, Bond 2007). Cryptic coloration is characterized by slow diversification rates compared to conspicuous coloration. Despite this, crypsis appears to be the more evolutionarily favorable state, as there are numerous instances of conspicuous lineages reverting to crypsis, but relatively few of the opposite (Arbuckle and Speed 2015). Furthermore, polytypic conspicuous lineages appear to be more likely to revert to crypsis than the reverse, supporting the idea that variation is evolutionarily favorable in cryptic species (Arbuckle and Speed 2015)

Aposematism

Aposematism is a widespread strategy in which a warning signal (e.g., conspicuous coloration) is coupled with a secondary defense (i.e., noxious smell, distasteful secretions, poisonous compounds; Poulton 1890) to warn predators about prey unprofitability. Thus, this defensive strategy relies on predator learning and/or biases towards conspicuous signals to protect prey species (Endler and Mappes 2004, Mappes et al. 2005). Notably, social animals such as birds, many of which feed on amphibians, have been shown to be capable of learning to avoid aposematic signals through social cues and observation without direct experience (Thorogood et al. 2018, Hämäläinen et al. 2020).

Among amphibians, frogs are the most well-studied in terms of understanding the role of aposematic coloration. Aposematic signaling has evolved multiple times in a number of frog families including, but not limited to, Dendrobatidae (Summers and Clough 2001, Fig. 3A, E), Mantellidae (Vences et al. 2003, Fig. 3D), Bufonidae (Bordignon et al. 2018, Fig. 3B-C), and Myobatrachidae (Lawrence et al. 2018, Fig. 3F). Warning signals are widespread in these families, but other families have evolved aposematic signals as well, though usually among individual lineages (e.g., *Eleutherodactylus iberia*; Rodríguez et al. 2011). Perhaps the best known and most studied example of aposematism in frogs is in the poison frogs in the family Dendrobatidae (e.g., Santos et al. 2003, Saporito et al. 2007, Rojas et al. 2014a, b, Casas-Cardona et al. 2018). There are more than 300 species in the family and the vast majority of species display aposematic signaling. With colors that span the entire visible spectrum, dendrobatids have received disproportionate attention to understanding the evolution and ecology of aposematic signaling. However, numerous other groups are notable. Frogs in the genus *Mantella* are restricted to the island of Madagascar and have evolved conspicuous coloration independently. *Mantella* and dendrobatids offer perhaps one of the best examples of convergent evolution in the animal kingdom (Vences et al. 2003). These two taxa shared a common ancestor approximately 170 MYA (Feng et al. 2017), yet are convergent in numerous traits. They evolved the same methods of sequestering defensive alkaloids from invertebrate prey (including many of the same types of toxins; Clark et al. 2005, Moskowitz et al. 2018); they have convergent phenotypes within the taxon (Chiari et al. 2004, Stuckert et al. 2014) as well as show considerable diversity in aposematic phenotypes (Santos et al. 2003, Klonoski et al. 2019); and they have evolved similar sizes and diurnality, as well as similar locomotion (hopping, as opposed to jumping). Furthermore, some species in both groups exhibit egg-feeding behavior, through which mothers pass on toxins to their offspring (Stynoski et al. 2014, Fischer et al. 2019). While dendrobatids are well represented in the literature, *Mantella* are less well-studied, largely due to dwindling populations (Harper et al. 2007, Randrianavelona et al. 2010).

Many species of aposematic amphibians display considerable variation in aposematic signaling within (polymorphism; Rojas and Endler 2013, Beukema et al. 2016) and among populations (polytypy; Summers et al. 2003, Twomey et al. 2016, Klonoski et al. 2019). This seems perplexing, as aposematic signaling is thought to be maintained via positive frequency-dependent selection (Endler 1988, Mallet and Joron 1999, Rönkä et al. 2020). Under the assumptions of positive frequency-dependent selection, novel signals will be removed by predators unfamiliar with new signals (Mallet and Barton 1989). Polymorphism and polytypy are, however, common in aposematic frogs, with numerous species displaying variation in color patterns. The best-studied example of such variation are dendrobatid frogs. Species in this group derive their defensive skin secretions from invertebrate prey, so as invertebrate communities vary across geographic space, so too do defensive compounds of these frogs (Saporito et al. 2007, Saporito et al. 2006). Thus, honest signaling, where signal is predictive of severity of secondary defense (Sherratt and Beatty 2003, Blount et al. 2009, Summers et al. 2015, White and Umbers 2021), has been suggested as driving regional phenotypic variation in this group. However,

Figure 3. Aposematism found in amphibians. (A) *Eleutherodactylus iberia;* (B) *Atelopus* sp.-dorsal; (C) *Atelopus sp.*-ventral; (D) *Mantella baroni;* (E) *Dendrobates tinctorius;* (F) *Pseudophryne corroboree;* (G) *Taricha granulosa;* (H) *Tylototriton shanjing;* (I) *Salamandra salamandra;* (J) *Notophthalmus viridescens;* (K) *Pseudotriton ruber.* Image Credits: Ariel Rodríguez A; Daniela Rößler B, C; J.P. Lawrence D, E, F, G, J, K; Jodi Rowley H; Christoph Leeb I.

experimental evidence of honest signaling has resulted in a convoluted picture. In some species, more conspicuous colors are more toxic (i.e., *Oophaga pumilio*; Maan and Cummings 2012), but in others, the inverse is true (i.e., *O. granulifera*; Wang 2011). Importantly, this toxicity has been measured via injections on mice and/or via quantification of the amount of alkaloids in the frogs' skin, which may yield different results than when palatability of defensive alkaloids is examined

(Bolton et al. 2017, Lawrence et al. 2019). Thus, these patterns of alkaloid variation and predator response warrant further investigation.

Aposematism is less well-studied among salamanders, and most examples of warning signals come from species in the family Salamandridae. Newts and salamandrid salamanders are notable for their potent skin secretions, which in some genera (i.e., *Taricha, Tylototriton;* Fig. 3G-H), are among the most toxic substances naturally produced by animals (Brodie 1968, Brodie et al. 1974, Lorentz et al. 2016). Furthermore, some of these salamanders have an interesting ontogenetic color change throughout their lives. The eastern newt (*Notophthalmus viridescens*) is notable for having a terrestrial subadult phase called an "eft." Efts are bright red and could be considered a classic example of aposematism (Shure et al. 1989; Fig. 3J). This terrestrial phase allows them to disperse across the landscape. When newts leave this subadult phase, they change to a green color (see the Color Change through Ontogeny section above about the proximate mechanisms underlying this ontogenetic color change). The only vestige of the conspicuous coloration of their previous stage are small red spots along the dorsum. While the adults are still toxic, their toxicity is reduced compared to their eft phase. Therefore, it is unclear how much their subdued coloration protects them (Brodie 1968), but the low number of reports of predation of adults suggests that they are protected, too. Aposematic signaling in other salamander families is likely, although much less studied. Many plethodontid salamanders generate sticky or distasteful skin secretions (particularly on their tails) that are known to dissuade predation (Brodie 1977, von Byern et al. 2017), and many species do appear to have conspicuous coloration (von Byern et al. 2017). Future research in salamanders should focus on the evolutionary ecology of coloration in these and other salamander species to better understand the function of conspicuous signals in the group.

Among the three amphibian orders, the ecology, natural history and behavior of gymnophionans (i.e., caecilians) is least understood. This includes, of course, understanding the use of coloration in these animals. Some species do appear to display conspicuous coloration (i.e., *Ichthyophis kohtaoensis, Rhinatrema bivittatum*; Fig. 1I), but the function of this coloration is unclear, partly because these and most species of caecilians are fossorial; understanding why fossorial species would display conspicuous coloration is conceptually difficult (Wollenberg and Measey 2009). However, given that most species of caecilians have reduced eyes and likely have poor vision, it is tempting to assume that their conspicuous coloration, when present, is likely used in antipredator defense as opposed to conspecific signaling. Indeed, current research suggests that conspicuous coloration in caecilians is largely driven by natural selection for when they are above surface, suggesting a mix of aposematic and cryptic defenses (Wollenberg and Measey 2009). Some species, such as *Siphonops annulatus*, do have enlarged poison glands reminiscent of macroglands found in frogs and salamanders, which suggests chemical defense (Arun et al. 2020). However, this is evidently a research area which warrants further investigation.

One interesting aspect of aposematism is that it has been suggested to provide benefits to prey beyond protection from predators, for example allowing prey to gather resources while wandering around freely in their environment (Speed et al. 2010). Using ancestral state reconstruction and multivariate phylogenetic analyses,

Carvajal-Castro et al. (2021) tested this idea and demonstrated that warning coloration and the possession of skin toxins allowed dendrobatid poison frogs to start using phytotelmata instead of larger bodies of water, thus diversifying their tadpole deposition sites. Despite the importance of *both* coloration and chemical defenses in the success of this antipredator strategy, most studies on aposematism continue to focus on coloration in isolation, and to neglect selective pressures other than predation (i.e., sexual selection) in its evolution (Rojas et al. 2018).

Mimicry

Mimicry, which can be broadly classified as either Batesian (defended model, undefended mimic) or Müllerian (defended model, defended mimic), allows species to capitalize on predator experience to improve fitness (Speed 1999). In amphibians, evidence of mimicry is rather sparse. Perhaps the best example of Müllerian mimicry in amphibians is with the imitator poison frog (*Ranitomeya imitator;* Chouteau et al. 2011, Stuckert et al. 2014, Twomey et al. 2016), a species which shows considerable polytypy in the Peruvian Amazon. Phenotypes of this species covary with a number of other congeners with impressive fidelity to phenotype.

Batesian mimicry, however, appears to be less common among amphibians. In salamanders, two potential examples exist. First, highly toxic eft phases of the eastern newt (*Notophthalmus viridescens;* Fig. 3J) have a number of putative mimics that co-occur with this species including red salamanders (*Pseudotriton ruber;* Howard and Brodie 1973, Huheey and Brandon 1974; Fig. 3K) and red-backed salamander (*Plethodon cinereus;* Kraemer and Adams 2014, Kraemer et al. 2015). Notably, erythristic populations exist of the red-backed salamander, and it has been hypothesized that such a population may be a mimic of sympatric newts (Tilley et al. 1982). The other putative example of Batesian mimicry in salamanders is the red-cheeked salamander (*Plethodon jordani*) and red-legged salamander (*Plethodon shermani*) and their potential mimic, the imitator salamander (*Desmognathus imitator, sensu lato D. ochrophaeus;* Tilley et al. 1978). Neither *Plethodon* species has overlapping ranges with one another, but where *D. imitator* is sympatric with either species, it either has an orange cheek patch, mimicking *P. jordani* or orange legs, mimicking *P. shermani* (Labanick and Brandon 1981). On the west coast of North America, *Ensatina eschscholtzii* forms a ring species complex (Moritz et al. 1992). Some of the subspecies of *E. eschscholtzii* (i.e., *E. e. xanthopicta*) have been implicated as part of a mimicry complex with sympatric rough-skinned newts (*Taricha*; Kuchta 2005, Kuchta et al. 2008). By and large, however, little work has been recently conducted on mimicry in salamanders.

Among frogs, Batesian mimicry is poorly understood, too. Perhaps the only example thought to illustrate that phenomenon is that of *Ameerega picta* and *Leptodactylus lineatus* (Fig. 1F). These two species are relatively similar in phenotype and, while *A. picta* does produce defensive compounds (Mebs et al. 2010), *L. lineatus* was long assumed to be nontoxic. However, recent evidence suggests that *L. lineatus* may be defended (Prates et al. 2012), which would make this system an(other) example of Müllerian mimicry. This, and other putative mimicry systems, are yet to be confirmed with rigorous experiments where predator response to the species involved is assessed (e.g., Bolton et al. 2017, Winters et al. 2018, Lawrence et al. 2019).

Mimicry in amphibians, as a whole, is an area of considerable opportunity for novel research. Convergent phenotypes are common across taxa, particularly conspicuous taxa (e.g., *Mantella baroni*/*M. madagascarensis*, *Notophthalmus viridescens* efts/other red salamander species) but, with a few exceptions discussed above, little work has been conducted to determine if such convergent phenotypes are mimetic. Mimicry confers a fitness advantage on mimics that breaks down when models are in low numbers or absent (Valkonen and Mappes 2014). While phenotypic similarity does suggest mimetic complexes, further research is necessary to determine whether it is predation or other selective pressures driving phenotypic similarity.

Deimatic Displays

Deimatic displays, or startle displays, are those that use bright or contrasting coloration to startle a predator, thus giving the prey species opportunity for escape (Umbers et al. 2015, Umbers et al. 2017). Such displays are well documented in insects (Umbers and Mappes 2015, O'Hanlon et al. 2018), but only beginning to be understood in amphibians. Perhaps the best-known putative example of a deimatic display in amphibians is the "unken" reflex, first described in the fire-bellied toads (*Bombina*) of Europe (Löhner 1919, Bajger 1980). While normally cryptic above, fire-bellied toads are characterized by highly conspicuous color blotches across their venter. According to the reports (Bajger 1980), this coloration is normally hidden until the toad is disturbed, at which point it raises its head and reads up with the palms of its feet facing upward, exposing the conspicuous ventral coloration to potential predators. As *Bombina* produce toxic peptides (Michl and Kaiser 1963, Csordas and Michl 1969), this may allow them to capitalize on aposematic coloration that predators can learn from (Kang et al. 2017) while avoiding excessive predation due to the cryptic dorsal coloration. This behavior has also been reported in species of the South American genus *Melanophryniscus* (Bordignon et al. 2018) and Australian *Pseudophryne* (Williams et al. 2000).

Other amphibians also display the so-called unken reflex, particularly salamandrid salamanders. The rough-skinned newts (*Taricha*) are characterized by dull reddish-brown dorsal coloration and bright orange ventral coloration (Johnson and Brodie 1975). Other salamandrids, such as the eastern newt (*Notophthalmus viridescens*), lift their head and tail up to expose the bright coloration when disturbed, presumably to advertise the potent tetrodotoxins that they harbor in their skin (Brodie 1983, Fig. 4A). Other salamanders (i.e., the slimy salamander [*Plethodon glutinosus*] complex) will show similar behavior while excreting sticky white secretions from their tails (von Byern et al. 2017).

In insects, some of the most common deimatic displays involve using false eyespots (Stevens 2005). Some amphibians appear to employ similar tactics, although the effects of putative eyespots as an antipredator defense are poorly understood. In insects, false eyespots are thought to mimic predator eyes, thus startling predators into thinking that the prey may in turn be their predator (De Bona et al. 2015).

Figure 4. Putative deimatism in amphibians. (A) *Notophthalmus viridescens*; (B) *Edalorhina perezi*; (C) *Phyllomedusa* sp.; (D) *Pseudophryne dendyi*; (E) *Pseudophryne semimarmorata*; (F) *Melanophryniscus admirabilis*; (G) *Litoria adelaidensis*; (H) *Dendropsophus marmoratus*; (I) *Pleurodema brachyops*. Image Credits: J.P. Lawrence A, B, D, E, G, H; Daniela Rößler C; Natália D. Vargas F; Luis Alberto Rueda I.

Several species of frogs are known to have eye spots in their groin which are made visible when the frog inflates itself in defense and raises the inferior part of its body. The best examples of these putative deimatic displays using eye spots seem to occur in Leiuperinae, with a number of species displaying coevolution of lumbar glands and eye-like coloration (Ferraro et al. 2021). *Physalaemus nattereri* has black eye spots that correspond with poison glands (Ferraro et al. 2021). Likewise, *Pleurodema brachyops* (as well as other *Pleurodema*) have lumbar glands with contrasting colors resembling eye spots (Fig. 4I, Martins 1989). *Edalorhina perezi*, notably, has conspicuous black spots in its groin surrounded by yellow which could serve as a reasonable mimic for eyes (Fig. 4B), and thus may protect them from predation. While most research into eye spots in *E. perezi* and other frogs suggests defensive purposes, little empirical work has been conducted to examine the effect of these displays on predators.

Visual Illusions

Visual illusions, where pattern (often coupled with movement) causes difficulty in identification of prey species or their trajectory, are another form of protection against predators. Many snakes, for example, display longitudinal stripes which, when moving through vegetation like grass, may hamper predators' ability to identify the head of the snake (Murali and Kodandaramaiah 2016). Furthermore, many snakes have latitudinal bands which may produce a flicker fusion effect when in movement (Jackson et al. 1976); this illusion can confuse predators, too (Titcomb et al. 2014).

While such patterns are not common in amphibians, they may still employ visual illusions. The dyeing poison frog (*Dendrobates tinctorius*, Fig. 3E), while aposematic, has a highly variable pattern, from longitudinal stripes to interrupted markings, which appear to be associated to different types of movement (i.e., frogs with longitudinal elongated patterns move directionally and at a higher speed than frogs with blotched patterns; Rojas et al. 2014a). This association has been suggested to be able to generate the so-called motion dazzle, which is thought to create an illusion of immobility or reduced speed that makes difficult for aerial predators to strike a successful attack on the moving prey (Stevens et al. 2011). While a recent experimental approach did not find support for striped patterns to get increased protection from predators in comparison to interrupted patterns, the authors found that patterned prey are harder to catch than their pattern-less counterparts (Hämäläinen et al. 2015). Similarly, the brightly-colored yellow-and-black striped southern corroboree frog (*Pseudophryne corroboree*) was originally thought to be aposematic, but recent evidence suggests that the pattern may actually be cryptic, particularly from a distance, in the alpine grasses in which they are typically found (Umbers et al. 2020). Notably, Umbers et al. (2020) proposed the potential crypsis of corroboree frog (Fig. 3F) patterns using stationary models. Given the longitudinal dorsal stripes of this species, however, it could be that such a pattern allows for a visual illusion. Among the different types of anti-predator coloration discussed, visual illusions in amphibians is likely least known, though may be most fruitful for research given the broad array of color and pattern found in amphibians.

Sexual Selection on Amphibian Coloration

Sexual selection is perhaps the least-studied aspect in relation to amphibian coloration, possibly because of the prevalence of other sensory modalities (e.g., chemical, acoustic; see Chapters 10 and 11) in mating interactions among amphibians. To date, some of the studies addressing the potential link between colour patterns and sexual selection in urodeles have demonstrated that colors are used in interactions between males and females (Rojas 2017). For example, Secondi et al. (2012) showed the role of UV reflectance in mate choice in the newt *Lissotriton vulgaris*, whereby males appeared less attractive in the absence of UV (Secondi et al. 2012). Eastern red-backed salamanders (*Plethodon.cinereus*) are polymorphic for color patterns (striped and unstriped) throughout part of their range. Field and laboratory experiments suggest that striped males are more attractive to females, which may contribute to positive assortative mating in the wild (Acord et al. 2013).

While acoustic signals are known to be the prominent signal type used during anuran mating (Chapter 11), there is growing evidence that colors may be involved in mating, too, particularly in diurnal species. Among diurnal frogs, *Oophaga pumilio* is inarguably the species in which visual mate choice has stimulated most research (Fig. 5F). In choice experiments done in the laboratory under controlled conditions, *O. pumilio* females have been shown to prefer males with higher dorsal brightness (Maan and Cummings 2009), suggesting that females pay attention to males' dorsal coloration when choosing a mate (Summers et al. 1999, Maan and Cummings 2008, Dreher et al. 2017). Given their extensive geographic variation in coloration, a few studies have also directly addressed whether this variation is maintained via assortative mating for colors. When presented with males from two different color morphs, representing two different populations, females were more likely to spend time with the male from their own population (Summers et al. 1999, Reynolds and Fitzpatrick 2007). A series of studies addressing similar questions using molecular pedigrees (Richards-Zawacki et al. 2012) or intensive field observations (Yang et al. 2019b) indicate that the assortative mating found in the laboratory does not necessarily reflect the patterns seen in the field, and that assortative mate preferences might occur among (some) populations, but not within the same population (Yang et al. 2016, Dreher et al. 2017, but see Gade et al. 2016 for an example of color assortative mating operating in the wild, in a mainland population). This has highlighted the need for more field studies to get a better idea of the role of color patterns in mating interactions under natural conditions. Interestingly, it has been recently shown that *Oophaga pumilio* males and females can imprint on coloration, which has important and previously unexplored implications in potential speciation processes (Yang et al. 2019a).

Colors have also been shown to play a role in mate choice of non-diurnal frog species. This is not as surprising as it intuitively seems, as frogs have been shown to use a dual rod system to discriminate colors at the visual threshold (Yovanovich et al. 2017, Donner and Yovanovich 2020), and to respond to some color cues under low light (Gomez et al. 2010). For instance, Gomez et al. (2009) used video playbacks of frogs emitting identical calls but differing in the color and brightness of their vocal sac to demonstrate that female *Hyla arborea* are more attracted to males with a more colorful sac and salient lateral stripe (Fig. 5G); these two traits are thought to enhance male conspicuousness when combined. Interestingly, the authors hypothesized that carotenoid-based colors in the vocal sac may provide information about male quality, given that carotenoid production has proven to be costly (Gomez et al. 2009). Female *Hyla squirrella* appear to be more attracted to males with larger lateral stripes when presented with two models differing in the size of lateral stripes but emitting identical calls (Taylor et al. 2007) and in *Scaphiopus couchii*, male color patterns are reliable

Figure 5. Amphibian species which display sexual dichromatism or engage in sexual selection based on color. (A) *Hoplobatrachus tigerinus*; (B-C) *Rana arvalis*; (D) *Incilius luetkenii*; (E) *Ambystoma opacum*; (F) Pair of *Oophaga pumilio* in courtship; (G) *Hyla arborea*. Image Credits: Doris Preininger A; Christoph Leeb B, C; Daniel Mennill D; J.P. Lawrence E; Andrius Pašukonis F; Susi Stückler G.

indicators of body size, which may in turn be used as an indicator of male quality by females. Indeed, females have been shown to prefer brighter males (Vásquez and Pfennig 2007). In the Puerto Rican cave-dwelling frog, *Eleutherodactylus cooki*, males with larger proportions of their venter covered with yellow had the highest reproductive success (Burrowes 2000). Interestingly, those males with the highest amount of yellow coloration suffer, too, the highest parasitization by ticks (Longo et al. 2020). For some other species the role of color patterns is not as clear cut, but there is increasing evidence pointing at an important role in species recognition, female choice and male-male competition in species such as *Agalychnis callidryas* (Jacobs et al. 2017) and, possibly, other phyllomedusines (Robertson and Greene 2017).

Interesting cases highlighting the role of coloration in sexual selection involve species with so-called dynamic sexual dichromatism, a term first coined by Doucet and Mennill (2010). These species, which are often explosive breeders (i.e., species that breed during just one or a few days during the year, following torrential rain), exhibit some of the most fascinating behavioral and phenotypic changes to ensure either species or sex recognition, or both (e.g., Ries et al. 2008, Sztatecsny et al. 2012). The Neotropical toad *Incilius luetkenii* (Fig. 5D), for example, forms breeding aggregations around temporary ponds at the start of the rainy season. Males call to attract females and, along with their vocalizations, they change their color from a dull brown to bright yellow (Doucet and Mennill 2010). Using frog models with brown or yellow coloration, the authors demonstrated that, regardless of the time of the day (i.e., both during the day and during the night), males were more likely to attempt amplecting brown models than their brown counterparts, suggesting that color plays a crucial role in sex discrimination in the large aggregations formed during their brief breeding period (Rehberg-Besler et al. 2015). Females, on the other hand, do not seem to show any preference for either yellow or brown males, indicating the female choice is not a driver of male color change (Gardner et al. 2021).

Sex differences in coloration have been reported for numerous amphibian species (e.g., Bell and Zamudio 2012, Bell et al. 2017) but, despite an increasing number of studies, the function, evolution and mechanisms of maintenance of amphibian sexual dichromatism are still poorly understood. Darwin's frogs (*Rhinoderma darwinii*), for example, are sexually dimorphic for color patterns, such that males, particularly at the brooding stage, are greener. Females, on the other hand, are mainly brown (Bourke et al. 2011). Reed frogs (Hyperoliidae) have striking sexual dimorphism in coloration, which is linked to rapid diversification and, presumably, speciation by sexual selection (Portik et al. 2019), and *Epidalea calamita* toads (Zamora-Camacho and Comas 2019) exhibit a sexual dichromatism in which differences in coloration also seem to be linked to habitat, morphology and locomotor performance. However, the function of these differences in coloration between males and females remain unknown. Interestingly, in the golden rocket frog (*Anomaloglossus beebei*), females are the colorful sex, exhibiting a bright golden coloration that contrasts with the brownish coloration of males (Engelbrecht-Wiggans and Tumulty 2019). While the authors suggest that these colors may serve intraspecific communication purposes, no evidence has been provided yet to confirm their hypothesis.

That is also the case in urodeles where, despite the occurrence of sexual dichromatism, conclusive evidence on the significance of such differences between the sexes is, for most part, still lacking. In the spotted salamander, *Ambystoma maculatum*, males have a larger spotted dorsal area and less bright yellow than females (Morgan et al. 2014), while in the marbled salamander, *A. opacum*, males exhibit whiter and a higher proportion of dorsal white coloration (Todd and Davis 2007, Pokhrel et al. 2013, Fig. 5E). Male fire salamanders, *Salamandra salamandra*, exhibit significantly larger yellow spots (Balogová and Uhrin 2015) and a larger yellow-to-black ratio than females, but no differences between the sexes have been found in color traits such as hue, saturation or brightness (Balogová and Uhrin 2015).

Intra-specific Competition in Relation to Coloration

Colors can influence the outcome of agonistic interactions in several taxa, where they are known to be reliable signals of status (e.g., Gerald 2001, Pryke et al. 2001, Whiting et al. 2006). Among amphibians, this aspect has been less studied than in other groups, but there are a few examples going in the same direction. Again, possibly the most studied species in that respect is *Oophaga pumilio*, whose coloration (brightness, in particular) appears to be a reliable indicator of aggressiveness that influences the outcome of agonistic encounters (Crothers et al. 2011, Galeano and Harms 2016, Crothers and Cummings 2015). However, field observations seem to point at factors (e.g., prior residency status) other than color as the determinants of the outcome of agonistic interactions (Yang et al. 2020). Also, populations with individuals conspicuously colored tend to have higher levels of aggressiveness than populations with duller individuals (Rudh et al. 2013). Interestingly, these trends about the role of color in agonistic encounters do not apply only to males. A recent review on frogs of the genus *Mannophryne* reported the females of all known species have a yellow throat with a dark collar whereas in eight of them males change their color to black when vocalizing, showcasing interesting instances of both ontogenetic and dynamic color change, which warrants further investigation (Greener et al. 2020). In *Mannophryne trinitatis*, for example, agonistic encounters between females involve a female adopting an upright posture and pulsating her yellow-colored throat facing an intruder (Wells 1980). However, the role of coloration on female aggressive behavior remains an understudied topic, particularly considering the many instances of reported agonistic interactions between females in species of colorful frogs (e.g., *Dendrobates auratus*, Summers 1989; *Oophaga pumilio*, Meuche et al. 2011; *D. tinctorius*, Rojas and Pašukonis 2019).

Areas of Future Research

Amphibians use coloration in a number of different ways, many of which are very well studied, but we would like to highlight a few areas deserving further research, as well as caution for inferences without further research. Biofluorescent animals have been increasingly discovered in recent years, garnering headlines in scientific and popular media (e.g., platypus, Anich et al. 2021; geckos, Prötzel et al. 2021). Amphibians joined the ranks of biofluorescent animals when researchers discovered that the polkadot tree frog (*Boana punctata*, Fig. 1C) fluoresces under moonlight conditions (Taboada et al. 2017a; see section Physiological Chlorosis and Fluorescence). Further work has shown that fluorescence is widespread across amphibians, at least under artificial UV stimulation (Lamb and Davis 2020). These discoveries are intriguing and novel, and open promising future research avenues on both the organismal and ecological functions of fluorescence in amphibians. For example, work to determine how much fluorescent light is emitted under biologically relevant ambient light conditions, or whether animals interacting with fluorescent signals alter behavior based on the presence or absence of such signals could provide insights on whether amphibian biofluorescence is involved in inter- or intraspecific communication.

Notably, understanding coloration in amphibians where color is not obviously advantageous (i.e., fossorial or nocturnal species) appears to be particularly challenging. For example, the most fossorial group of amphibians, the caecilians, are poorly understood in this regard, which as noted above, is likely due to their secretive nature. Like many caecilians, a number of fossorial frogs (i.e., many myobatrachids and microhylids) display distinct coloration that could potentially serve important functionality in their survival. Conspicuous coloration in fossorial, or even nocturnal, species is a puzzling phenomenon ripe for further research on how amphibians use coloration. Likewise, salamanders show remarkable diversity in coloration, but with the exception of a few notable species (i.e., *Taricha granulosa, Plethodon cinereus*), relatively little study has been conducted on the wide multitude of salamander species, most of whom are fossorial or nocturnal.

As mentioned above, an emerging area of research elsewhere in the field of color biology is that of deimatic displays. These remarkably effective displays are common in other taxa, particularly insects, but have received relatively little attention in amphibians, offering an exciting avenue of research that can provide insight into the dynamic function of coloration.

Our understanding of color, its use and evolution in amphibians, is derived from a relatively small sampling of one of the largest animal clades, which has a particular overrepresentation in frogs. Indeed, even this has been instrumental in providing understanding of everything from cell function and ontogeny to evolution of coloration across the animal kingdom. Amphibians, with their broad array of habits and lifestyles, are frequently models in understanding biology. No different should this be for understanding coloration, its development, evolution, and function.

Acknowledgements

We thank Gregorio Moreno-Rueda and Mar Comas for the invitation to write this chapter. BR is grateful for funding from the Academy of Finland (Academy Research Fellowship, decision No. 318404). RM is supported by the Michigan Society of Fellows and award No. IOS-1827333 from the USA's National Science Foundation. We are indebted to the many researchers that agreed to share their photographs with us (see credits in each figure legend).

References

Acord, M.A., C.D. Anthony and C.-A.M. Hickerson. 2013. Assortative mating in a polymorphic salamander. Copeia 2013: 676–683.

Allen, B.M. 1916. The results of extirpation of the anterior lobe of the hypophysis and of the thyroid of *Rana pipiens* larvae. Science 44: 755–758.

Altincicek, B., J.L. Kovacs and N.M. Gerardo. 2012. Horizontally transferred fungal carotenoid genes in the two-spotted spider mite *Tetranychus urticae*. Biol. Lett. 8: 253–257.

Andrade, P., C. Pinho, G.P. i de Lanuza, S. Afonso, J. Brejcha, C.J. Rubin et al. 2019. Regulatory changes in pterin and carotenoid genes underlie balanced color polymorphisms in the wall lizard. PNAS 116: 5633–5642.

Anich, P.S., S. Anthony, M. Carlson, A. Gunnelson, A.M. Kohler, J.G. Martin et al. 2021. Biofluorescence in the platypus (*Ornithorhynchus anatinus*). Mammalia 85: 179–181.

Arbuckle, K. and M.P. Speed. 2015. Antipredator defenses predict diversification rates. PNAS 112: 13597–13602.

Arun, D., S. Sandhya, M.A. Akbarsha, O.V. Oommen and L. Divya. 2020. An insight into the skin glands, dermal scales and secretions of the caecilian amphibian *Ichthyophis beddomei*. Saudi J. Biol. Sci. 27: 2683–2690.

Aspengren, S., D. Hedberg, H.N. Sköld and M. Wallin. 2009. New insights into melanosome transport in vertebrate pigment cells. Int. Rev. Cell Mol. Biol. 272: 245–302.

Bagnara, J.T. 1966. Cytology and cytophysiology of non-melanophore pigment cells. Int. Rev. Cytol. 20: 173–205.

Bagnara, J.T., J.D. Taylor and M.E. Hadley. 1968. The dermal chromatophore unit. J. Cell Biol. 38: 67–79.

Bagnara, J.T., J.D. Taylor and G. Prota. 1973. Color changes, unusual melanosomes, and a new pigment from leaf frogs. Science 182: 1034–1035. Bagnara, J.T., S.K. Frost and J. Matsumoto. 1978. On the development of pigment patterns in amphibians. Am. Zool. 18: 301–312.

Bagnara, J.T. 2003. Enigmas of pterorhodin, a red melanosomal pigment of tree frogs. Pigm. Cell Res. 16: 510–516.

Bagnara, J.T. and J. Matsumoto. 2006. Comparative anatomy and physiology of pigment cells in nonmammalian tissues. pp. 11–59. *In:* Nordlund, J.J., R.E. Boissy, V.J. Hearing, R.A. King, W.S. Oetting and J. Ortonne [eds.]. The Pigmentary System. Blackwell Publishing Ltd, Oxford, UK.

Bagnara, J.T., P.J. Fernandez and R. Fujii. 2007. On the blue coloration of vertebrates. Pigm. Cell Res. 20: 14–26.

Bajger, J. 1980. Diversity of defensive responses in populations of fire toads (*Bombina bombina* and *Bombina variegata*). Herpetologica 36: 133–137.

Balogová, M. and M. Uhrin. 2015. Sex-biased dorsal spotted patterns in the fire salamander (*Salamandra salamandra*). Salamandra 51: 12–18.

Barnett, J.B., C. Michalis, N.E. Scott-Samuel and I.C. Cuthill. 2018. Distance-dependent defensive coloration in the poison frog *Dendrobates tinctorius*, Dendrobatidae. PNAS 115: 6416–6421.

Barnett, J.B., C. Michalis, N.E. Scott-Samuel and I.C. Cuthill. 2021. Colour pattern variation forms local background matching camouflage in a leaf-mimicking toad. J. Evol. Biol. 34: 1531–1540.

Barrio, A. 1965. Cloricia fisiológica en batracios anuros. Physis 25: 137–142.

Beckerle M.C. and K.R. Porter. 1983. Analysis of the role of microtubules and actin in erythrophore intracellular motility. J. Cell Biol. 96:354–62.

Bell, R.C. and K.R. Zamudio. 2012. Sexual dichromatism in frogs: natural selection, sexual selection and unexpected diversity. Proc. R. Soc. B 279: 4687–4693.

Bell, R.C., G.N. Webster and M.J. Whiting. 2017. Breeding biology and the evolution of dynamic sexual dichromatism in frogs. J. Evol. Biol. 30: 2104–2115.

Belt, T. 1874. The naturalist in Nicaragua: a narrative of a residence at the gold mines of Chontales; journeys in the savannahs and forests. With observations on animals and plants in reference to the theory of evolution of living forms. J. Murray, London, UK.

Bennett, D.C. and M.L. Lamoreux. 2003. The color loci of mice--a genetic century. Pigm. Cell Res. 16: 333–344.

Beukema, W., A.G. Nicieza, A. Lourenço and G. Velo-Antón. 2016. Colour polymorphism in *Salamandra salamandra* (Amphibia: Urodela), revealed by a lack of genetic and environmental differentiation between distinct phenotypes. J. Zool. Syst. Evol. Res. 54: 127–136.

Blount, J.D., M.P. Speed, G.D. Ruxton and P.A. Stephens. 2009. Warning displays may function as honest signals of toxicity. Proc. R. Soc. B 276: 871–877.

Bond, A.B. and A.C. Kamil. 2002. Visual predators select for crypticity and polymorphism in virtual prey. Nature 415: 609–613.

Bond, A.B. and A.C. Kamil. 2006. Spatial heterogeneity, predator cognition, and the evolution of color polymorphism in virtual prey. PNAS 103: 3214–3219.

Bond, A.B. 2007. The evolution of color polymorphism: Crypticity searching images, and apostatic selection. Annu. Rev. Ecol. Evol. Syst. 38: 489–514.

Bolton, S.K., K. Dickerson and R. Saporito. 2017. Variable alkaloid defenses in the dendrobatid poison frog *Oophaga pumilio* are perceived as differences in palatability to arthropods. J. Chem. Ecol. 43: 273–289.

Bonett, R.M. and A.L. Blair. 2017. Evidence for complex life cycle constraints on salamander body form diversification. PNAS 114: 9936–9941

Bordignon, D.W., V.Z. Caorsi, P. Colombo, M. Abadie, I.V. Brack, B.T. Dasoler et al. 2018. Are the unken reflex and the aposematic colouration of Red-Bellied Toads efficient against bird predation? PloS One 13: e0193551.

Bourke, J., K.C. Busse and T.M. Bakker. 2011. Sex differences in polymorphic body coloration and dorsal pattern in Darwin's frogs (*Rhinoderma darwinii*). Herpetol. J. 21: 227–234.

Brandon, R.A. and J.E. Huheey. 1975. Diurnal activity, avian predation, and the question of warning coloration and cryptic coloration in salamanders. Herpetologica 31: 252–255.

Brandon, R.A., G.M. Labanick and J.E. Huheey. 1979a. Learned avoidance of brown efts, *Notophthalmus viridescens louisianensis* (Amphibia, Urodela, Salamandridae), by chickens. J. Herpetol. 13: 171–176.

Brandon, R.A., G.M. Labanick and J.E. Huheey. 1979b. Relative palatability, defensive behavior, and mimetic relationships of Red Salamanders (*Pseudotriton ruber*), Mud Salamanders (*Pseudotriton montanus*), and Red Efts (*Notophthalmus viridescens*). Herpetologica 35: 289–303.

Brodie, E.D. 1968. Investigations on the skin toxin of the red-spotted newt, *Notophthalmus viridescens viridescens*. Am. Midl. Nat. 80: 276–280.

Brodie, Jr., E.D., J.L. Hensel Jr. and J.A. Johnson. 1974. Toxicity of the urodele amphibians *Taricha*, *Notophthalmus*, *Cynops* and *Paramesotriton* (Salamandridae). Copeia 506–511.

Brodie, Jr., E.D. 1977. Salamander antipredator postures. Copeia 523–535.

Brodie, E.D., Jr. and E.D. Brodie 3rd. 1980. Differential avoidance of mimetic salamanders by free-ranging birds. Science 208: 181–182.

Brodie, E.D. 1983. Antipredator adaptations of salamanders: evolution and convergence among terrestrial species. pp. 109–133. In Adaptations to terrestrial environments . Springer, Boston, MA, USA.

Bueno-Villafane, D., A. Caballlero-Gini, M. Ferreira, F. Netto, D.F. Rios and F. Brusquetti. 2020. Ontogenetic changes in the ventral colouration of post metamorphic *Elachistocleis haroi* Pereyra, Akmentins, Laufer, Vaira, 2013 (Anura: Microhylidae). Amph.- Rept. 41: 191–200.

Bulbert, M.W., T.E. White, R.A. Saporito and F. Kraus. 2018. Ontogenetic colour change in *Oreophryne ezra* (Anura: Microhylidae) reflects an unusual shift from conspicuousness to crypsis but not in toxicity. Biol. J. Linn. Soc. 123: 12–20.

Burgon, J.D., D.R. Vieites, A. Jacobs, S.K. Weidt, H.M. Gunter, S. Steinfartz, K. Burgess, B.K. Mable and K.R. Elmer. 2020. Functional colour genes and signals of selection in colour-polymorphic salamanders. Mol. Ecol. 29(7): 1284–1299.

Burrowes, P.A. 2000. Parental care and sexual selection in the Puerto Rican cave-dwelling frog, *Eleutherodactylus cooki*. Herpetologica 56: 375–386.

Byers, H.R. and K.R. Porter. 1977. Transformations in the structure of the cytoplasmic ground substance in erythrophores during pigment aggregation and dispersion: I. a study using whole-cell preparations in stereo high voltage electron microscopy. J. Cell Biol. 75: 541–558.

Cabello Ruz, J. 1943. Biliverdinemia del sapo. Rev. Soc. Arg. Biol. 19: 81–93.

Carvajal-Castro, J.D., F. Vargas-Salinas, S. Casas-Cardona, B. Rojas and J.C. Santos. 2021. Aposematism facilitates the diversification of parental care strategies in poison frogs. Sci. Rep. 11: 19047.

Casas-Cardona, S., R. Márquez and F. Vargas-Salinas. 2018. Different colour morphs of the poison frog *Andinobates bombetes* (Dendrobatidae) are similarly effective visual predator deterrents. Ethology 124: 245–255.

Castroviejo-Fisher, S., J.M. Guayasamin, A. Gonzalez-Voyer and C. Vilà. 2014. Neotropical diversification seen through glassfrogs. J. Biogeogr. 41: 66–80.

Chiari, Y., M. Vences, D.R. Vieites, F. Rabemananjara, P. Bora, O. Ramilijaona et al. 2004. New evidence for parallel evolution of colour patterns in Malagasy poison frogs (*Mantella*). Mol. Ecol. 13: 3763–3774.

Chouteau, M., K. Summers, V. Morales and B. Angers. 2011. Advergence in Mullerian mimicry: the case of the poison dart frogs of Northern Peru revisited. Biol. Lett. 7: 796–800.

Clark, V.C., C.J. Raxworthy, V. Rakotomalala, P. Sierwald and B.L. Fisher. 2005. Convergent evolution of chemical defense in poison frogs and arthropod prey between Madagascar and the Neotropics. PNAS 102: 11617–11622.

Cobbs, C., J. Heath, J.O. Stireman III and P. Abbot. 2013. Carotenoids in unexpected places: Gall midges, lateral gene transfer, and carotenoid biosynthesis in animals. Mol. Phylogen. Evol. 68(2): 221–228.

Colleran, E. and P. O'Carra. 1977. Enzymology and comparative physiology of biliverdin reduction. pp. 69–80. *In:* Berk, P.D. and N.I. Berlin [eds.]. Chemistry and Physiology of Bile Pigments. Dept. of Health, Education, and Welfare, Public Health Service, National Institutes of Health, Bethesda, Md, U.S.

Cornelius, C.E. 1991. Bile pigments in fishes: A review. Vet. Clin. Pathol. 20: 106–116.

Cox, C.L., J. Bowers, C. Obialo, J.D. Curlis and J.W. Streicher. 2020. Spatial and temporal dynamics of exuberant colour polymorphism in the southern cricket frog. J. Nat. Hist. 54: 2249–2264.

Crothers, L., E. Gering and M. Cummings. 2011. Aposematic signal variation predicts male–male interactions in a polymorphic poison frog. Evolution 65: 599–605.

Crothers, L., R.A. Saporito, J. Yeager, K. Lynch, C. Friesen, C.L. Richards-Zawacki et al. 2016. Warning signal properties covary with toxicity but not testosterone or aggregate carotenoids in a poison frog. Evol. Ecol. 30: 601–621.

Crothers, L.R. and M.E. Cummings. 2013. Warning signal brightness variation: sexual selection may work under the radar of natural selection in populations of a polytypic poison frog. Am. Nat. 181: E116–E124.

Crothers, L.R. and M.E. Cummings. 2015. A multifunctional warning signal behaves as an agonistic status signal in a poison frog. Behav. Ecol. 26: 560–568.

Csordas, A. and H. Michl. 1969. Primary structure of two oligopeptides of the toxin of *Bombina variegata* L. Toxicon 7: 103–108.

D'Alba, L. and M.D. Shawkey. 2019. Melanosomes: Biogenesis, properties, and evolution of an ancient organelle. Physiol. Rev. 99: 1–19.

d'Ischia, M., K. Wakamatsu, F. Cicoira, E. Di Mauro, J.C. Garcia-Borron, S. Commo et al. 2015. Melanins and melanogenesis: From pigment cells to human health and technological applications. Pigm. Cell Melan. Res. 28: 520–544.

De Bona, S., J.K. Valkonen, A. López-Sepulcre and J. Mappes. 2015. Predator mimicry, not conspicuousness, explains the efficacy of butterfly eyespots. Proc. R. Soc. B 282: 20150202.

Donner, K. and C.A. Yovanovich. 2020. A frog's eye view: Foundational revelations and future promises. Semin. Cell Dev. Biol. 106: 72–85.

Doucet, S.M. and D.J. Mennill. 2010. Dynamic sexual dichromatism in an explosively breeding Neotropical toad. Biol. Lett. 6: 63–66.

Dreher, C.E., A. Rodríguez, M.E. Cummings and H. Pröhl. 2017. Mating status correlates with dorsal brightness in some but not all poison frog populations. Ecol. Evol. 7: 10503–10512.

Duellman, W.E. and P.M. Ruiz-Carranza. 1986. Ontogenetic polychromatism in marsupial frogs (Anura: Hylidae). Caldasia 15: 617–627.

Dunlap, D.G. 1955. Inter-and intraspecific variation in Oregon frogs of the genus *Rana*. Am. Midl. Nat. 54: 314–331.

DuShane, G.P. 1935. An experimental study of the origin of pigment cells in Amphibia. J. Exp. Zool. 72: 1–31.

DuShane, G.P. 1943. The embryology of vertebrate pigment cells. Part I. Amphibia. Q. Rev. Biol. 18: 109–127.

Ehrmann, S. 1892. Beitrag zur Physiologie der Pigmentzellen nach Versuchen am Farbenwechsel der Amphibien. Archiv Dermatol. Syph. 24: 519–539.

Endler, J.A. 1988. Frequency-dependent predation, crypsis and aposematic coloration. Phil. Trans. R. Soc. Lond. B 319: 505–523.

Endler, J.A. and J. Mappes. 2004. Predator mixes and the conspicuousness of aposematic signals. Am. Nat. 163: 532–547.

Engelbrecht-Wiggans, E. and J.P. Tumulty. 2019. "Reverse" sexual dichromatism in a Neotropical frog. Ethology 125: 957–964.

Eom, D.S., S. Inoue, L.B. Patterson, T.N. Gordon, R. Slingwine, S. Kondo et al. 2012. Melanophore migration and survival during zebrafish adult pigment stripe development require the immunoglobulin superfamily adhesion molecule Igsf11. PLoS Genetics 8: e1002899.

Eom, D.S., E.J. Bain, L.B. Patterson, M.E. Grout and D.M. Parichy. 2015. Long-distance communication by specialized cellular projections during pigment pattern development and evolution. eLife 4: 1–25.

Everts, R.E., J. Rothuizen and B.A. van Oost. 2000. Identification of a premature stop codon in the melanocyte-stimulating hormone receptor gene (MC1R) in Labrador and Golden retrievers with yellow coat colour. Anim. Gen. 31: 194–199.

Eycleshymer, A.C. 1906. The development of chromatophores in *Necturus*. Am. J. Anat. 5: 309–313.

Feng, Y.-J., D.C. Blackburn, D. Liang, D.M. Hillis, D.B. Wake, D.C. Cannatella et al. 2017. Phylogenomics reveals rapid, simultaneous diversification of three major clades of Gondwanan frogs at the Cretaceous–Paleogene boundary. PNAS 114: E5864–E5870.

Fernandez, P.J. and J.P. Collins. 1988. Effect of environment and ontogeny on color pattern variation in Arizona tiger salamanders (*Ambystoma tigrinum nebulosum* Hallowell). Copeia 1988: 928–938.

Fernandez, P.J. and J.T. Bagnara. 1991. Effect of background color and low temperature on skin color and circulating α-MSH in two species of leopard frog. Gen. Comp. Endocrinol. 83: 132–141.

Ferraro, D.P., M.O. Pereyra, P.E. Topa and J. Faivovich. 2021. Evolution of macroglands and defensive mechanisms in Leiuperinae (Anura: Leptodactylidae). Zool. J. Linn. Soc. 193: 388–412.

Fischer, E.K., A.B. Roland, N.A. Moskowitz, C. Vidoudez, N. Ranaivorazo, E.E. Tapia et al. 2019. Mechanisms of convergent egg provisioning in poison frogs. Curr. Biol. 29: 4145–4151.

Fogleman, J.C., P.S. Corn and D. Pettus. 1980. The genetic basis of a dorsal color polymorphism in *Rana pipiens*. J. Hered. 71: 439–440.

Frohnhöfer, H.G., J. Krauss, H.M. Maischein and C. Nüsslein-Volhard. 2013. Iridophores and their interactions with other chromatophores are required for stripe formation in zebrafish. Development 140: 2997–3007.

Frost, S.K. and J.T. Bagnara. 1979. Allopurinol-induced melanism in the tiger salamander (*Ambystoma tigrinum nebulosum*). J. Exp. Zool. 209: 455–465.

Fujii, R. 1993. Cytophysiology of fish chromatophores. pp. 191–255. *In:* Jeon, K.W., M. Friedlander and J. Jarvik [eds.]. International Review of Cytology (Vol. 143). Academic Press, Cambridge, MA, USA.

Fukuzawa, T., P. Samaraweera, F.T. Mangano, J.H. Law and J.T. Bagnara. 1995. Evidence that MIF plays a role in the development of pigmentation patterns in the frog. Dev. Biol. 167: 148–158.

Fukuzawa, T. and H. Ide. 1988. A ventrally localized inhibitor of melanization in *Xenopus laevis* skin. Dev. Biol. 129: 25–36.

Gade, M.R., M. Hill and R.A. Saporito. 2016. Color assortative mating in a mainland population of the poison frog *Oophaga pumilio*. Ethology 122: 851–858.

Gage, S.H. 1891. Life-history of the Vermilion-Spotted Newt (*Diemyctylus viridescens* Raf.). Am. Nat. 25: 1084–1110.

Galeano, S.P. and K.E. Harms. 2016. Coloration in the polymorphic frog *Oophaga pumilio* associates with level of aggressiveness in intraspecific and interspecific behavioral interactions. Behav. Ecol. Sociobiol. 70: 83–97.

Gally, M., J. Zina, C.V. de Mira-Mendes and M. Solé. 2014. Legs-interweaving: An unusual defense behaviour of anurans displayed by *Agalychnis aspera* (Peters, 1983). Herpetol. Notes 7: 623–625.

Gardner, K.M., D.J. Mennill, L.M. Savi, N.E. Shangi and S.M. Doucet. 2021. Sexual selection in a tropical toad: Do female toads choose brighter males in a species with rapid colour change? Ethology 127: 475–483.

Gerald, M.S. 2001. Primate colour predicts social status and aggressive outcome. Anim. Behav. 61: 559–566.

Goin C.J. 1950. Color pattern inheritance in some frogs of the genus *Eleutherodactylus*. Bull. Chic. Acad. Sci. 9: 1–15.

Gomez, D., C. Richardson, T. Lengagne, S. Plenet, P. Joly, J.P. Léna et al. 2009. The role of nocturnal vision in mate choice: females prefer conspicuous males in the European tree frog (*Hyla arborea*). Proc. R. Soc. Lond. B 276: 2351–2358.

Gomez, D., C. Richardson, T. Lengagne, M. Derex, S. Plenet, P. Joly et al. 2010. Support for a role of colour vision in mate choice in the nocturnal European treefrog (*Hyla arborea*). Behaviour 147: 1753–1768.

Goodwin, T.W. 1984. The Biochemistry of the Carotenoids (Vol. II: Animal). Springer, Dordrecht, Netherlands.

Goutte, S., I. Hariyani, K.D. Utzinger, Y. Bourgeois and S. Boissinot. 2021. Coloring inside the lines: genomic architecture and evolution of a widespread color pattern in frogs. BioRxiv, 2021.10.28.466315.

Goutte, S., M.J. Mason, M.M. Antoniazzi, C. Jared, D. Merle, L. Cazes et al. 2019. Intense bone fluorescence reveals hidden patterns in pumpkin toadlets. Sci. Rep. 9: 428.

Greener, M.S., E. Hutton, C.J. Pollock, A. Wilson, C.Y. Lam, M. Nokhbatolfoghahai et al. 2020. Sexual dichromatism in the neotropical genus *Mannophryne* (Anura: Aromobatidae). PLoS One 15: e0223080.

Gross, J.B., R. Borowsky and C.J. Tabin. 2009. A novel role for Mc1r in the parallel evolution of depigmentation in independent populations of the cavefish *Astyanax mexicanus*. PLoS Genetics 5: e1000326.

Guayasamin, J.M., D.F. Cisneros-Heredia, R.W. McDiarmid, P. Peña and C.R. Hutter. 2020. Glassfrogs of Ecuador: Diversity, evolution, and conservation. Diversity 12: 222.

Guimond, R.W. and V.H. Hutchison. 1973. Trimodal gas exchange in the large aquatic salamander, *Siren lacertina* (Linnaeus). Comp. Biochem. Physiol. A 46: 249–268.

Hairston, N.G. and C.H. Pope. 1948. Geographic variation and speciation in Appalachian salamanders (*Plethodon jordani* group). Evolution 2: 266–278.

Hämäläinen, L., J. Valkonen, J. Mappes and B. Rojas. 2015. Visual illusions in predator–prey interactions: birds find moving patterned prey harder to catch. Anim. Cogn. 18: 1059–1068.

Hämäläinen, L., J. Mappes, H.M. Rowland, M. Teichmann and R. Thorogood. 2020. Social learning within and across predator species reduces attacks on novel aposematic prey. J. Anim. Ecol. 89: 1153–1164.

Hargitt, C.W. 1912. Behavior and color changes of tree frogs. J. Anim. Behav. 2: 51–78.

Harper, G.J., M.K. Steininger, C.J. Tucker, D. Juhn and F. Hawkins. 2007. Fifty years of deforestation and forest fragmentation in Madagascar. Environ. Conserv. 34: 325–333.

Hayes, T.B. and K.P. Menendez. 1999. The effect of sex steroids on primary and secondary sex differentiation in the sexually dichromatic reedfrog (*Hyperolius argus*: Hyperolidae) from the Arabuko Sokoke forest of Kenya. Gen. Comp. Endocrinol. 115: 188–199.

Heatwole, H. and G.T. Barthalmus. 1995. Amphibian Biology, vol. 1, The Integument. Surrey, Beatty and Sons, Norton, New South Wales, Australia.

Hoekstra, H.E. 2006. Genetics, development and evolution of adaptive pigmentation in vertebrates. Heredity 97: 222–234.

Hoffman, E.A. and M.S. Blouin. 2000. A review of colour and pattern polymorphisms in anurans. Biol. J. Linn. Soc. 70: 633–665.

Hofreiter, M. and T. Schöneberg. 2010. The genetic and evolutionary basis of colour variation in vertebrates. Cell. Mol. Life Sci. 67: 2591–2603.

Hooper, D.M., S.C. Griffith and T.D. Price. 2019. Sex chromosome inversions enforce reproductive isolation across an avian hybrid zone. Mol. Ecol. 28: 1246–1262.

Houssay, B.A. 1925. Factores que regulan la pigmentación de la rana "*Leptodactylus ocellatus*" (L) Gir. Imprenta de la Universidad.

Howard, R.R. and E.D. Brodie, Jr. 1973. A Batesian mimetic complex in salamanders: responses of avian predators. Herpetologica 29: 33–41.

Hubbard, J.K., J.A.C. Uy, M.E. Hauber, H.E. Hoekstra and R.J. Safran. 2010. Vertebrate pigmentation: from underlying genes to adaptive function. Trends Genet. 26: 231–239.

Huheey, J.E. 1960. Mimicry in the color pattern of certain Appalachian salamanders. J. Elisha Mitchell Sci. Soc. 76: 246–251.

Huheey, J.E. and R.A. Brandon. 1974. Studies in warning coloration and mimicry. VI. Comments on the warning coloration of Red Efts and their presumed mimicry by Red Salamanders. Herpetologica 30: 149–155.

Hurlbert, S.H. 1970. Predator responses to the Vermilion-Spotted Newt (*Notophthalmus viridescens*). J. Herpetol. 4: 47–55.

Irion, U., A.P. Singh and C. Nüsslein-Volhard. 2016. The developmental genetics of vertebrate color pattern formation: lessons from zebrafish. Curr. Top. Dev. Biol. 117: 141–169.

Ito, S. and K. Wakamatsu. 2008. Chemistry of mixed melanogenesis—pivotal roles of dopaquinone. Photochem. Photobiol. 84(3): 582–592.

Jackson, J.F., W. Ingram III and H.W. Campbell. 1976. The dorsal pigmentation pattern of snakes as an antipredator strategy: a multivariate approach. Am. Nat. 110: 1029–1053.

Jacobs, L.E., A. Vega, S. Dudgeon, K. Kaiser and J.M. Robertson. 2017. Local not vocal: assortative female choice in divergent populations of red-eyed treefrogs, *Agalychnis callidryas* (Hylidae: Phyllomedusinae). Biol. J. Linn. Soc. 120: 171–178.

Johnson, J.A. and E.D. Brodie, Jr. 1975. The selective advantage of the defensive posture of the newt, *Taricha granulosa*. Am. Midl. Nat. 93: 139–148.

Kang, C., T.N. Sherratt, Y.E. Kim, Y. Shin, J. Moon, U. Song et al. 2017. Differential predation drives the geographical divergence in multiple traits in aposematic frogs. Behav. Ecol. 28: 1122–1130.

Kelly, H.A. 1878. Identity of *Diemictylus miniatus* with *Diemictylus viridescens*. Am. Nat. 12: 399.

Kemp, N.E. 1961. Replacement of the larval basement lamella by adult-type basement membrane in anuran skin during metamorphosis. Dev. Biol. 3: 391–410.

Kindermann, C. and J.M. Hero. 2016. Pigment cell distribution in a rapid colour changing amphibian (*Litoria wilcoxii*). Zoomorphology 135: 197–203.

Klonoski, K., K. Bi and E.B. Rosenblum. 2019. Phenotypic and genetic diversity in aposematic Malagasy poison frogs (genus *Mantella*). Ecol. Evol. 9: 2725–2742.

Kraemer, A.C. and D.C. Adams. 2014. Predator perception of Batesian mimicry and conspicuousness in a salamander. Evolution 68: 1197–1206.

Kraemer, A.C., J.M. Serb and D.C. Adams. 2015. Batesian mimics influence the evolution of conspicuousness in an aposematic salamander. J. Evol. Biol. 28: 1016–1023.

Kraus, F. and A. Allison. 2009. A remarkable ontogenetic change in color pattern in a new species of *Oreophryne* (Anura: Microhylidae) from Papua New Guinea. Copeia 2009: 690–697.

Kronforst, M.R., G.S. Barsh, A. Kopp, J. Mallet, A. Monteiro, S.P. Mullen et al. 2012. Unraveling the thread of nature's tapestry: the genetics of diversity and convergence in animal pigmentation. Pigm. Cell Melan. Res. 25: 411–433.

Kuchta, S.R. 2005. Experimental support for aposematic coloration in the salamander *Ensatina eschscholtzii xanthoptica*: implications for mimicry of Pacific newts. Copeia 2005: 265–271.

Kuchta, S.R., A.H. Krakauer and B. Sinervo. 2008. Why does the Yellow-eyed Ensatina have yellow eyes? Batesian mimicry of Pacific newts (genus *Taricha*) by the salamander *Ensatina eschscholtzii xanthoptica*. Evolution 62: 984–990.

Kuriyama, T., A. Murakami, M. Brandley and M. Hasegawa. 2020. Blue, black, and stripes: Evolution and development of color production and pattern formation in lizards and snakes. Front. Ecol. Evol. 8: 232.

Labanick, G.M. and R.A. Brandon. 1981. An experimental study of Batesian mimicry between the salamanders *Plethodon jordani* and *Desmognathus ochrophaeus*. J. Herpetol. 15: 275–281.

Lamb, J.Y. and M.P. Davis. 2020. Salamanders and other amphibians are aglow with biofluorescence. Sci. Rep. 10: 2821.

Land, M.F. 1972. The physics and biology of animal reflectors. Prog. Biophys. Mol. Biol. 24: 75–106.

Lawrence, J.P., M. Mahony and B.P. Noonan. 2018. Differential responses of avian and mammalian predators to phenotypic variation in Australian Brood Frogs. PloS One 13: e0195446.

Lawrence, J.P., B. Rojas, A. Fouquet, J. Mappes, A. Blanchette, R.A. Saporito et al. 2019. Weak warning signals can persist in the absence of gene flow. PNAS 116: 19037–19045.

Lechaire, J.P. and J.P. Denefle. 1991. Dermal breaks: Migratory cell pathways opened in anuran tadpole skin at climactic metamorphosis. Tiss. Cell 23: 307–315.

Ligon, R.A. and K.L. McCartney. 2016. Biochemical regulation of pigment motility in vertebrate chromatophores: A review of physiological color change mechanisms. Curr. Zool. 62: 237–252.

Löhner, L. 1919. Über einen eigentümlichen Reflex der Feuerunken nebst Bemerkungen über die "tierische Hypnose". Pflüger's Archiv für die gesamte Physiologie des Menschen und der Tiere 174: 324–351.

Longo, A.V., C.A. Rodríguez-Gómez, J.P. Zegarra, O. Monzón, H.J. Claudio-Hernández, R.L. Joglar et al. 2020. Tick parasitism as a cost of sexual selection and male parental care in a Neotropical frog. Ecosphere 11: e03010.

Lopes, R.J.J., J.D. Johnson, M.B.B. Toomey, M.S.S. Ferreira, P.M.M. Araujo, J. Melo-Ferreira et al. 2016. Genetic basis for red coloration in birds. Curr. Biol. 26: 1427–1434.

Lorentz, M.N., A.N. Stokes, D.C. Rößler and S. Lötters. 2016. Tetrodotoxin. Curr. Biol. 26: R870–R872.

Maan, M.E. and M.E. Cummings. 2008. Female preferences for aposematic signal components in a polymorphic poison frog. Evolution 62: 2334–2345.

Maan, M.E. and M.E. Cummings. 2009. Sexual dimorphism and directional sexual selection on aposematic signals in a poison frog. PNAS 106: 19072–19077.

Maan, M.E. and M.E. Cummings. 2012. Poison frog colors are honest signals of toxicity, particularly for bird predators. Am. Nat. 179: E1–E14.

Maderspacher, F. and C. Nüsslein-Volhard. 2003. Formation of the adult pigment pattern in zebrafish requires leopard and obelix dependent cell interactions. Development 130: 3447–3457.

Mallet, J. and N. H. Barton. 1989. Strong natural selection in a warning-color hybrid zone. Evolution 43: 421–431.

Mallet, J. and M. Joron. 1999. Evolution of diversity in warning color and mimicry: Polymorphisms, shifting balance, and speciation. Annu. Rev. Ecol. Syst. 30: 201–233.

Mappes, J., N. Marples and J.A. Endler. 2005. The complex business of survival by aposematism. Trends Ecol. Evol. 20: 598–603.

Márquez, R., T.P. Lindereth, D. Mejía-Vargas, R. Nielsen, A. Amézquita and M.R. Kronforst. 2020. Divergence, gene flow, and the origin of leapfrog geographic distributions: The history of colour pattern variation in *Phyllobates* poison-dart frogs. Mol. Ecol. 29: 3702–3719.

Martins, M. 1989. Deimatic behavior in *Pleurodema brachyops*. J. Herpetol. 23: 305–307.

McCollum, S.A. and J. Van Buskirk. 1996. Costs and benefits of a predator-induced polyphenism in the gray treefrog *Hyla chrysoscelis*. Evolution 50(2): 583–593.

McCollum, S.A. and J.D. Leimberger. 1997. Predator-induced morphological changes in an amphibian: Predation by dragonflies affects tadpole shape and color. Oecologia 109(4): 615–621.

McNamara, M.E., V. Rossi, T.S. Slater, C.S. Rogers, A.L. Ducrest, S. Dubey et al. 2021. Decoding the evolution of melanin in vertebrates. Trends Ecol. Evol. 36: 430–443.

Mebs, D., M. Jansen, G. Köhler, W. Pogoda and G. Kauert. 2010. Myrmecophagy and alkaloid sequestration in amphibians: a study on *Ameerega picta* (Dendrobatidae) and *Elachistocleis* sp. (Microhylidae) frogs. Salamandra 46: 11–15.

Meuche, I., K.E. Linsenmair and H. Pröhl. 2011. Female territoriality in the strawberry poison frog (*Oophaga pumilio*). Copeia 2011(3): 351–356.

Michl, H. and E. Kaiser. 1963. Chemie und biochemie der amphibiengifte. Toxicon 1: 175–228.

Millar, S.E., M.W. Miller, M.E. Stevens and G.S. Barsh. 1995. Expression and transgenic studies of the mouse agouti gene provide insight into the mechanisms by which mammalian coat color patterns are generated. Development 121: 3223–3232.

Mills, M.G. and L.B. Patterson. 2009. Not just black and white: Pigment pattern development and evolution in vertebrates. Semin. Cell Dev. Biol. 20: 72–81.

Moran, N.A. and T. Jarvik. 2010. Lateral transfer of genes from fungi underlies carotenoid production in aphids. Science 328: 624–627.

Morgan, S.K., M.W. Pugh, M.M. Gangloff and L. Siefferman. 2014. The Spots of the spotted salamander are sexually dimorphic. Copeia 2014: 251–256.

Moritz, C., C.J. Schneider and D.B. Wake. 1992. Evolutionary relationships within the *Ensatina eschscholtzii* complex confirm the ring species interpretation. Syst. Biol. 41: 273–291.

Moriwaki, T. 1953. The inheritance of the dorso-median stripe in *Rana limnocharis* Wiegmann. J. Sci. Hiroshima Univ. (Zoology) 14: 159–164.

Moskowitz, N.A., A.B. Roland, E.K. Fischer, N. Ranaivorazo, C. Vidoudez, M.T. Aguilar et al. 2018. Seasonal changes in diet and chemical defense in the Climbing Mantella frog (*Mantella laevigata*). PloS One 13: e0207940.

Mundy, N.I., J. Stapley, C. Bennison, R. Tucker, H. Twyman, K.W. Kim et al. 2016. Red carotenoid coloration in the zebra finch is controlled by a cytochrome P450 gene cluster. Curr. Biol. 26: 1435–1440.

Murali, G. and U. Kodandaramaiah 2016. Deceived by stripes: conspicuous patterning on vital anterior body parts can redirect predatory strikes to expendable posterior organs. R. Soc. Open Sci. 3: 160057.

Myers, C.W., J.W. Daly and B. Malkin. 1978. A dangerously toxic new frog (*Phyllobates*) used by Emberá Indians of western Colombia, with discussion of blowgun fabrication and dart poisoning. Bull. Am. Mus. Nat. Hist. 161: 309–365.

Nascimento, A.A., J.T. Roland and V.I. Gelfand. 2003. Pigment cells: a model for the study of organelle transport. Annu. Rev. Cell Dev. Biol. 19: 469–491.

Nielsen, H.I. 1978. Ultrastructural changes in the dermal chromatophore unit of *Hyla arborea* during color change. Cell Tiss. Res. 194: 405–418.

Nielsen, H.I. and J. Dyck. 1978. Adaptation of the tree frog, *Hyla cinerea*, to colored backgrounds, and the role of the three chromatophore types. J. Exp. Zool. 205: 79–94.

Nokelainen, O. and M. Stevens. 2016. Camouflage. Curr. Biol. 26: R654–R656.

Obika, M. and J.T. Bagnara. 1964. Pteridines as pigments in amphibians. Science 143: 485–487.

O'Hanlon, J.C., D.N. Rathnayake, K.L. Barry and K.D.L. Umbers. 2018. Post-attack defensive displays in three praying mantis species. Behav. Ecol. Sociobiol. 72: 1–7.

O'Neill, E.M. and K.H. Beard. 2010. Genetic basis of a color pattern polymorphism in the coqui frog *Eleutherodactylus coqui*. J. Hered. 101: 703–709.

O'Quin C.T., A.C. Drilea, M.A. Conte and T.D. Kocher. 2013. Mapping of pigmentation QTL on an anchored genome assembly of the cichlid fish, *Metriaclima zebra*. BMC Genomics. 14: 287.

Parichy, D.M. 1996a. Pigment patterns of larval salamanders (Ambystomatidae, Salamandridae): the role of the lateral line sensory system and the evolution of pattern-forming mechanisms. Dev. Biol. 175: 265–282.

Parichy, D.M. 1996b. When neural crest and placodes collide: Interactions between melanophores and the lateral lines that generate stripes in the salamander *Ambystoma tigrinum tigrinum* (Ambystomatidae). Dev. Biol. 175: 283–300.

Parichy, D.M. 1998. Experimental analysis of character coupling across a complex life cycle: pigment pattern metamorphosis in the tiger salamander, *Ambystoma tigrinum tigrinum*. J. Morphol. 237: 53–67.

Parichy, D.M., M.V. Reedy and C.A. Erickson. 2006. Regulation of melanoblast migration and differentiation. pp. 108–139. *In:* Nordlund, J.J., R.E. Boissy, V.J. Hearing, R.A. King, W.S. Oetting and J.-P. Ortonne [eds.]. The Pigmentary System and its Disorders (2nd Edition). Oxford University Press, Oxford, UK.

Parker, G.H. 1948. Animal colour changes and their neurohumours: a survey of investigations 1910–1943. Q. Rev. Biol. 23: 368–369.

Patterson, L.B. and D.M. Parichy. 2019. Zebrafish pigment pattern formation: insights into the development and evolution of adult form. Annu. Rev. Genet. 53: 505–530.

Pokhrel, L.R., I. Karsai, M.K. Hamed and T.F. Laughlin. 2013. Dorsal body pigmentation and sexual dimorphism in the marbled salamander (*Ambystoma opacum*). Ethol. Ecol. Evol. 25: 214–226.

Portik, D.M., R.C. Bell, D.C. Blackburn, A.M. Bauer, C.D. Barratt, W.R. Branch et al. 2019. Sexual dichromatism drives diversification within a major radiation of African amphibians. Syst. Biol. 68: 859–875.

Posso-Terranova, A. and J. Andrés. 2017. Diversification and convergence of aposematic phenotypes: truncated receptors and cellular arrangements mediate rapid evolution of coloration in harlequin poison frogs. Evolution 71: 2677–2692.

Poulton, E.B. 1890. The Colours of Animals: their Meaning and Use, especially Considered in the Case of Insects. Kegan Paul, Trench, Trubner, UK.

Prates, I., M.M. Antoniazzi, J.M. Sciani, D.C. Pimenta, L.F. Toledo, C.F.B Haddad et al. 2012. Skin glands, poison and mimicry in dendrobatid and leptodactylid amphibians. J. Morphol. 273: 279–290.

Prötzel, D., M. Heß, M. Schwager, F. Glaw and M.D. Scherz. 2021. Neon-green fluorescence in the desert gecko *Pachydactylus rangei* caused by iridophores. Sci. Rep. 11: 297.

Pryke, S.R., M.J. Lawes and S. Andersson. 2001. Agonistic carotenoid signalling in male red-collared widowbirds: aggression related to the colour signal of both the territory owner and model intruder. Anim. Behav. 62: 695–704.

Rada, M., J.J. Ospina-Sarria and J.M. Guayasamin. 2017. A taxonomic review of Tan-Brown Glassfrogs (Anura: Centrolenidae), with the description of a new species from southwestern Colombia. South Am. J. Herpetol. 12(2): 136–156.

Rafinesque, C.S. 1820. Annals of nature or annual synopsis of new genera and species of animals, plants, etc., discovered in North America. Thomas Smith, Lexington, KY, USA.

Randrianavelona, R., H. Rakotonoely, J. Ratsimbazafy and R.K.B. Jenkins. 2010. Conservation assessment of the critically endangered frog *Mantella aurantiaca* in Madagascar. Afr. J. Herpetol. 59: 65–78.

Rehberg-Besler, N., D.J. Mennill and S.M. Doucet. 2015. Dynamic sexual dichromatism produces a sex signal in an explosively breeding Neotropical toad: a model presentation experiment. Behav. Proc. 121: 74–79.

Reynolds, R.G. and B.M. Fitzpatrick. 2007. Assortative mating in poison-dart frogs based on an ecologically important trait. Evolution 61: 2253-2259.

Richards, C.M. 1976. The development of color dimorphism in *Hyperolius v. viridiflavus*, a reed frog from Kenya. Copeia 1976: 65–70.

Richards, C.M. 1982. The alteration of chromatophore expression by sex hormones in the Kenyanreed frog, *Hyperolius viridiflavus*. Gen. Comp. Endocrinol. 46: 59–67.

Richards, C.M. and G.W. Nace. 1983. Dark pigment variants in anurans: classification, new descriptions, color changes and inheritance. Copeia 1983: 979–990.

Richards-Zawacki, C.L., I.J. Wang and K. Summers. 2012. Mate choice and the genetic basis for colour variation in a polymorphic dart frog: inferences from a wild pedigree. Mol. Ecol. 21: 3879–3892.

Ries, C., J. Spaethe, M. Sztatecsny, C. Strondl and W. Hödl. 2008. Turning blue and ultraviolet: sex-specific colour change during the mating season in the Balkan moor frog. J. Zool. 276: 229–236.

Robertson, J.M. and H.W. Greene. 2017. Bright colour patterns as social signals in nocturnal frogs. Biol. J. Linn. Soc. 121: 849–857.

Rodríguez, A., D. Poth, S. Schulz and M. Vences. 2011. Discovery of skin alkaloids in a miniaturized eleutherodactylid frog from Cuba. Biol. Lett. 7: 414–418.

Rodríguez, A., N.I. Mundy, R. Ibáñez and H. Pröhl. 2020. Being red, blue and green: the genetic basis of coloration differences in the strawberry poison frog (*Oophaga pumilio*). BMC Genomics 21: 301.

Rogers, R.L., L. Zhou, C. Chu, R. Márquez, A. Corl, T. Linderoth et al. 2018. Genomic takeover by transposable elements in the Strawberry poison frog. Mol. Biol. Evol. 35: 2913–2927.

Rojas, B. and J.A. Endler. 2013. Sexual dimorphism and intra-populational colour pattern variation in the aposematic frog Dendrobates tinctorius. Evol. Ecol. 27: 739–753.

Rojas, B., P. Rautiala and J. Mappes. 2014a. Differential detectability of polymorphic warning signals under varying light environments. Behav. Proc. 109: 164–172.

Rojas, B., J. Devillechabrolle and J.A. Endler. 2014b. Paradox lost: colour pattern and movement are associated in an aposematic frog. Biol. Lett. 10: 20140193.

Rojas, B. 2017. Behavioural, ecological, and evolutionary aspects of diversity in frog colour patterns. Biol. Rev. 92: 1059–1080.

Rojas, B., E. Burdfield-Steel, S.P. Gordon, C. De Pasqual, L. Hernández, J. Mappes et al. 2018. Multimodal aposematic signals and their emerging role in mate attraction. Front. Ecol. Evol. 9: 93.

Rojas, B. and A. Pašukonis. 2019. From habitat use to social behavior: natural history of a voiceless poison frog, *Dendrobates tinctorius*. PeerJ 7: e7648.

Rönkä, K., J. Valkonen, O. Nokelainen, B. Rojas, S.P. Gordon, E. Burdfield-Steel et al. 2020. Geographic mosaic of selection by avian predators on hindwing warning colour of a polymorphic aposematic moth. Ecol. Lett. 23: 1654–1663.

Rudh, A. and A. Qvarnström. 2013. Adaptive colouration in amphibians. Semin. Cell Dev. Biol. 24: 553–561.

Rudh, A., M.F. Breed and A. Qvarnström. 2013. Does aggression and explorative behaviour decrease with lost warning coloration? Biol. J. Linn. Soc. 108: 116–126.

Ruxton, G.D., W.L. Allen, T.N. Sherratt and M.P. Speed. 2018. Avoiding Attack. The Evolutionary Ecology of Crypsis, Aposematism, and Mimicry. Oxford University Press, Oxford, UK.

Sanchez, E., H. Pröhl, T. Lüddecke, S. Schulz, S. Steinfartz and M. Vences. 2019. The conspicuous postmetamorphic coloration of fire salamanders, but not their toxicity, is affected by larval background albedo. J. Exp. Zool. B 332: 26–35.

Santos, J.C., L.A. Coloma and D.C. Cannatella. 2003. Multiple, recurring origins of aposematism and diet specialization in poison frogs. PNAS 100: 12792–12797.

Santos, J.C., M. Baquero, C.L. Barrio-Amorós, L.A. Coloma, L.K. Erdtmann, A.P. Lima et al. 2014. Aposematism increases acoustic diversification and speciation in poison frogs. Proc. R. Soc. B 281: 20141761.

Saporito, R.A., M.A. Donnelly, H.M. Garraffo, T.F. Spande and J.W. Daly. 2006. Geographic and seasonal variation in alkaloid-based chemical defenses of *Dendrobates pumilio* from Bocas del Toro, Panama. J. Chem. Ecol. 32: 795–814.

Saporito, R.A., R. Zuercher, M. Roberts, K.G. Gerow and M.A. Donnelly. 2007. Experimental evidence for aposematism in the dendrobatid poison frog *Oophaga pumilio*. Copeia 2007: 1006–1011.

Saunders, L.M., A.K. Mishra, A.J. Aman, V.M. Lewis, M.B. Toomey, J.S. Packer et al. 2019. Thyroid hormone regulates distinct paths to maturation in pigment cell lineages. eLife 8: e45181.

Schallreuter, K.U. 2007. Advances in melanocyte basic science research. Demartol. Clin. 25(3): 283–291.

Schiøtz, A. 1967. The treefrogs (Rhacophoridae) of West Africa. Spolia Zool. Musei Haunienses 25: 1–346.

Secondi, J., V. Lepetz and M. Théry. 2012. Male attractiveness is influenced by UV wavelengths in a newt species but not in its close relative. PloS One 7: e30391.

Sherratt, T.N. and C.D. Beatty. 2003. The evolution of warning signals as reliable indicators of prey defense. Am. Nat. 162: 377–389.

Shure, D.J., L.A. Wilson and C. Hochwender. 1989. Predation on aposematic efts of *Notophthalmus viridescens*. J. Herpetol. 23: 437–439.

Silverstone, P.A. 1976. A revision of the poison-arrow frogs of the genus *Phyllobates* Bibron in Sagra (Family Dendrobatidae). Nat. Hist. Mus. Los Angeles County Sci. Bull. 27: 1–53.

Skelhorn, J., H.M. Rowland, M.P. Speed and G.D. Ruxton. 2010. Masquerade: camouflage without crypsis. Science 327: 51–51.

Sköld, H.N., S. Aspengren and M. Wallin. 2013. Rapid color change in fish and amphibians–function, regulation, and emerging applications. Pigm. Cell Melan. Res. 26(1): 29–38.

Smith, P.E. 1916. Experimental ablation of the hypophysis in the frog embryo. Science 44: 280–282.

Speed, M.P. 1999. Batesian, quasi-Batesian or Müllerian mimicry? Theory and data in mimicry research. Evol. Ecol. 13: 755–776.

Speed, M.P., M.A. Brockhurst and G.D. Ruxton. 2010. The dual benefits of aposematism: predator avoidance and enhanced resource collection. Evolution 64: 1622–1633.

Stackhouse, H.L. 1966. Some aspects of pteridine biosynthesis in amphibians. Comp. Biochem. Physiol. 17: 219–226.

Stearner, S.P. 1946. Pigmentation studies in salamanders, with special reference to the changes at metamorphosis. Physiol. Zool. 19: 375–404.

Stevens, M. 2005. The role of eyespots as anti-predator mechanisms, principally demonstrated in the Lepidoptera. Biol. Rev. 80: 573–588.

Stevens, M., W.T.L. Searle, J.E. Seymour, K.L. Marshall and G.D. Ruxton. 2011. Motion dazzle and camouflage as distinct anti-predator defenses. BMC Biol. 9: 1–11.

Stuckert, A.M.M., P.J. Venegas and K. Summers. 2014. Experimental evidence for predator learning and Müllerian mimicry in Peruvian poison frogs (*Ranitomeya*, Dendrobatidae). Evol. Ecol. 28: 413–426.

Stuckert, A.M.M., E. Moore, K.P. Coyle, I. Davison, M.D. MacManes, R. Roberts et al. 2019. Variation in pigmentation gene expression is associated with distinct aposematic color morphs in the poison frog *Dendrobates auratus*. BMC Evol. Biol. 19: 1–15.

Stuckert, A.M.M., M. Chouteau, M. McClure, T.M. LaPolice, T. Linderoth, R. Nielsen et al. 2021. The genomics of mimicry: gene expression throughout development provides insights into convergent and divergent phenotypes in a Müllerian mimicry system. Mol. Ecol. 30: 4039–4061.

Sturm, R.A. 2009. Molecular genetics of human pigmentation diversity. Hum. Mol. Genet. 18: R9–R17.

Stynoski, J.L., Y. Torres-Mendoza, M. Sasa-Marin and R.A. Saporito. 2014. Evidence of maternal provisioning of alkaloid-based chemical defenses in the strawberry poison frog *Oophaga pumilio*. Ecology 95: 587–593.

Summers, K. 1989. Sexual selection and intra-femalecompetition in the green poison-dart frog, *Dendrobates auratus*. Anim. Behav. 37: 797–805.

Summers, K. and M.E. Clough. 2001. The evolution of coloration and toxicity in the poison frog family (Dendrobatidae). PNAS 98: 6227–6232.

Summers, K., R. Symula, M. Clough and T. Cronin. 1999. Visual mate choice in poison frogs. Proc. R. Soc. B 266: 2141–2145.

Summers, K., T.W. Cronin and T. Kennedy. 2003. Variation in spectral reflectance among populations of *Dendrobates pumilio*, the strawberry poison frog, in the Bocas del Toro Archipelago, Panama. J. Biogeogr. 30: 35–53.

Summers, K., T.W. Cronin and T. Kennedy. 2004. Cross-breeding of distinct color morphs of the strawberry poison frog (*Dendrobates pumilio*) from the Bocas del Toro Archipelago, Panama. J. Herpetol. 38: 1–8.

Summers, K., M.P. Speed, J.D. Blount and A.M.M. Stuckert. 2015. Are aposematic signals honest? A review. J. Evol. Biol. 28: 1583–1599.

Sun, Y.B., Y. Zhang and K. Wang. 2020. Perspectives on studying molecular adaptations of amphibians in the genomic era. Zool. Res. 41: 351–364.

Sztatecsny, M., D. Preininger, A. Freudmann, M.-C. Loretto, F. Maier and W. Hödl. 2012. Don't get the blues: conspicuous nuptial colouration of male moor frogs (*Rana arvalis*) supports visual mate recognition during scramble competition in large breeding aggregations. Behav. Ecol. Sociobiol. 66: 1587–1593.

Taboada, C., A.E. Brunetti, F.N. Pedron, F. Carnevale Neto, D.A. Estrin, S.E. Bari et al. 2017a. Naturally occurring fluorescence in frogs. PNAS 114: 3672–3677.

Taboada, C., A.E. Brunetti, C. Alexandre, M.G. Lagorio and J. Faivovich. 2017b. Fluorescent frogs: a herpetological perspective. South Am. J. Herpetol. 12: 1–13.

Taboada, C., A.E. Brunetti, M.L. Lyra, R.R. Fitak, A.F. Soverna, S.R. Ron et al. 2020. Multiple origins of green coloration in frogs mediated by a novel biliverdin-binding serpin. PNAS 117: 18574–18581.

Taylor, R.C., B.W. Buchanan and J.L. Doherty. 2007. Sexual selection in the squirrel treefrog *Hyla squirella*: the role of multimodal cue assessment in female choice. Anim. Behav. 74: 1753–1763.

Teyssier, J., S.V. Saenko, D. Van Der Marel and M.C. Milinkovitch. 2015. Photonic crystals cause active colour change in chameleons. Nat. Comm. 6: 1–7.

Thibaudeau, G. and R. Altig. 2012. Coloration of anuran tadpoles (Amphibia): development, dynamics, function, and hypotheses. Internat. Schol. Res. Notices 2012: 725203.

Tilley, S.G., R.B. Merritt, B. Wu and R. Highton. 1978. Genetic differentiation in salamanders of the *Desmognathus ochrophaeus* complex (Plethodontidae). Evolution 32: 3–115.

Tilley, S.G., B.L. Lundrigan and L.P. Brower. 1982. Erythrism and mimicry in the salamander *Plethodon cinereus*. Herpetologica 38: 409–417.

Titcomb, G.C., D.W. Kikuchi and D.W. Pfennig. 2014. More than mimicry? Evaluating scope for flicker-fusion as a defensive strategy in coral snake mimics. Curr. Zool. 60: 123–130.

Thorogood, R., H. Kokko and J. Mappes. 2018. Social transmission of avoidance among predators facilitates the spread of novel prey. Nat. Ecol. Evol. 2: 254–261.

Todd, B.D. and A.K. Davis. 2007. Sexual dichromatism in the marbled salamander, *Ambystoma opacum*. Can. J. Zool. 85: 1008–1013.

Toews, D.P.L., N.R. Hofmeister and S.A. Taylor. 2017. The evolution and genetics of carotenoid processing in animals. Trends Genet. 3: 171–182.

Toledo, L.F. and C.F.B. Haddad. 2009. Colors and some morphological traits as defensive mechanisms in anurans. Int. J. Zool. 2009: 910892.

Tuma, M.C. and V.I. Gelfand. 1999. Molecular mechanisms of pigment transport in melanophores. Pigm. Cell Res. 12: 283–294.

Twomey, E., J.S. Vestergaard, P.J. Venegas and K. Summers. 2016. Mimetic divergence and the speciation continuum in the mimic poison frog *Ranitomeya imitator*. Am. Nat. 187: 205–224.

Twomey, E., J.D. Johnson, S. Castroviejo-Fisher and I. Van Bocxlaer. 2020a. A ketocarotenoid-based colour polymorphism in the Sira poison frog *Ranitomeya sirensis* indicates novel gene interactions underlying aposematic signal variation. Mol. Ecol. 29: 2004–2015.

Twomey, E., M. Kain, M. Claeys, K. Summers, S. Castroviejo-Fisher and I. Van Bocxlaer. 2020b. Mechanisms for color convergence in a mimetic radiation of poison frogs. Am. Nat. 195: E132–E149.

Twyman, H., M. Prager, N.I. Mundy and S. Andersson. 2018. Expression of a carotenoid-modifying gene and evolution of red coloration in weaverbirds (Ploceidae). Mol. Ecol. 27: 449–458.

Umbers, K.D., J. Lehtonen and J. Mappes. 2015. Deimatic displays. Curr. Biol. 25: R58–R59.

Umbers, K.D. and J. Mappes. 2015. Postattack deimatic display in the mountain katydid, *Acripeza reticulata*. Anim. Behav. 100: 68–73.

Umbers, K.D., S. De Bona, T.E. White, J. Lehtonen, J. Mappes and J.A. Endler. 2017. Deimatism: a neglected component of antipredator defence. Biol. Lett. 13: 20160936.

Umbers, K.D., J.L. Riley, M.B. Kelly, G. Taylor-Dalton, J.P. Lawrence and P.G. Byrne. 2020. Educating the enemy: Harnessing learned avoidance behavior in wild predators to increase survival of reintroduced southern corroboree frogs. Cons. Sci. Pract. 2: e139.

Vaissi, S., P. Parto and M. Sharifi. 2018. Ontogenetic changes in spot configuration (numbers, circularity, size and asymmetry) and lateral line in *Neurergus microspilotus* (Caudata: Salamandridae). Acta Zool. 99: 9–19.

Valkonen, J.K. and J. Mappes. 2014. Resembling a viper: implications of mimicry for conservation of the endangered smooth snake. Cons. Biol. 28: 1568–1574.

Vásquez, T. and K.S. Pfennig. 2007. Looking on the bright side: females prefer coloration indicative of male size and condition in the sexually dichromatic spadefoot toad, *Scaphiopus couchii*. Behav. Ecol. Sociobiol. 62: 127–135.

Vences, M., J. Kosuch, R. Boistel, C.F.B Haddad, E. La Marca, S. Lötters et al. 2003. Convergent evolution of aposematic coloration in Neotropical poison frogs: a molecular phylogenetic perspective. Org. Div. Evol. 3: 215–226.

Vestergaard, J.S., E. Twomey, R. Larsen, K. Summers and R. Nielsen. 2015. Number of genes controlling a quantitative trait in a hybrid zone of the aposematic frog *Ranitomeya imitator*. Proc. R. Soc. B 282: 20141950.

von Byern, J., C. Müller, K. Voigtländer, V. Dorrer, M. Marchetti-Deschmann, P. Flammang et al. 2017. Examples of bioadhesives for defence and predation. pp. 141–191. *In:* Gorb, S.N. and E.V. Gorb [eds.]. Functional Surfaces in Biology III. Springer, Cham, Switzerland.

Wang, I.J. and H.B. Shaffer. 2008. Rapid color evolution in an aposematic species: a phylogenetic analysis of color variation in the strikingly polymorphic strawberry poison-dart frog. Evolution 62: 2742–2759.

Wang, I.J. 2011. Inversely related aposematic traits: reduced conspicuousness evolves with increased toxicity in a polymorphic poison-dart frog. Evolution 65: 1637–1649.

Watanabe, M., M. Iwashita, M. Ishii, Y. Kurachi, A. Kawakami, S. Kondo et al. 2006. Spot pattern of leopard *Danio* is caused by mutation in the zebrafish connexin41.8 gene. EMBO Reports 7: 893–897.

Weiner, L., R. Han, B.M. Scicchitano, J. Li, K. Hasegawa, M. Grossi et al. 2007. Dedicated epithelial recipient cells determine pigmentation patterns. Cell 130: 932–942.

Wells, K.D. 1980. Social behavior and communication of a dendrobatid frog (*Colostethus trinitatis*). Herpetologica 36: 189–199.

White, T.E. and K.D.L. Umbers. 2021. Meta-analytic evidence for quantitative honesty in aposematic signals. Proc. R. Soc. B 288: 20210679.

Whiting, M.J., D.M. Stuart-Fox, D. O'Connor, D. Firth, N.C. Bennett and S.P. Blomberg. 2006. Ultraviolet signals ultra-aggression in a lizard. Anim. Behav. 72: 353–363.

Williams, C.R., E.D. Brodie Jr., M.J. Tyler and S.J. Walker. 2000. Antipredator mechanisms of Australian frogs. J. Herpetol. 34: 431–443.

Winters, A.E., N.G. Wilson, C.P. van den Berg, M.J. How, J.A. Endler, N.J. Marshall et al. 2018. Toxicity and taste: unequal chemical defences in a mimicry ring. Proc. R. Soc. B 285: 20180457.

Wollenberg, K.C. and C.J. Measey. 2009. Why colour in subterranean vertebrates? Exploring the evolution of colour patterns in caecilian amphibians. J. Evol. Biol. 22: 1046–1056.

Wolnicka-Glubisz, A., A. Pecio, D. Podkowa, L.M. Kolodziejczyk and P.M. Plonka. 2012. Pheomelanin in the skin of *Hymenochirus boettgeri* (Amphibia: Anura: Pipidae). Exp. Dermatol. 21: 537–540.

Xiao, N., H. Li, L. Shafique, S. Zhao, X. Su, Y. Zhang et al. 2019. A novel pale-yellow coat color of rabbits generated via MC1R mutation with CRISPR/Cas9 system. Front. Genet. 10: 875.

Yang, Y., C.L. Richards-Zawacki, A. Devar and M.B. Dugas. 2016. Poison frog color morphs express assortative mate preferences in allopatry but not sympatry. Evolution 70: 2778–2788.

Yang, Y., M.R. Servedio and C.L. Richards-Zawacki. 2019a. Imprinting sets the stage for speciation. Nature 574: 99–102.

Yang, Y., S. Blomenkamp, M.B. Dugas, C.L. Richards-Zawacki and H. Pröhl. 2019b. Mate choice versus mate preference: inferences about color-assortative mating differ between field and lab assays of poison frog behavior. Am. Nat. 193: 598–607.

Yang, Y., V. Prémel and C.L. Richards-Zawacki. 2020. Prior residence effect determines success of male–male territorial competition in a color polymorphic poison frog. Ethology 126: 1131–1140.

Yasutomi, M. 1987. Migration of epidermal melanophores to the dermis through the basement membrane during metamorphosis in the frog, *Rana japonica*. Pigm. Cell Res. 1: 181–187.

Yovanovich, C.A., S.M. Koskela, N. Nevala, S.L. Kondrashev, A. Kelber and K. Donner. 2017. The dual rod system of amphibians supports colour discrimination at the absolute visual threshold. Phil. Trans. R. Soc. Lond. B 372: 20160066.

Zamora-Camacho, F.J. and M. Comas. 2019. Beyond sexual dimorphism and habitat boundaries: coloration correlates with morphology, age, and locomotor performance in a toad. Evol. Biol. 46: 60–70.

Zimova, M., K. Hackländer, J.M. Good, J. Melo-Ferreira, P.C. Alves and L.S. Mills. 2018. Function and underlying mechanisms of seasonal colour moulting in mammals and birds: what keeps them changing in a warming world? Biol. Rev. 93: 1478–1498.

Zuasti, A. 2002. Melanization stimulating factor (MSF) and melanization inhibiting factor (MIF) in the integument of fish. Microsc. Res. Tech. 58: 488–495.

Index

For Product Safety Concerns and Information please contact our EU
representative GPSR@taylorandfrancis.com
Taylor & Francis Verlag GmbH, Kaufingerstraße 24, 80331 München, Germany

www.ingramcontent.com/pod-product-compliance
Lightning Source LLC
Chambersburg PA
CBHW060352220326
41598CB00023B/2896

* 9 7 8 0 3 6 7 5 5 3 9 7 5 *